Developments in Electrochemistry

Developments in Electrochemistry

Science Inspired by Martin Fleischmann

Editors

DEREK PLETCHER

Chemistry, University of Southampton, UK

ZHONG-QUN TIAN

*State Key Laboratory of Physical Chemistry of Solid Surfaces,
Xiamen University, China*

DAVID E. WILLIAMS

School of Chemical Sciences, University of Auckland, New Zealand

WILEY

Contents

8   Applications of Electrochemical Surface-Enhanced Raman
      Spectroscopy (EC-SERS)   **137**
      Marco Musiani, Jun-Yang Liu and Zhong-Qun Tian

9   *In-Situ* Scanning Probe Microscopies: Imaging and Beyond   **163**
      Bing-Wei Mao

List of Contributors

Morteza Y. Abyaneh, Department of Philosophy, University of Uppsala, Sweden

Philip N. Bartlett, Chemistry, University of Southampton, UK

Alan M. Bond, School of Chemistry, Monash University, Australia

Salvatore Daniele, Dipartimento Scienze Molecolari e Nanosistemi, University of Venice, Italy

Guy Denuault, Chemistry, University of Southampton, UK

Claude Gabrielli, LISE, Université Pierre et Marie Curie, France

Hubert H. Girault, Department of Chemistry and Chemical Engineering, Ecole Polytechnique Fédérale de Lausanne, Switzerland

Jun-Yang Liu, State Key Laboratory of Physical Chemistry of Solid Surfaces, Xiamen University, China

Digby D. Macdonald, Departments of Materials Science and Engineering and Nuclear Engineering, University of California at Berkeley, USA

Bing-Wei Mao, State Key Laboratory of Physical Chemistry of Solid Surfaces, Xiamen University, China

Elena A. Mashkina, School of Chemistry, Monash University, Australia

Michael C.H. McKubre, Energy Research, SRI International, USA

Melvin H. Miles, Department of Chemistry, University of LaVerne, USA

Jorge Mostany, Departamento de Química, Universidad Simón Bolívar, Venezuela

Marco Musiani, IENI-CNR, Padova, Italy

Richard J. Nichols, Department of Chemistry, University of Liverpool, UK

Laurence Peter, Department of Chemistry, University of Bath, UK

Derek Pletcher, Chemistry, University of Southampton, UK

Stephen W.T. Price, Diamond Light Source Ltd, Harwell, UK

Andrea E. Russell, Chemistry, University of Southampton, UK

Benjamin R. Scharifker, Departamento de Química, Venezuela and Universidad Metropolitana, Rectorado, Universidad Simón Bolívar, Venezuela

Samin Sharifi-Asl, Departments of Materials Science and Engineering and Nuclear Engineering, University of California at Berkeley, USA

Alexandr N. Simonov, School of Chemistry, Monash University, Australia

Stephen J. Thompson, Chemistry, University of Southampton, UK

Zhong-Qun Tian, State Key Laboratory of Physical Chemistry of Solid Surfaces, Xiamen University, China

Frank C. Walsh, Engineering Sciences, University of Southampton, UK

David E. Williams, School of Chemical Sciences, University of Auckland, New Zealand

Robert J.K. Wood, Engineering Sciences, University of Southampton, UK

Xue-Min Zhang, State Key Laboratory of Physical Chemistry of Solid Surfaces, Xiamen University, China

1

Martin Fleischmann[1] – The Scientist and the Person

A group of electrochemists whose lives were enhanced by their contacts with Martin Fleischmann have joined together to produce this book; it is intended to celebrate the legacy that he has left to modern electrochemistry. Martin was an outstanding scientist with a great vision that allowed him to initiate a number of fields of activity. His detailed grasp

---

[1] Figure reproduced from http://en.wikipedia.org/wiki/File:Fleischmann-cf.jpg

Developments in Electrochemistry: Science Inspired by Martin Fleischmann, First Edition.
Edited by Derek Pletcher, Zhong-Qun Tian and David E. Williams.
© 2014 John Wiley & Sons, Ltd. Published 2014 by John Wiley & Sons, Ltd.

of chemistry, physics and mathematics provided a background for a continuous flow of new approaches and experiments. Martin was essentially "an ideas man." Indeed, often his ideas were ahead of the ability of equipment to carry out the experiments, and it was only a few years later that the ideas came to fruition and it became possible to obtain high-quality experimental data. As can be seen by the authorship of the following chapters, this ability to be ahead of "the state-of-the-art," combined with inspirational leadership, made him a reliable stepping stone to successful careers for many of his coworkers. His enthusiasm for science, combined with a very warm personality and a lifetime's interest in the arts, skiing, food and wine, led him to have a large group of friends, ex-students and other coworkers, throughout the world. Stories about Martin abound, and a few of these are set out below. Indeed, the affection with which Martin is held can be seen in all the following chapters. All authors have, however, been asked to concentrate on the developments from the work of Fleischmann that are important now, and hence to produce a book that is relevant to "Electrochemistry in 2014." This would surely have been the wish of Martin Fleischmann.

Martin Fleischmann FRS was born in Karlsbad, Czechoslovakia in March 1927 to a wealthy, German-speaking family. His father was a well-known lawyer and his mother the daughter of a senior Austrian civil servant whose family traced its roots back to the thirteenth century in Prague. In Martin's own words, he was born into a castle with a fantastic collection of paintings. All this was to change dramatically, however. His parents were vocal opponents of the Nazi regime and, inevitably, they were forced to flee their home and leave behind all their possessions. They arrived in England after a lengthy and dangerous journey by taxi and train through Germany and Holland with a total of £1.30 in their purse! Following a period living in a "chicken hut," and the death of his father resulting from injuries received during a period of imprisonment by the Nazis, the family circumstances began to improve. Support from a refugee committee led to the lease of a cottage in Rustington (Sussex), where his mother was to start a business making dolls (that was to continue for more than 30 years, http://www.oldcottagedolls.co.uk) and Martin went back to education at Worthing High School for Boys. During the war he served in the Czech Air Force Training Unit. Martin was both an Undergraduate and Postgraduate in the Department of Chemistry at Imperial College London. During these student days he courted – and married – Sheila, who was to be his wife and support for 62 years. Together, they brought up three children, Nicholas, Vanessa and Charlotte, and Martin was always a devoted and stimulating father. He died on August 3rd, 2012 at the age of 85 after an extended illness.

His introduction to electrochemistry was as a PhD student with Professor Herrington at Imperial College. His own project concerned the diffusion of electrogenerated hydrogen through thin palladium foils! Importantly to his later career, he became part of a larger group that included John Bockris, Brian Conway and Roger Parsons, all to become leading figures in the world of electrochemistry. These contacts led to a stimulating environment for discussion and catalyzed broad interests in electrochemistry. After graduation in 1951, Martin went to the University of Newcastle where he was to interact with Lord Wynne-Jones, Reg Thirsk, Alan Bewick, Ron Armstrong and Frank Goodridge, amongst others. He was quickly promoted to a Readership before, in 1967, being appointed to the Faraday Chair of Chemistry at the University of Southampton where, with the support of Graham Hills, he was to establish a large Electrochemistry Group that soon had a worldwide reputation and still flourishes today. Key colleagues included Alan Bewick, Pat Hendra, Bob Jannson,

Laurie Peter, Derek Pletcher, Jim Robinson and David Schiffrin. His work in Newcastle and Southampton led to numerous contributions in:

- Electrochemical nucleation and phase growth
- Surface-enhanced Raman spectroscopy
- In-situ X-ray techniques
- Potentiostat design
- Microelectrodes
- Theory and development of electroanalytical techniques
- Organic electrochemistry
- Electrolytic cell design and electrochemical engineering
- Corrosion
- Electrodes in biological science.

Martin was a consummate mathematician and liked nothing better than a model leading either to "back-of-the-envelope calculations" or many pages of equations; those who worked with him were regularly presented with 20 pages of mathematics, scribbled the evening before and often requiring one to learn about new mathematical transforms or functions! The idea was always to fit experimental data to the resulting equations, and hence to gain insight into the fundamentals of the electrode reaction mechanism. Martin already had the interpretation and conclusions fully worked out and ready for discussion!

During the late 1960s and throughout the 1970s, Southampton was an exciting place for electrochemists. Lectures and longer visits by the world's most distinguished electrochemists were frequent, while Martin was always full of ideas for new experiments and would discuss them energetically. The Electrochemistry Laboratory was bigger than many entire Chemistry Departments at the time, and it had many diverse projects. The atmosphere at Southampton at the time is captured in Jim McQuillan's recollection: "From June 1972, I was a postdoctoral fellow at Southampton with Martin Fleischmann and Pat Hendra. Both Martin and Pat were innovative scientists who enjoyed competing with each other in scientific brainstorming and both were excited by the prospect of audacious experiments. I well remember those sessions when ideas were flying."

Pat Hendra's view was that Martin used him as an intellectual "punch bag." Pat particularly remembers one morning (and there were many like it) when he was giving a tutorial to a small group of undergraduates. Suddenly, the door crashed open, unseating a secretary whose desk was behind the door, and in advanced the "Great Man," as Pat always called him. With the oh so familiar words, "I've had an idea," he started to outline it! He was, of course, bearing a coffee cup in his left hand and spilling some on the floor! Several minutes later, after repeated reassurances that Pat would find him after the end of teaching, Martin left to acquire another coffee while Pat returned to his students. No more tutorial – they were speechless. "Who was THAT?" Pat was left to explain that they had been privileged to see a genius at work!

Martin was involved in the early years of the International Society of Electrochemistry, and served as both its Secretary/Treasurer (1964–1967) and President (1973–1974). He was for a period Head of Chemistry in Southampton, and also served on Research Councils and National Committees, duties that he carried out in his own particular style. Again, his lasting contributions were ideas. He was not a detailed administrator; Derek Pletcher describes how Martin's office was always covered with stacks of reports/correspondence and so on, and if your particular interest dropped below a certain level in the piles you were wise to sneak in and return it to the top of the stack. Martin's then secretary, Kate, had a system where piles were regularly moved to boxes in a cupboard and then destroyed, if MF had not noticed, in two years! Derek also commented that he used to tease Martin: "The only admin that you do efficiently is to book your skiing holidays." Despite these shortcomings, Martin was an effective leader with a great talent for inspiring novel research activity.

Martin's work led to a large number of publications in scientific journals (see the list below), many plenary lectures at conferences, and also invitations to visit laboratories throughout the world. Recognition peaked with the election to Fellowship of the Royal Society in 1985. He was also awarded several medals, perhaps the most prestigious being the Electrochemistry and Thermodynamics Medal (1979) of the Royal Society of Chemistry, and the Olin Palladium Medal (1986) of the US Electrochemical Society.

Martin Fleischmann took early retirement from Southampton in 1983 but, despite some serious health problems, he was to remain a very active scientist for a further 25 years. He continued to collaborate with colleagues in Southampton but spent extended periods at Harwell, the University of Utah, and the Laboratories of IMRA (part of Toyota) in the South of France. David Williams remembers first meeting Martin on a staircase during a scientific meeting and asking whether he would like to think about applying stochastic modeling to the problem of pitting corrosion; this topic piqued Martin's interest and led to a longstanding collaboration. The period with Stan Pons in Utah led to a large volume of publications related to the theory and practice of microelectrodes. Also, of course, Salt Lake City saw the early experiments that led to the birth of "Cold Fusion," and these continued in France. Later, Martin became a more itinerant scientist working with several laboratories, especially in the USA and Italy; he remained a focus for work on "cold fusion" but his interest in nucleation, biological systems and microelectrodes was undimmed. He also developed an enthusiasm for the applications of quantum electrodynamics to explaining scientific observations, but because of ill-health many of these ideas never matured into the published literature.

Details of all his scientific endeavors can be seen both in the following chapters and the list of his publications. Here, we will only summarize some highlights. During the early 1950s, Martin Fleischmann recognized the importance of "potential" in determining the rate of electrode reactions, and set out to design instrumentation that was capable of potential control and variation in a programmed way. These potentiostats and function generators were large, unreliable and often temperamental (literally, sparks could fly!) but, when working, they allowed new experiments. Later generations of such instrumentation remain essential to all electrochemical experiments in laboratories throughout the world. At the same time, Martin started to study the early stages of the deposition of conducting materials on electrode surfaces. The theory of nucleation and growth of such phases remained an interest throughout his life, and this is reflected in recent papers. Martin was one of the first to recognize the need for spectroscopic methods capable of interrogating the interface between electrodes and solutions. In particular, he developed surface-enhanced Raman spectroscopy(SERS), and in 1974 published the first such spectrum. Later, it was shown that SERS could provide new insights into many systems. The application of ultra-microelectrodes was another topic developed in his laboratory that was to become a routine laboratory technique. Martin also championed the use of electrolysis for the manufacture of chemicals and started an electrochemical engineering group charged with the development and study of novel flow cell designs, including cells with three-dimensional electrodes.

Martin had a career-long interest in the palladium/hydrogen system, piqued by his PhD studies and stimulated by further results from experiments performed by one of his students during the late 1960s that he could not "fully explain." This eventually led to the experiments that were to claim "excess heat" and the birth of "cold fusion." The concept promised a final solution to meeting the need for energy generation, thereby creating enormous interest. The concept of "cold fusion" was, however, totally contrary to accepted wisdom, and the experiments could only be reproduced in some laboratories. Overall, the response of most scientists was extremely hostile and often very personal, and not helped by the unfortunate way that the effect was announced at a news conference in Salt Lake City. This was all a great sadness to Martin, and contributed to his poor health. Certainly, to the end Martin remained willing to defend the underlying concepts as well as his experiments; he believed that there was an unusual phenomenon that deserved further study. It is inevitable and appropriate that this book contains a chapter on cold fusion that takes a positive view. Whatever one's opinion about cold fusion, however, it should not be allowed to dominate our view of Martin Fleischmann as a remarkable and outstanding scientist. Even those who were amongst his critics would agree. David Williams, heavily critical of cold fusion, remembers Martin as a kind personality full of energy and enthusiasm as well as a quirky humor and, most importantly, with a deep insight into scientific problems.

Laurence Peter recalls: "Martin was a real European intellectual with broad interests in the arts (and wine) as well as science. I first met him in 1966; needless to say I was absolutely captivated by Martin, his Central European accent and dynamic personality and this led me to a career in electrochemistry." Those who have worked with him all agree that Martin was a formative influence on a whole generation of electrochemists. We remember those wonderful ideas sessions in and around the laboratory; Martin taught that science is great fun, and his love of skiing, food, good wine and a good joke were never far from the surface. He was also a generous and kind man. Marco Musiani remembers being a visitor to Southampton for a summer and reporting to Martin that his family's apartment had been burgled. The immediate response was that Marco's wife and daughter could not stay in the flat; Martin's house and car were offered as Martin and his wife were to be in the USA. In consequence, the Musiani family were to enjoy a memorable month in an English country house and village.

Another abiding memory of Martin Fleischmann is the ending of a BBC documentary on "Cold Fusion." He appears purchasing cream cakes from a patisserie in the South of France with the accompaniment of Edith Piaf singing "Je ne regrette rien"!

The Publications of Martin Fleischmann

Electrochemical Nucleation and Phase Growth

(1) Fleischmann, M. and Thirsk, H.R. (1955) An investigation of electrochemical kinetics at constant overpotential – the behaviour of the lead dioxide electrode. *Transactions of the Faraday Society*, **51**, 71.

(2) Fleischmann, M. and Liler, M. (1958) The anodic oxidation of solution of plumbous salts. *Transactions of the Faraday Society*, **54**, 1370.

(3) Fleischmann, M., Bone, S.J. and Wynne-Jones, W.F.K. (1959) The exchange of β-lead dioxide with plumbous ions in solution. *Transactions of the Faraday Society*, **55**, 1783.

(4) Fleischmann, M. and Thirsk, H.R. (1958) The rate and nature of phase changes at electrode surfaces. Proceedings of the 1st International Symposium on Batteries, Bournemouth.

(5) Fleischmann, M. and Thirsk, H.R. (1959) The potentiostatic study of deposits on electrodes. *Electrochimica Acta*, **1**, 146.

(6) Fleischmann, M. and Sanghi, I. (1959) Some potentiostatic studies on zinc. *Electrochimica Acta*, **1**, 161.

(7) Fleischmann, M. and Thirsk, H.R. (1960) Anodic electrocrystallisation. *Electrochimica Acta*, **2**, 22.

(8) Fleischmann, M. and Thirsk, H.R. (1960) Some problems in electrocrystallisation. *The Electrochemical Society of Japan*, **28**, 175.

(9) Dugdale, I., Fleischmann, M. and Wynne-Jones, W.K.F. (1962) The anodic oxidation of silver sulfate to silver oxide at constant potential. *Electrochimica Acta*, **5**, 229.

(10) Bewick, A., Fleischmann, M. and Thirsk, H.R. (1962) Kinetics of the electrocrystallisation of thin films of calomel. *Transactions of the Faraday Society*, **58**, 2200.

(11) Fleischmann, M., Tordesillas, I.M. and Thirsk, H.R. (1962) Kinetics of electrodeposition of γ-manganese dioxide. *Transactions of the Faraday Society*, **58**, 1865.

(12) Fleischmann, M., Rajogopalan, K.S. and Thirsk, H.R. (1963) The growth of thin films of cadmium hydroxide an cadmium amalgam. *Transactions of the Faraday Society*, **59**, 741.

(13) Fleischmann, M. and Thirsk, H.R. (1963) The growth of thin passivating layers on metals. *Journal of The Electrochemical Society*, **110**, 688.

(14) Fleischmann, M. and Thirsk, H.R. (1964) Electrochemical kinetics of formation of monolayers of solid layers. *Electrochimica Acta*, **9**, 757.

(15) Fleischmann, M., Pattison, J. and Thirsk, H.R. (1965) Electrocrystallisation of thin films of thallous chloride on thallium amalgam. *Transactions of the Faraday Society*, **61**, 1256.

(16) Briggs, G.W.D., Fleischmann, M. and Thirsk, H.R. (1965) Behaviour of electrodes of the second kind. Proceedings of the 4th International Symposium on Batteries, Brighton, p. 167.

(17) Armstrong, R.D., Fleischmann, M. and Thirsk, H.R. (1965) Anodic behaviour of mercury in chloride media. *Transactions of the Faraday Society*, **61**, 2238.

(18) Fleischmann, M., Harrison, J.A. and Thirsk, H.R. (1965) Electrocrystallisation of thin films of nickel. *Transactions of the Faraday Society*, **61**, 2742.

(19) Armstrong, R.D. and Fleischmann, M. (1965) The formation of thin layers of solid phases on liquid electrodes. *Journal of the Polarographic Society*, **11**, 31.

(20) Armstrong, R.D., Fleischmann, M. and Koryta, J. (1965) Anodic polarographic waves involving insoluble mercury salt formation. *Collection of Czechoslovak Chemical Communications*, **30**, 4324.

(21) Armstrong, R.D., Fleischmann, M. and Thirsk, H.R. (1966) Anodic behaviour of mercury in hydroxide ion solutions. *Journal of Electroanalytical Chemistry*, **11**, 208.

(22) Fleischmann, M. and Harrison, J.A. (1966) The calculation of current-time curves for the electrocrystallisation of metals. *Journal of Electroanalytical Chemistry*, **12**, 183.

(23) Fleischmann, M. and Harrison, J.A. (1966) The relative roles of adatoms and of ions in solution near step lines during electrocrystallisation reactions. *Electrochimica Acta*, **11**, 749.

(24) Briggs, G.W.D. and Fleischmann, M. (1966) The anodic growth of NiOOH from nickel acetate at constant potential. *Transactions of the Faraday Society*, **62**, 3217.

(25) Armstrong, R.D. and Fleischmann, M. (1966) The anodic formation of mercury oxalate. *Zeitschrift für Physikalische Chemie*, **52**, 131.

(26) Fleischmann, M., Koryta, J. and Thirsk, H.R. (1967) Kinetics of electrodeposition and catalytic activity of thin films of ruthenium. *Transactions of the Faraday Society*, **63**, 1261.

(27) Fleischmann, M., Rangarajan, S.K. and Thirsk, H.R. (1967) Effects of diffusion through solutions and along surfaces on electrocrystallisation processes, part 1. *Transactions of the Faraday Society*, **63**, 1240.

(28) Fleischmann, M., Rangarajan, S.K. and Thirsk, H.R. (1967) Effects of diffusion through solutions and along surfaces on electrocrystallisation processes, part 2. *Transactions of the Faraday Society*, **63**, 1251.

(29) Fleischmann, M., Rangarajan, S.K. and Thirsk, H.R. (1967) Effects of diffusion through solutions and along surfaces on electrocrystallisation processes, part 3. *Transactions of the Faraday Society*, **63**, 1256.

(30) Armstrong, R.D., Fleischmann, M. and Oldfield, D.W. (1967) The anodic formation of phosphates of mercury. *Journal of Electroanalytical Chemistry*, **14**, 235.

(31) Briggs, G.W.D., Fleischmann, M., Lax, D.J. and Thirsk, H.R. (1968) Texture, growth and orientation of anodically formed silver oxides. *Transactions of the Faraday Society*, **64**, 3120.

(32) Fleischmann, M., Lax, D.J. and Thirsk, H.R. (1968) Kinetic studies of the formation of silver oxide. *Transactions of the Faraday Society*, **64**, 3128.

(33) Fleischmann, M., Lax, D.J. and Thirsk, H.R. (1968) Electrochemical studies of the Ag_2O/AgO phase change in alkaline solutions. *Transactions of the Faraday Society*, **64**, 3137.

(34) Fleischmann, M., Armstrong, R.D. and Oldfield, J.W. (1969) Properties of anodically formed mercury barbiturates. *Transactions of the Faraday Society*, **65**, 3053.

(35) Briggs, G.W.D. and Fleischmann, M. (1971) Oxidation and reduction of nickel hydroxide at constant potential. *Transactions of the Faraday Society*, **67**, 2397.

(36) Fleischmann, M. and Grenness, M. (1972) Electrocrystallisation of ruthenium and electrocatalysis of hydrogen evolution. *Journal of the Chemical Society, Faraday Transactions I*, **68**, 2305.

(37) Bindra, P., Fleischmann, M., Oldfield, J.W. and Singleton, D. (1973) Nucleation. *Discussions of the Faraday Society*, **56**, 180.

(38) Armstrong, R.D., Bewick, A. and Fleischmann, M. (1979) Comment on the application of the Avrami equation to electrocrystallisation. *Journal of Electroanalytical Chemistry*, **99**, 375.

(39) Manandhar, K., Pletcher, D. and Fleischmann, M. (1979) Further studies of the mechanism for the formation of mercury barbiturate films on a mercury anode. *Journal of Electroanalytical Chemistry*, **98**, 241.

(40) Abyaneh, M. and Fleischmann, M. (1980) The electrocrystallisation of nickel. *Transactions of the Institute of Metal Finishing*, **58**, 91.

(41) Abyaneh, M. and Fleischmann, M. (1981) The electrocrystallisation of nickel. Part I. Generalised models of electrocrystallisation. *Journal of Electroanalytical Chemistry*, **119**, 187.

(42) Abyaneh, M. and Fleischmann, M. (1981) The electrocrystallisation of nickel. Part II. Comparison of models with experimental data. *Journal of Electroanalytical Chemistry*, **119**, 197.

(43) Abyaneh, M. and Fleischmann, M. (1982) The role of nucleation and of overlap in electrocrystallisation reactions. *Electrochimica Acta*, **27**, 1513.

(44) Abyaneh, M., Berkem, M. and Fleischmann, M. (1982) The electrocrystallisation of nickel – the effects of additives. *Transactions of the Institute of Metal Finishing*, **60**, 114.

(45) Budeski, E., Fleischmann, M., Gabrielli, C. and Labram, M. (1983) Statistical analysis of the 2-D nucleation and electrocrystallisation of silver. *Electrochimica Acta*, **28**, 925.

(46) Fleischmann, M. and Saraby-Reintjes, A. (1984) The simultaneous deposition of nickel and hydrogen on vitreous carbon. *Electrochimica Acta*, **29**, 69.

(47) Li, L.J., Fleischmann, M. and Peter, L.M. (1989) Molecular level measurements of the kinetics of nucleation of α-lead dioxide on carbon microelectrodes. *Electrochimica Acta*, **34**, 475.

(48) Abyaneh, M. and Fleischmann, M. (1991) General models for surface nucleation and three dimensional growth; the effects of concurrent redox reactions and of diffusion. *Journal of The Electrochemical Society*, **138**, 2491.

(49) Abyaneh, M. and Fleischmann, M. (1991) Two-dimensional phase transformations on substrates of infinite and finite size. *Journal of The Electrochemical Society*, **138**, 2485.

(50) Abyaneh, M. and Fleischmann, M. (1991) The modelling of electrocrystallisation processes in battery systems. *Proceedings of the Electrochemical Society Conference*, **91–10**, 96.

(51) Abyaneh, M. and Fleischmann, M. (2002) Extracting nucleation rates from current-time transients, part I. The choice of growth models. *Journal of Electroanalytical Chemistry*, **530**, 82.

(52) Abyaneh, M. and Fleischmann, M. (2002) Extracting nucleation rates from current-time transients, part II. Comparing the computer fit and pre-pulse method. *Journal of Electroanalytical Chemistry*, **530**, 89.

(53) Abyaneh, M. and Fleischmann, M. (2002) Extracting nucleation rates from current-time transients; comments on the criticisms of Fletcher. *Journal of Electroanalytical Chemistry*, **530**, 108.

(54) Abyaneh, M. and Fleischmann, M. (2002) Extracting nucleation rates from current-time transients: further comments. *Journal of Electroanalytical Chemistry*, **530**, 123.

(55) Abyaneh, M., Fleischmann, M., Del Giudice, E. and Vitiello, G. (2009) The investigation of nucleation using microelectrodes, part I. The ensemble averages of the times of birth of the first nucleus. *Electrochimica Acta*, **54**, 879.

(56) Abyaneh, M., Fleischmann, M., Del Giudice, E. and Vitiello, G. (2009) The investigation of nucleation using microelectrodes, part II. The second moment of the times of birth of the first nucleus. *Electrochimica Acta*, **54**, 888.

Surface-Enhanced Raman Spectroscopy

(1) Fleischmann, M., Hendra, P.J. and McQuillan, A.J. (1973) Raman spectra from electrode surfaces. *Chemical Communications*, **9**, 80.

(2) Fleischmann, M., Hendra, P.J. and McQuillan, A.J. (1974) Adsorption of pyridine at the surface of silver electrodes. *Chemical Physics Letters*, **26**, 163.

(3) Fleischmann, M., Hendra, P.J. and McQuillan, A.J. (1975) Raman spectroscopic investigation of silver electrodes. *Journal of Electroanalytical Chemistry*, **65**, 933.

(4) Fleischmann, M., Hendra, P.J., McQuillan, A.J. and Paul, R.L. (1975) Raman spectroscopy at the copper electrode surface. *Journal of Electroanalytical Chemistry*, **66** 248.

(5) Fleischmann, M., Hendra, P.J., McQuillan, A.J. *et al.* (1976) Raman spectra at electrode–electrolyte interfaces. *Journal of Raman Spectroscopy*, **4**, 269.

(6) Fleischmann, M., Hendra, P.J., Cooney, R. and Reid, E.S. (1977) The Raman spectrum of adsorbed iodine on a Pt electrode surface. *Journal of Raman Spectroscopy*, **6**, 264.

(7) Fleischmann, M., Hendra, P.J., Cooney, R. and Reid, E.S. (1977) A Raman spectroscopic study of corrosion of lead electrodes in chloride media. *Journal of Electroanalytical Chemistry*, **80**, 405.

(8) Fleischmann, M., Hendra, P.J., Cooney, R. and Reid, E.S. (1977) Thiocyanate adsorption at silver electrodes. A Raman spectroscopic study. *Journal of the Chemical Society, Faraday Transactions I*, 1691.

(9) Fleischmann, M., Hendra, P.J. and Cooney, R. (1977) The Raman spectra of carbon monoxide on a Pt surface. *Chemical Communications*, **7**, 235.

(10) Fleischmann, M., Hendra, P.J. and Cooney, R. (1977) Raman spectra from adsorbed iodine species on an unroughened Pt electrode surface. *Journal of Raman Spectroscopy*, **6**, 264.

(11) Fleischmann, M., Hendra, P.J., Hill, I.R. and Pemble, M.E. (1981) Enhanced Raman spectra for species formed by the co-adsorption of halide ions and water molecules on silver electrodes. *Journal of Electroanalytical Chemistry*, **117**, 243.

(12) Fleischmann, M., Hill, I.R. and Pemble, M.E. (1981) Surface-enhanced Raman spectroscopy of $^{12}CN^-$ and $^{13}CN^-$ adsorbed at silver electrodes. *Journal of Electroanalytical Chemistry*, **136**, 361.

(13) Fleischmann, M. and Hill, I.R. (1983) Surface-enhanced Raman scattering from silver electrodes: formation and photolysis of chemisorbed pyridine species. *Journal of Electroanalytical Chemistry*, **146**, 353.

(14) Mengoli, G., Musiani, M., Pelli, B. and Fleischmann, M. (1983) The effect of triton on the electropolymerisation of phenol; An investigation of the adhesion of coatings using SERS. *Electrochimica Acta*, **28**, 1733.

(15) Fleischmann, M., Hill, I.R., Mengoli, G. and Musiani, M. (1983) A Raman spectroscopic study of the electropolymerisation of phenol on silver electrodes. *Electrochimica Acta*, **28**, 1545.

(16) Fleischmann, M. and Hill, I.R. (1983) The observation of solvated metal ions in the double layer region at silver electrodes using surface-enhanced Raman scattering. *Journal of Electroanalytical Chemistry*, **146**, 367.

(17) Fleischmann, M., Graves, P.R., Hill, I.R. and Robinson, J. (1983) Raman spectroscopy of pyridine adsorbed on roughened β-palladium hydride electrodes. *Chemical Physics Letters*, **95**, 322.

(18) Fleischmann, M., McCreery, R.L. and Hendra, P.J. (1983) Fiber optic probe for remote Raman spectroscopy. *Analytical Chemistry*, **55**, 146.

(19) Fleischmann, M., Hill, I.R. and Robinson, J. (1983) Surface-enhanced Raman scattering from silver electrodes; Potential and cation dependences of the very low frequency mode. *Chemical Physics Letters*, **97**, 441.

(20) Fleischmann, M., Graves, P.R., Hill, I.R. and Robinson, J. (1983) Raman spectroscopy of pyridine adsorbed on roughened β-palladium hydride electrodes. *Chemical Physics Letters*, **95**, 322.

(21) Fleischmann, M., Graves, P.R., Hill, I.R. and Robinson, J. (1983) Simultaneous Raman spectroscopy and differential double layer capacitance measurement of pyridine adsorbed on roughened silver electrodes. *Chemical Physics Letters*, **98**, 503.

(22) Fleischmann, M., Hill, I.R. and Sundholm, G. (1983) A Raman spectroscopic study of thiourea adsorbed on silver and copper electrodes. *Journal of Electroanalytical Chemistry*, **157**, 359.

(23) Fleischmann, M., Hill, I.R. and Sundholm, G. (1983) A Raman spectroscopic study of quinoline and isoquinoline adsorbed on silver and copper electrodes. *Journal of Electroanalytical Chemistry*, **158**, 153.

(24) Abrantes, L.M., Fleischmann, M., Hill, I.R. *et al.* (1984) An investigation of copper acetylide films on copper electrodes. Part I electrochemical and Raman spectroscopic analysis of the film formation. *Journal of Electroanalytical Chemistry*, **164**, 177.

(25) Fleischmann, M., Graves, P.R. and Robinson, J. (1985) Enhanced and normal Raman scattering from pyridine adsorbed on rough and smooth silver electrodes. *Journal of Electroanalytical Chemistry*, **182**, 73.

(26) Fleischmann, M., Sundholm, G. and Tian, Z.Q. (1986) A SERS study of silver electrodeposition from thiourea- and cyanide-containing solutions. *Electrochimica Acta*, **31**, 907.

(27) Mengoli, G., Musiani, M.M., Fleischmann, M. *et al.* (1987) Enhanced Raman scattering from iron electrodes. *Electrochimica Acta*, **32**, 1239.

(28) Fleischmann, M., Tian, Z.Q. and Li, L.J. (1987) Raman spectroscopy of adsorbates on thin film electrodes deposited on silver substrates. *Journal of Electroanalytical Chemistry*, **217**, 397.

(29) Fleischmann, M. and Tian, Z.Q. (1987) The effects of the underpotential and overpotential deposition of lead and thallium on silver on the Raman spectra of adsorbates. *Journal of Electroanalytical Chemistry*, **217**, 385.

(30) Fleischmann, M. and Tian, Z.Q. (1987) The induction of SERS on smooth silver by the deposition of nickel and cobalt. *Journal of Electroanalytical Chemistry*, **217**, 411.

(31) Korzeniewski, C., Severson, M.W., Schmidt, P.P. *et al.* (1987) Theoretical analysis of the vibrational spectra of ferricyanide and ferrocyanide adsorbed on metal electrodes. *The Journal of Physical Chemistry*, **91**, 5568.

(32) Crookell, A., Fleischmann, M., Hanniet, M. and Hendra, P.J. (1988) Surface enhanced Fourier transform Raman spectroscopy in the near infrared. *Chemical Physics Letters*, **149**, 123.

(33) Tian, Z.Q., Lian, Y.Z. and Fleischmann, M. (1990) In situ Raman spectroscopic studies on coadsorption of thiourea with anions at silver electrodes. *Electrochimica Acta*, **35**, 879.

(34) Fleischmann, M., Sockalingum, D. and Musiani, M.M. (1990) The use of near-infrared Fourier transform techniques in the study of surface-enhanced Raman spectra. *Spectrochimica Acta Part A*, **46A**, 285.

(35) Sockalingum, D., Fleischmann, M. and Musiani, M.M. (1991) Near-infrared transform SERS scattering of azole copper corrosion inhibitors in aqueous chloride media. *Spectrochimica Acta Part A*, **47A**, 1475.

In-Situ NMR

(1) Newmark, R.D., Fleischmann, M. and Pons, S. (1988) The observation of surface species on platinum colloids using ^{195}Pt NMR. *Journal of Electroanalytical Chemistry*, **255**, 235.

In-Situ X-Ray Techniques

(1) Fleischmann, M., Hendra, P.J. and Robinson, J. (1980) X-ray diffraction from adsorbed iodine on graphite. *Nature*, **288**, 152.
(2) Fleischmann, M., Graves, P., Hill, I.R. *et al.* (1983) Raman spectroscopic and X-ray diffraction studies of electrode–solution interfaces. *Journal of Electroanalytical Chemistry*, **150**, 33.
(3) Fleischmann, M., Oliver, A. and Robinson, J. (1986) In situ X-ray diffraction studies on electrode solution interfaces. *Electrochimica Acta*, **31**, 899.
(4) Fleischmann, M. and Mao, B.W. (1987) In situ X-ray diffraction studies of platinum electrode/solution interfaces. *Journal of Electroanalytical Chemistry*, **229**, 125.
(5) Mao, B.W. and Fleischmann, M. (1988) In-situ X-ray diffraction measurements of the surface structure of platinum in the presence of weakly adsorbed hydrogen. *Journal of Electroanalytical Chemistry*, **247**, 311.
(6) Mao, B.W. and Fleischmann, M. (1988) In-situ X-ray diffraction investigations of the UPD of thallium and lead on silver and gold electrodes. *Journal of Electroanalytical Chemistry*, **247**, 297.

Equipment Design

(1) Bewick, A., Fleischmann, M. and Liler, M. (1959) Some factors in potentiostat design. *Electrochimica Acta*, **1**, 83.
(2) Bewick, A. and Fleischmann, M. (1963) The design and performance of potentiostats. *Electrochimica Acta*, **8**, 89.
(3) Bewick, A. and Fleischmann, M. (1966) The design of potentiostats for use at very short times. *Electrochimica Acta*, **11**, 1397.

Microelectrodes

(1) Fleischmann, M., Lasserre, F., Robinson, J. and Swan, D. (1984) The application of microelectrodes to the study of homogeneous processes coupled to electrode reactions. Part I EC′ and EC reactions. *Journal of Electroanalytical Chemistry*, **177**, 97.
(2) Fleischmann, M., Lasserre, F. and Robinson, J. (1984) The application of microelectrodes to the study of homogeneous processes coupled to electrode reactions. Part II ECE and DISP 1 reactions. *Journal of Electroanalytical Chemistry*, **177**, 115.
(3) Bond, A.M., Fleischmann, M. and Robinson, J. (1984) Voltammetric measurements using microelectrodes in highly dilute solutions. Theoretical considerations. *Journal of Electroanalytical Chemistry*, **172**, 11.
(4) Bond, A.M., Fleischmann, M. and Robinson, J. (1984) Electrochemistry in organic solvents without supporting electrolyte using Pt microelectrodes. *Journal of Electroanalytical Chemistry*, **168**, 291.
(5) Bond, A.M., Fleischmann, M. and Robinson, J. (1984) The use of Pt microelectrodes for electrochemical investigations in low temperature glasses of non-aqueous solvents. *Journal of Electroanalytical Chemistry*, **180**, 257.
(6) Fleischmann, M., Bandyopadhyay, S. and Pons, S. (1985) The behaviour of microring electrodes. *The Journal of Physical Chemistry*, **89**, 5537.
(7) Fleischmann, M., Ghoroghchian, J. and Pons, S. (1985) Electrochemical behaviour of dispersions of spherical ultramicroelectrodes, part I. Theoretical considerations. *The Journal of Physical Chemistry*, **89**, 5530.
(8) Cassidy, J., Khoo, S.B., Pons, S. and Fleischmann, M. (1985) Electrochemistry at very high potentials: the use of ultramicroelectrodes in the anodic oxidation of short chain alkanes. *The Journal of Physical Chemistry*, **89**, 3933.
(9) Bond, A.M., Fleischmann, M., Khoo, S.B. *et al.* (1986) The construction and behaviour of ultramicroelectrodes; investigations of novel electrochemical systems. *Indian Journal of Technology*, **24**, 492.

(10) Bixler, J.W., Bond, A.M., Lay, P.A. *et al.* (1986) Instrumental configurations for the determination of sub-micromolar concentrations with carbon, gold and platinum microdisc electrodes in static and flow-through cells. *Analytica Chimica Acta*, **187**, 67.
(11) Ghoroghchian, J., Sarfarazi, F., Dibble, T. *et al.* (1986) Electrochemistry in the gas phase. Use of ultramicroelectrodes for the analysis of electroactive species in gas mixtures. *Analytical Chemistry*, **58**, 2278.
(12) Russell, A.E., Repka, K., Dibble, T. *et al.* (1986) Determination of electrochemical heterogeneous electron transfer rates from steady state measurements at ultramicroelectrodes. *Analytical Chemistry*, **58**, 2961.
(13) Fleischmann, M., Ghoroghchian, J., Rolison, D. and Pons, S. (1986) Electrochemical behaviour of dispersions of spherical ultramicroelectrodes. *The Journal of Physical Chemistry*, **90**, 6392.
(14) Dibble, T., Bandyopadhyaya, S., Ghoroghchian, J. *et al.* (1986) Electrochemistry at high potentials: oxidation of rare gases and other gases in non-aqueous solvents at ultramicroelectrodes. *The Journal of Physical Chemistry*, **90**, 5275.
(15) Li, L.J., Fleischmann, M. and Peter, L.M. (1987) In situ measurements of lead(II) concentration in lead-acid battery using mercury ultramicroelectrode. *Electrochimica Acta*, **32**, 1585.
(16) Pena, M.J., Fleischmann, M. and Garrard, N. (1987) Voltammetric measurements with microelectrodes in low conductivity systems. *Journal of Electroanalytical Chemistry*, **220**, 31.
(17) Fleischmann, M. and Pons, S. (1987) The behaviour of microdisc and microring electrodes. *Journal of Electroanalytical Chemistry*, **222**, 107.
(18) Pons, J.W., Daschbach, J., Pons, S. and Fleischmann, M. (1988) The behaviour of the mercury ultramicroelectrode. *Journal of Electroanalytical Chemistry*, **239**, 427.
(19) Brina, R., Pons, S. and Fleischmann, M. (1988) Ultramicroelectrode sensors and detectors. Considerations of the stability, sensitivity, reproducibility and mechanism of ion transport in gas-phase chromatography and in high-performance liquid-phase chromatography. *Journal of Electroanalytical Chemistry*, **244**, 244.
(20) Fleischmann, M. and Pons, S. (1988) The behaviour of microdisc and microring electrodes. Mass transport to the disc in the unsteady state. Chronopotentiometry. *Journal of Electroanalytical Chemistry*, **250**, 257.
(21) Fleischmann, M., Daschbach, J. and Pons, S. (1988) The behaviour of microdisc and microring electrodes. Mass transport to the disc in the unsteady state. Chronoamperometry. *Journal of Electroanalytical Chemistry*, **250**, 269.
(22) Fleischmann, M. and Pons, S. (1988) The behaviour of microdisc and microring electrodes. Mass transport to the disc in the unsteady state. The AC response. *Journal of Electroanalytical Chemistry*, **250**, 277.
(23) Fleischmann, M. and Pons, S. (1988) The behaviour of microdisc and microring electrodes. Mass transport to the disc in the unsteady state. The effects of coupled chemical reactions. The CE mechanism. *Journal of Electroanalytical Chemistry*, **250**, 285.
(24) Abrantes, L.M., Fleischmann, M., Peter, L.M. *et al.* (1988) On the diffusional impedance of microdisc electrodes. *Journal of Electroanalytical Chemistry*, **256**, 229.
(25) Li, L.J., Hawkins, M., Pons, J.W. *et al.* (1989) The behaviour of microdisc and microring electrodes. Prediction of the chronoamperometric response at microdisc and microring electrodes. *Journal of Electroanalytical Chemistry*, **262**, 45.
(26) Abrantes, L.M., Fleischmann, M., Li, L.J. *et al.* (1989) Behaviour of microdisc electrodes – chronopotentiometry and linear sweep voltammetric experiments. *Journal of Electroanalytical Chemistry*, **262**, 55.
(27) Fleischmann, M., Daschback, J. and Pons, S. (1989) The behaviour of microdisc and microring electrodes. Application of Neumann's integral theorem to the prediction of steady state responses. *Journal of Electroanalytical Chemistry*, **263**, 189.
(28) Daschbach, J., Pons, S. and Fleischmann, M. (1989) The behaviour of microdisc and microring electrodes. Application of Neumann's integral theorem to the prediction of steady state responses. Numerical illustrations. *Journal of Electroanalytical Chemistry*, **263**, 205.
(29) Fleischmann, M., Denuault, G., Pletcher, D. *et al.* (1989) The prediction of the chronoamperometric response at microdisc electrodes of the steady state for EC and CE catalytic reactions via Neumann's theorem. *Journal of Electroanalytical Chemistry*, **263**, 225.

(30) Li, L.J., Fleischmann, M. and Peter, L.M. (1989) Microelectrode studies of lead-acid battery electrochemistry. *Electrochimica Acta*, **34**, 459.

(31) Fleischmann, M., Denuault, G., Pletcher, D. and Tutty, O. (1990) Development of the theory for the interpretation of steady-state limiting currents at a microelectrode: EC' reactions. *Journal of Electroanalytical Chemistry*, **280**, 243.

(32) Fleischmann, M., Denuault, G. and Pletcher, D. (1990) A microelectrode study of the mechanism of the reactions of silver (II) with manganese (II) and chromium (III) in sulphuric acid. *Journal of Electroanalytical Chemistry*, **280**, 255.

(33) Ghoroghchian, J., Pons, S. and Fleischmann, M. (1991) Gas-phase electrochemistry on dispersions of microelectrodes. *Journal of Electroanalytical Chemistry*, **317**, 101.

(34) Fleischmann, M., Pons, S. and Daschbach, J. (1991) The AC impedance of spherical, disc and ring microelectrodes. *Journal of Electroanalytical Chemistry*, **317**, 1.

(35) Sousa, J.P., Pons, S. and Fleischmann, M. (1993) A novel approach to the silver electrodeposition mechanism. *Electrochimica Acta*, **11**, 265.

(36) Fleischmann, M., Pons, S., Sousa, J.P. and Ghoroghchian, J. (1994) Electrodeposition and electrocatalysis; the deposition and dissolution of single catalyst centres. *Journal of Electroanalytical Chemistry*, **366**, 171.

(37) Sousa, J.P., Pons, S. and Fleischmann, M. (1994) Studies of silver nucleation onto carbon microelectrodes. *Journal of the Chemical Society, Faraday Transactions I*, **90**, 1923.

Theory and Development of Other Electroanalytical Techniques

(1) Bewick, A., Fleischmann, M. and Huddleston, J.N. (1966) The determination of the kinetics of fast reactions in solution by electrochemical methods. Proceedings of the 3rd International Conference on Polarography, Southampton, p. 57.

(2) Fleischmann, M., Mansfield, J.R., Thirsk, H.R. *et al.* (1967) The investigation of the kinetics of electrode reactions by the application of repetitive square pulses of potential. *Electrochimica Acta*, **12**, 967.

(3) Fleischmann, M. and Oldfield, J.W. (1970) Generation- recombination noise in weak electrolytes. *Journal of Electroanalytical Chemistry*, **27**, 207.

(4) Fleischmann, M., Pletcher, D. and Rafinski, A. (1972) Homogeneous kinetics using cells with a large ratio of electrode area to solution volume; Part I. *Journal of Electroanalytical Chemistry*, **38**, 323.

(5) Fleischmann, M., Pletcher, D. and Rafinski, A. (1972) Homogeneous kinetics using cells with a large ratio of electrode area to solution volume; Part II. *Journal of Electroanalytical Chemistry*, **38**, 329.

(6) Brown, A.P., Fleischmann, M. and Pletcher, D. (1974) A new design of thin layer cell – application to the study of the reactions of silver(II). *Journal of Electroanalytical Chemistry*, **50**, 65.

(7) Fleischmann, M., Joslin, T. and Pletcher, D. (1974) A further study of diffusion and its role in electrode reactions. The development of a novel double potential step technique. *Electrochimica Acta*, **19**, 511.

(8) Bindra, P., Brown, A.P., Fleischmann, M. and Pletcher, D. (1975) The determination of the kinetics of very fast reactions by means of a quasi-steady state method. The mercury(I)/mercury system, part I. Theory. *Journal of Electroanalytical Chemistry*, **58**, 31.

(9) Bindra, P., Brown, A.P., Fleischmann, M. and Pletcher, D. (1975) The determination of the kinetics of very fast reactions by means of a quasi-steady state method. The mercury(I)/mercury system, part II. Experimental results. *Journal of Electroanalytical Chemistry*, **58**, 39.

Organic Electrochemistry

(1) Fleischmann, M., Petrov, I.N. and Wynne-Jones, W.F.K. (1963) The investigation of the kinetics of the electrode reactions of organic compounds by potentiostatic methods. Proceedings of the 1st Australian Conference on Electrochemistry, p. 500.

(2) Fleischmann, M., Mansfield, J.R. and Wynne-Jones, W.K.F. (1965) The anodic oxidation of acetate ions at smooth Pt electrodes. Part 1. The non-steady state oxidation of Pt. *Journal of Electroanalytical Chemistry*, **10**, 511.

(3) Fleischmann, M., Mansfield, J.R. and Wynne-Jones, W.K.F. (1965) The anodic oxidation of acetate ions at smooth Pt electrodes. Part 2. The non-steady state of the Kolbe synthesis of ethane. *Journal of Electroanalytical Chemistry*, **10**, 522.

(4) Atherton, G., Fleischmann, M. and Goodridge, F. (1967) Kinetic study of the Hofer-Moest reaction. *Transactions of the Faraday Society*, **63**, 1468.

(5) Fleischmann, M. and Goodridge, F. (1968) Anodic oxidation under pulse conditions. *Discussions of the Faraday Society*, **45**, 254.

(6) Fleischmann, M. and Pletcher, D. (1968) The electrochemical oxidation of aliphatic hydrocarbons in acetonitrile. *Tetrahedron Letters*, 6255.

(7) Fleischmann, M., Pletcher, D. and Race, G.M. (1969) The electrochemical oxidation of mercury(II)-propene complexes. *Journal of Electroanalytical Chemistry*, **23**, 369.

(8) Fleischmann, M., Pletcher, D. and Race, G.M. (1970) The electrochemical oxidation of mercury(II)-olefin complexes. *Journal of the Chemical Society B*, 1746.

(9) Fleischmann, M., Pletcher, D. and Faita, G. (1970) The electrochemical oxidation of cyclohexene/chloride ion mixtures in acetonitrile. *Journal of Electroanalytical Chemistry*, **25**, 455.

(10) Fleischmann, M. and Pletcher, D. (1970) The electrochemistry of polynuclear aromatic hydrocarbons in the molten salt system, AlCl₃-NaCl-KCl. *Journal of Electroanalytical Chemistry*, **25**, 449.

(11) Fleischmann, M., Pletcher, D. and Rafinski, A. (1971) The kinetics of the silver(I)/silver(II) couple in acid solution. *Journal of Applied Electrochemistry*, **1**, 1.

(12) Fleischmann, M., Pletcher, D. and Vance, C.J. (1971) The reduction of alkyl halides at a lead cathode in dimethylformamide. *Journal of Electroanalytical Chemistry*, **29**, 325.

(13) Bertram, J., Fleischmann, M. and Pletcher, D. (1971) The anodic oxidation of alkanes in fluorosulfonic acid; a novel synthesis of αβ-unsaturated ketones. *Tetrahedron Letters*, 349.

(14) Fleishmann, M., Korinek, K. and Pletcher, D. (1971) The oxidation of organic compounds on a nickel anode in alkaline solutions. *Journal of Electroanalytical Chemistry*, **31**, 39.

(15) Fleischmann, M., Pletcher, D. and Sundholm, G. (1971) The anodic oxidation of organomercury compounds. *Journal of Electroanalytical Chemistry*, **31**, 51.

(16) Fleishmann, M., Korinek, K. and Pletcher, D. (1972) The oxidation of organic compounds on a cobalt anode in alkaline solutions. *Journal of Electroanalytical Chemistry*, **33**, 478.

(17) Clark, D.B., Fleischmann, M. and Pletcher, D. (1972) The partial anodic oxidation of propylene in acetonitrile. *Journal of Electroanalytical Chemistry*, **36**, 137.

(18) Fleischmann, M. and Pletcher, D. (1972) The partial anodic oxidation of aliphatic hydrocarbons. *Chemie Ingenieur Technik*, **44**, 187.

(19) Fleischmann, M., Pletcher, D. and Vance, C.J. (1972) The anodic oxidation of ethylaluminium and ethylmagnesium compounds at a lead electrode. *Journal of Organometallic Chemistry*, **40**, 1.

(20) Fleischmann, M., Korinek, K. and Pletcher, D. (1972) The kinetics and mechanism of the oxidation of amines and alcohols at oxide covered nickel, silver, copper and cobalt electrodes. *Journal of the Chemical Society, Perkin Transactions 2*, **31**, 1396.

(21) Fleischmann, M., Mengoli, G. and Pletcher, D. (1973) The reduction of simple alkyl iodides at tin cathodes in dimethylformamide. *Electrochimica Acta*, **18**, 231.

(22) Fleischmann, M., Mengoli, G. and Pletcher, D. (1973) The cathodic reduction of acetonitrile. A new synthesis of tin tetramethyl. *Journal of Electroanalytical Chemistry*, **43**, 308.

(23) Koch, V.R., Miller, L.L., Clark, D.B. *et al.* (1973) Anodic fluorination from fluoroborate electrolytes. *Journal of Electroanalytical Chemistry*, **43**, 318.

(24) Clark, D.B., Fleischmann, M. and Pletcher, D. (1973) The partial anodic oxidation of aliphatic hydrocarbons in aprotic solvents. *Journal of Electroanalytical Chemistry*, **42**, 133.

(25) Bertram, J., Coleman, J.P., Fleischmann, M. and Pletcher, D. (1973) The electrochemical behaviour of alkanes in fluorosulfonic acid. *Journal of the Chemical Society, Perkin Transactions 2*, 374.

(26) Coleman, J.P., Fleischmann, M. and Pletcher, D. (1973) Anodic oxidation of polyfluorinated aromatic hydrocarbons in fluorosulfonic acid – the production of stable cation radicals. *Electrochimica Acta*, **18**, 331.

(27) Clark, D.B., Fleischmann, M. and Pletcher, D. (1973) The partial anodic oxidation of n-alkanes in acetonitrile and trifluoroacetic acid. *Journal of the Chemical Society, Perkin Transactions 2*, 1578.

(28) Fleischmann, M., Gara, W.B. and Hills, G.J. (1975) Electrode reactions of organic compounds as a function of pressure. *Journal of Electroanalytical Chemistry*, **60**, 313.

(29) Chatten, L.G., Fleischmann, M. and Pletcher, D. (1979) The anodic oxidation of some tetracyclines. *Journal of Electroanalytical Chemistry*, **102**, 407.

(30) Bloom, J.A., Fleischmann, M. and Mellor, J.M. (1984) Indirect electrochemical nitration; novel nitroacetamidation of dienes. *Tetrahedron Letters*, **25**, 4971.

(31) Bloom, J.A., Fleischmann, M. and Mellor, J.M. (1984) Nitroacetamidation of styrenes. *Journal of the Chemical Society, Perkin Transactions 1*, 2357.

(32) Bloom, J.A., Fleischmann, M. and Mellor, J.M. (1986) Nitroacetamidation of dienes by indirect electrochemical nitration. *Journal of the Chemical Society, Perkin Transactions 1*, 79.

(33) Bloom, J.A., Fleischmann, M. and Mellor, J.M. (1987) Nitration procedures using electrogenerated reagents. *Electrochimica Acta*, **32**, 785.

Cell Design and Electrochemical Engineering

(1) Fleischmann, M., Backhurst, J.R., Coulson, J.M. *et al.* (1969) A preliminary investigation of fluidised bed electrodes. *Journal of The Electrochemical Society*, **116**, 1600.

(2) Fleischmann, M., Backhurst, J.R., Plimley, R.E. and Goodridge, F. (1969) Fluidised zinc/oxygen electrode system. *Nature*, **221**, 55.

(3) Fleischmann, M. and Oldfield, J.W. (1971) Fluidised bed electrodes: I. Polarisation predicted by simplified models. *Journal of Electroanalytical Chemistry*, **29**, 211.

(4) Fleischmann, M. and Oldfield, J.W. (1971) Fluidised bed electrodes: II. Effective resistivity of the discontinuous metal phase. *Journal of Electroanalytical Chemistry*, **29**, 231.

(5) Fleischmann, M., Oldfield, J.W. and Porter, D.F. (1971) Fluidised bed electrodes: III. The cathodic reduction of oxygen on silver in a fluidised bed electrode. *Journal of Electroanalytical Chemistry*, **29**, 241.

(6) Fleischmann, M., Oldfield, J.W. and Tennakoon, L. (1971) Electrodeposition of copper in a fluidised bed. *Journal of Applied Electrochemistry*, **1**, 103.

(7) Fleischmann, M. (1973) Some engineering aspects of electrosynthesis. *Pure and Applied Chemistry*, **5**, 1.

(8) Chu, A.K., Fleischmann, M. and Hills, G.J. (1974) Packed bed electrodes – the electrochemical extraction of copper ions from dilute aqueous solutions. *Journal of Applied Electrochemistry*, **4**, 323.

(9) Fleischmann, M. and Kelsall, G.H. (1975) An investigation of the local behaviour of a copper fluidised bed electrode. *Chemistry & Industry*, **8**, 320.

(10) Fleischmann, M. and Jansson, R.E.W. (1975) Characterisation of electrochemical reactors. *Chemical Engineering*, 603.

(11) Fleischmann, M. and Jansson, R.E.W. (1977) The characterisation and performance of three dimensional electrodes. *Chemie Ingenieur Technik*, **49**, 283.

(12) Fleischmann, M. and Jansson, R.E.W. (1979) Dispersion in electrochemical cells with radial flow between parallel electrodes part I. A dispersive plug flow mathematical model. *Journal of Applied Electrochemistry*, **9**, 427.

(13) Fleischmann, M., Ghoroghchian, J. and Jansson, R.E.W. (1979) Dispersion in electrochemical cells with radial flow between parallel electrodes part II. Experimental results for capillary gap cell and pump cell configurations. *Journal of Applied Electrochemistry*, **9**, 437.

(14) Fleischmann, M. and Jansson, R.E.W. (1979) Effect of cell design on selectivity and conversion in electroorganic processes. *AIChE Symposium Series 185*, **75**, 2.

(15) Fleischmann, M. and Jansson, R.E.W. (1979) Electrochemical pump cells. *Transactions of the SAEST*, **12**, 277.

(16) Fleischmann, M. and Jansson, R.E.W. (1979) Cells for electrosynthetic reactions. *Chemisch Weekblad*, 539.

(17) Fleischmann, M. and Justinijanovic, I. (1980) The characterisation of three dimensional electrodes using an electrochemical tracer method. *Journal of Applied Electrochemistry*, **10**, 143.

(18) Fleischmann, M. and Jansson, R.E.W. (1980) The application of the principles of reaction engineering to electrochemical cell design. *Journal of Chemical Technology and Biotechnology*, **30**, 351.

(19) Fleischmann, M. and Ibrisagic, Z. (1980) Electrochemical measurements in bipolar trickle tower reactors. *Journal of Applied Electrochemistry*, **10**, 151.

(20) Fleischmann, M. and Ibrisagic, Z. (1980) Examination of flow models for bipolar trickle tower reactors. *Journal of Applied Electrochemistry*, **10**, 157.

(21) Fleischmann, M. and Ibrisagic, Z. (1980) Performance of bipolar trickle tower reactors. *Journal of Applied Electrochemistry*, **10**, 169.

(22) Fleischmann, M., Ghoroghchian, J. and Jansson, R.E.W. (1981) Reduction of 2-nitropropane in an undivided, flooded bipolar cell. *Journal of Applied Electrochemistry*, **11**, 55.

(23) Edhaie, S., Fleischmann, M. and Jansson, R.E.W. (1982) Application of the trickle tower to problems of pollution control. Part I. The scavenging of metal ions. *Journal of Applied Electrochemistry*, **12**, 59.

(24) Fleischmann, M. and Jansson, R.E.W. (1982) The reaction engineering of electrochemical cells. Part I. Axial packed bed electrodes. *Electrochimica Acta*, **27**, 1023.

(25) Fleischmann, M. and Jansson, R.E.W. (1982) The reaction engineering of electrochemical cells. Part II. Packed bed electrodes with orthogonal flow of current and solution. *Electrochimica Acta*, **27**, 1029.

(26) Fleischmann, M., Tennakoon, L., Bampfield, A. and Williams, P.J. (1983) Electrosynthesis in systems of two immiscible liquids and a phase transfer catalyst, part III. Characterisation of thin film contactor cells characterisation of thin film contactor cells. *Journal of Applied Electrochemistry*, **13**, 593.

(27) Fleischmann, M., Tennakoon, L., Gough, P. *et al.* (1983) Electrosynthesis in systems of two immiscible liquids and a phase transfer catalyst, part IV. Electro/synthesis using thin film contactor cells. *Journal of Applied Electrochemistry*, **13**, 603.

(28) Fleischmann, M. and Kelsall, G.H. (1984) A parametric study of copper deposition in a fluidised bed electrode. *Journal of Applied Electrochemistry*, **14**, 269.

(29) Fleischmann, M. and Kelsall, G.H. (1984) A feasibility study of mercury deposition in a lead fluidised bed. *Journal of Applied Electrochemistry*, **14**, 277.

(30) Wu, W.S., Rangaiah, G.P. and Fleischmann, M. (1993) Effect of gas evolution on dispersion in an electrochemical reactor. *Journal of Applied Electrochemistry*, **23**, 113.

Corrosion

(1) Bignold, G.J. and Fleischmann, M. (1974) Identification of transient phenomena during the anodic polarisation of iron in dilute sulfuric acid. *Electrochimica Acta*, **19**, 363.

(2) Fleischmann, M., Hill, I.R., Mengoli, G. and Musiani, M.M. (1983) The synergetic effect of benzylamine on the corrosion inhibition of copper by benzotriazole. *Electrochimica Acta*, **28**, 1325.

(3) Williams, D.E., Westcott, C. and Fleischmann, M. (1983) A statistical approach to the study of localised corrosion. Proceedings of the 5th International Symposium in Passivity of Metals and Semiconductors, p. 217.

(4) Williams, D.E., Westcott, C. and Fleischmann, M. (1984) Stochastic models of the initiation of pitting corrosion on stainless steels. *Journal of Electroanalytical Chemistry*, **180**, 549.

(5) Williams, D.E., Westcott, C. and Fleischmann, M. (1985) Stochastic models of pitting corrosion of stainless steel part I. Modelling of the initiation and growth of pits at constant potential. *Journal of The Electrochemical Society*, **132**, 1796.

(6) Williams, D.E., Westcott, C. and Fleischmann, M. (1985) Stochastic models of pitting corrosion of stainless steel part II. Measurement and interpretation of data at constant potential. *Journal of The Electrochemical Society*, **132**, 1804.

(7) Fleischmann, M., Hill, I.R., Mengoli, G. *et al.* (1985) A comparative study of the efficiency of some organic inhibitors for the corrosion of copper in aqueous chloride media using electrochemical and Raman scattering techniques. *Electrochimica Acta*, **30**, 879.

(8) Fleischmann, M., Mengoli, G., Musiani, M.M. and Pagura, C. (1985) An electrochemical and Raman spectroscopic investigation of synergetic effects in the corrosion inhibition of copper. *Electrochimica Acta*, **30**, 1591.

(9) Williams, D.E., Fleischmann, M., Stewart, J. and Brooks, T. (1986) Some characteristics of the initiation phase of pitting corrosion of stainless steel. *Materials Science Forum*, **8**, 151.

(10) Fleischmann, M., Mengoli, G., Musiani, M.M. and Lowry, R.B. (1987) An electrochemical and SERS investigation of the influence of pH on the effectiveness of some corrosion inhibitors of copper. *Journal of Electroanalytical Chemistry*, **217**, 187.

Biological Science

(1) Fleischmann, M., Gabrielli, C., Labram, M.T.G. *et al.* (1980) Alamethicin-induced conductances in lipid bilayers, part I. data analysis and simple steady state model. *The Journal of Membrane Biology*, **55**, 9.

(2) Fleischmann, M., Labram, M.T.G., Gabrielli, C. and Sattar, A. (1980) The measurement and interpretation of stochastic effects in electrochemistry and bioelectrochemistry. *Surface Science*, **101**, 583.

(3) Fleischmann, M., Gabrielli, C. and Labram, M.T.G. (1983) Analysis of the voltage-gated alamethicin induced channel conductance in lipid bilayers as a triggered birth and death stochastic process. *Physical Chemistry of Transmembrane Ion Motions*, 367.

(4) Fleischmann, M., Labram, M.T.G. and Gabrielli, C. (1983) The measurement and interpretation of fluctuations in electrochemical systems. The electrocrystallisation of silver and the voltage-gated alamethicin-induced channel conductance in lipid bilayers. *Journal of Electroanalytical Chemistry*, **150**, 111.

(5) Fleischmann, M., Gabrielli, C., Labram, M.T.G. and Markvart, T. (1986) Alamethicin-induced conductances in lipid bilayers, part II. Decomposition of the observed (compound) process. *Journal of Electroanalytical Chemistry*, **214**, 427.

(6) Fleischmann, M., Gabrielli, C. and Labram, M.T.G. (1986) Alamethicin-induced conductances in lipid bilayers, part III. Derivation of the properties of the elementary process from the observed (compound) process. *Journal of Electroanalytical Chemistry*, **214**, 441.

(7) Cullane, V.J., Schiffrin, D.J., Fleischmann, M. *et al.* (1988) The kinetics of ion transfer across adsorbed phospholipid layers. *Journal of Electroanalytical Chemistry*, **243**, 455.

(8) Li, J.L., Pons, S. and Fleischmann, M. (1991) The adsorption of single molecules of DNA on carbon microelectrodes. *Journal of Electroanalytical Chemistry*, **316**, 255.

(9) Comisso, N., Del Giudice, E., De Ninno, A. *et al.* (2006) Dynamics of the ion cyclotron resonance effect on amino acid adsorbed at interfaces. *Bioelectromagnetics*, **27**, 16.

Cold Fusion

(1) Fleischmann, M., Pons, S. and Hawkins, M. (1989) Electrochemically induced nuclear fusion of deuterium. *Journal of Electroanalytical Chemistry*, **261**, 261.

(2) Fleischmann, M., Pons, S., Hawkins, M. and Hoffman, R.J. (1989) Measurement of γ-rays from cold fusion. *Nature*, **339**, 667.

(3) Fleischmann, M. and Pons, S. (1990) Calorimetry of the palladium/deuterium system. 1st Annual Cold Fusion Conference Proceedings, p. 1.

(4) Fleischmann, M. and Pons, S. (1990) Calorimetric measurements of the palladium/deuterium system: fact and fiction. *Fusion Technology*, **17**, 669.

(5) Fleischmann, M., Pons, S., Anderson, M.W. *et al.* (1990) Calorimetry of the palladium-deuterium-heavy water system. *Journal of Electroanalytical Chemistry*, **287**, 293.

(6) Fleischmann, M. and Pons, S. (1991) The calorimetry of electrode reactions and measurements of excess enthalpy generation in the electrolysis of heavy water using palladium based cathodes. Proceedings, Italian Physics Society Conference, vol. **33**, p. 349.

(7) Fleischmann, M. and Pons, S. (1992) Concerning the detection of neutrons and γ-rays from cells containing palladium cathodes polarised in heavy water. *Nuovo Cimento della Societa di Fisica A: Nuclei, Particles and Fields*, **105A**, 763.

(8) Fleischmann, M. and Pons, S. (1992) Some comments on the paper 'analysis of experiments on calorimetry of lithium deuteroxide/heavy water electrochemical cells'. *Journal of Electroanalytical Chemistry*, **332**, 33.

(9) Fleischmann, M. and Pons, S. (1993) Calorimetry of the palladium-heavy water system – from simplicity via complications to simplicity. *Frontiers of Science Series*, **4**, 47.

(10) Fleischmann, M. and Pons, S. (1993) Calorimetry of the palladium- heavy water system – from simplicity via complications to simplicity. *Physics Letters A*, **176**, 118.

(11) Fleischmann, M. and Pons, S. (1994) Reply to critique by Morrison 'Comments on Claims of Excess Enthalpy by Fleischmann and Pons Using Simple Cells Made to Boil'. *Physics Letters A*, **187**, 276.

(12) Pons, S. and Fleischmann, M. (1994) Heat after death. *Fusion Technology*, **21**, 87.

(13) Fleischmann, M., Pons, S., Le Roux, M. and Roulette, J. (1994) Calorimetry of the Pd-D_2O system; the search for simplicity and accuracy. *Fusion Technology*, **21**, 323.

(14) Fleischmann, M., Pons, S. and Preparata, G. (1994) Possible theories of cold fusion. *Nuovo Cimento della Societa di Fisica A: Nuclei, Particles and Fields*, **107A**, 143.

(15) Pons, S. and Fleischmann, M. (1996) Calibration of the Pd-D_2O system: effects of procedure and positive feedback. *Journal de Chimie Physique et de Physico-Chimie Biologique*, **93**, 711.

(16) Miles, M.H., Imam, M.A. and Fleischmann, M. (2000) 'Case studies' of two experiments carried out with the Icarus systems. Proceedings, Italian Physics Society Conference, vol. **70**, p. 105.

(17) Miles, M.H. and Fleischmann, M. (2001) Calorimetric analysis of a heavy water electrolysis experiment using a Pd-B alloy cathode. *Electrochemistry Society Proceedings*, **2001–13**, 194.

(18) Szpak, S., Mosier-Boss, P.A., Miles, M.H. and Fleischmann, M. (2004) Thermal behaviour of polarised Pd/H electrodes prepared by co-deposition. *Thermochimica Acta*, **410**, 101.

(19) Miles, M.H. and Fleischmann, M. (2008) Accuracy of isoperibolic calorimetry used in cold fusion control experiment. *ACS Symposium Series*, **198**, 153.

(20) Fleischmann, M. and Miles, M.H. (2006) The instrument function of isoperibolic calorimeters; Excess enthalpy generation due to the parasitic reduction of oxygen, in *Condensed Matter Nuclear Science, Proceedings 10th International Conference on Cold Fusion* (eds P.L. Hagerstein and S.R. Chubb), World Scientific, New Jersey, p. 247.

Quantum Electrodynamics

(1) Del Giudice, E., Fleischmann, M., Preparata, G. and Talpo, G. (2002) On the 'Unreasonable' effects of ELF magnetic fields upon a system of ions. *Bioelectromagnetics*, **23**, 522.

(2) Del Giudice, E., Preparata, G. and Fleischmann, M. (2000) QED coherence and electrolyte solutions. *Journal of Electroanalytical Chemistry*, **482**, 482.

(3) Del Giudice, E., De Ninno, A., Fleischmann, M. *et al.* (2005) Coherent quantum electrodynamics in living matter. *Bioelectromagnetics*, **24**, 199.

Other Investigations

(1) Fleischmann, M. and Thirsk, H.R. (1957) The electrochemical polarisation of silver single crystals in hydrochloric acid. *Transactions of the Faraday Society*, **53**, 91.

(2) Bewick, A., Fleischmann, M., Hiddleston, J.N. and Wynne-Jones, W.K.F. (1966) Examination of proton transfer reactions by temperature jump and electrochemical methods. *Discussions of the Faraday Society*, **39**, 149.

(3) Fleischmann, M. and Hiddleston, J.N. (1968) A palladium–hydrogen probe electrode for use as a microreference electrode. *Journal of Scientific Instruments*, **1**, 667.

(4) Fleischmann, M. and Sundholm, G. (1971) The influence of crystal growth on the polarographic behaviour of *cis*-dichlorodiamine Pt(II) complexes. *Journal of Electroanalytical Chemistry*, **30**, Appendix 4.

(5) Fleishmann, M., Korinek, K. and Pletcher, D. (1972) The oxidation of hydrazine at a nickel anode in alkaline solutions. *Journal of Electroanalytical Chemistry*, **34**, 494.

(6) Dandipani, B. and Fleischmann, M. (1972) Electrolytic separation factors on platinum under non-steady-state conditions. *Journal of Electroanalytical Chemistry*, **39**, 315.

(7) Dandipani, B. and Fleischmann, M. (1972) Electrolytic separation factors on palladium. *Journal of Electroanalytical Chemistry*, **39**, 323.

(8) Doughty, A.G., Fleischmann, M. and Pletcher, D. (1974) A study of fluorine evolution at a Pt electrode in anhydrous hydrogen fluoride. *Journal of Electroanalytical Chemistry*, **51**, 329.

(9) Doughty, A.G., Fleischmann, M. and Pletcher, D. (1974) Anode effects on reactions in anhydrous hydrogen fluoride. *Journal of Electroanalytical Chemistry*, **51**, 456.

(10) Fleischmann, M., Robinson, J. and Waser, R. (1981) An electrochemical study of the adsorption of pyridine and chloride ions on smooth and roughened silver electrodes. *Journal of Electroanalytical Chemistry*, **117**, 257.

(11) Fleischmann, M., Daolio, S. and Pletcher, D. (1981) A study of the deposition of molybdenum at a vitreous carbon cathode from an aqueous citrate bath. *Journal of Electroanalytical Chemistry*, **130**, 269.

(12) Gale, R.J., Sefaja, J. and Fleischmann, M. (1981) Modulated differential reflectance spectroscopy of lead dioxide films during growth. *Analytical Chemistry*, **53**, 1457.

(13) Mengoli, G., Musiani, M.M., Pletcher, D. and Fleischmann, M. (1984) Studies of pyrrole black as a possible battery positive electrode. *Journal of Applied Electrochemistry*, **14**, 285.

(14) Saraby-Reintjes, A. and Fleischmann, M. (1984) Kinetics of electrodeposition of nickel from Watts bath. *Electrochimica Acta*, **29**, 557.

(15) Abrantes, L.M., Castello, L.M., Fleischmann, M. *et al.* (1984) An investigation of copper acetylide films on copper electrodes. Part II photoelectrochemical studies of copper(I) phenylacetylide. *Journal of Electroanalytical Chemistry*, **177**, 129.

(16) Fleischmann, M., Graves, P.R. and Robinson, J. (1985) The Raman spectroscopy of the ferricyanide/ferrocyanide system at gold, *β*-palladium hydride and platinum electrodes. *Journal of Electroanalytical Chemistry*, **182**, 87.

(17) Zotti, G., Cattarin, S., Mengoli, G. *et al.* (1986) Photoelectrochemistry of copper(I) phenylacetylide films electrodeposited onto copper electrodes. *Journal of Electroanalytical Chemistry*, **200**, 341.

(18) Fleischmann, M., Mengoli, G., Musiani, M.M. and Garrard, N. (1987) Electrochemical synthesis and study of a poly(4-vinylpyridium hydrobromide perbromide) electrode. *Electrochimica Acta*, **32**, 55.

(19) Mao, B.W., Tian, Z.Q. and Fleischmann, M. (1992) Voltammetric studies of underpotential deposition of thallium from thallium(I) film confined to silver electrodes. *Electrochimica Acta*, **37**, 1767.

Reviews and Book Chapters

(1) Fleischmann, M. and Oldham, K.B. (1959) Electrochemistry. *Annual Report of the Chemical Society*, **55**, 67.

(2) Bewick, A. and Fleischmann, M. (1961) Fast reactions in solution. *Annual Report of the Chemical Society*, **57**, 90.

(3) Fleischmann, M. and Thirsk, H.R. (1963) Metal deposition and electrocrystallisation, in *Advances in Pure and Applied Electrochemistry*, vol. **3**, (ed. P. Delahay), Interscience, p. 123.

(4) Fleischmann, M. and Pletcher, D. (1969) Electrosynthesis of organic compounds. *Platinum Metals Review*, **13**, 46.

(5) Fleischmann, M. and Pletcher, D. (1969) Organic electrochemistry. *RIC Reviews*, **2**, 37.

(6) Fleischmann, M. and Pletcher, D. (1969) Organic electrochemistry. *Chemistry & Industry*, 1301.

(7) Fleischmann, M. and Pletcher, D. (1971) The electrode reactions of organic compounds, in *Reactions of Molecules on Electrodes* (ed. N. Hush), John Wiley, p. 347.

(8) Fleischmann, M. and Pletcher, D. (1973) Parameters for the control of electrosynthetic reactions, in *Advances in Physical and Organic Chemistry*, vol. **10** (ed. V. Gold), Academic Press, p. 155.

(9) Fleischmann, M. and Pletcher, D. (1975) Industrial electrosynthesis. *Chemistry in Britain*, **11**, 50.

(10) Bewick, A. and Fleischmann, M. (1978) Formation of surface compounds on electrodes, in *Topics in Surface Chemistry* (eds E. Kay and P.S. Bagus), Plenum, p. 45.

(11) Fleischmann, M. and Hendra, P.J. (1977) Raman spectroscopy at surfaces, in *Topics in Surface Chemistry* (eds E. Kay and P.S. Bagus), Plenum, p. 373.

(12) Bewick, A., Fleischmann, M. and Robinson, J. (1981) Structural investigations of electrode–solution interfaces. *Dechema Monographs*, **90**, 87.

(13) Fleischmann, M. and Pletcher, D. (1982) Progress in electrochemistry and electrochemical engineering, in *The Chemical Industry* (eds D.H. Sharpe and T.F. West), Ellis Horwood, Chichester, p. 581.

(14) Fleischmann, M. and Hill, I.R. (1982) Electrochemical effects in surface-enhanced Raman scattering, in *Surface Enhanced Raman Scattering* (eds R.K. Chang and T.E. Furtak), Plenum Press, New York, p. 275.

(15) Fleischmann, M. and Overstall, J. (1983) The status of electrochemical technology. *Chemical Engineering*, **393**, 24.

(16) Fleischmann, M. and Hill, I.R. (1984) Raman spectroscopy, in *Comprehensive Treatise of Electrochemistry*, vol. **8** (eds J.O'M. Bockris, B.R. Conway, E. Yeager and R.C. White), Plenum Press, New York, p. 373.

(17) Fleischmann, M. and Pletcher, D. (1985) Industrial applications of electrochemistry, in *Electrochemistry in Research and Development* (eds R. Kalvoda and R. Parsons), p. 261.

(18) Fleischmann, M. and Pons, S. (1987) The behaviour of microelectrodes. *Analytical Chemistry*, **59**, 1391.

(19) Fleischmann, M., Pons, S. and Daschbach, J. (1991) Non-steady state processes at microelectrodes, in *NATO ASI Series. Series E: Applied Science*, vol. **197** (eds M.I. Montenegro, M.A. Queiros and J. Daschbach), Kluwer, p. 51.

(20) Fleischmann, M. and Pons, S. (1991) Electrochemistry in the gas phase, in *NATO ASI Series. Series E: Applied Science*, vol. **197** (eds M.I. Montenegro, M.A. Queiros and J. Daschbach), Kluwer, p. 357.

(21) Fleischmann, M., Pons, S. and Daschbach, J. (1991) Adsorption and kinetics at a molecular level, in *NATO ASI Series. Series E: Applied Science*, vol. **197** (eds M.I. Montenegro, M.A. Queiros and J. Daschbach), Kluwer, p. 393.

(22) Fleischmann, M. (2006) Background to cold fusion; the genesis of the concept, in *Condensed Matter Nuclear Science, Proceedings 10th International Conference on Cold Fusion* (eds P.L. Hagerstein and S.R. Chubb), World Scientific, New Jersey, p. 1.

2

A Critical Review of the Methods Available for Quantitative Evaluation of Electrode Kinetics at Stationary Macrodisk Electrodes

Alan M. Bond, Elena A. Mashkina and Alexandr N. Simonov
Monash University, School of Chemistry, Australia

The evaluation of the thermodynamics and kinetics of an electron transfer reactions is frequently the objective of voltammetric studies. The reaction

$$\text{Red}_{\text{soln}} \xrightleftharpoons{E^0, k^0, \alpha} \text{Ox}_{\text{soln}} + e^- \tag{2.1}$$

may be fully characterized by determination of the standard potential (E^0), the heterogeneous charge transfer rate constant (k^0), and the charge transfer coefficient (α), if Butler–Volmer electron transfer theory is employed. In dynamic electrochemistry, electron transfer is coupled to mass transport, as an electroactive species must be able to move from the bulk solution phase to the electrode; alternatively, if generated at the electrode it must be able to move away from the electrode into the bulk solution. If voltammetry is carried out in a still solution and in the presence of excess electrolyte, then diffusion will be the dominant form of mass transport. Another key parameter determining the response to a dynamic electrochemical experiment is the uncompensated resistance (R_u); the contribution of this term is of course governed by Ohm's law. In transient voltammetric studies at macrodisk electrodes, the magnitude of double-layer capacitance (C_{dl}) also needs to be known as this

Developments in Electrochemistry: Science Inspired by Martin Fleischmann, First Edition.
Edited by Derek Pletcher, Zhong-Qun Tian and David E. Williams.
© 2014 John Wiley & Sons, Ltd. Published 2014 by John Wiley & Sons, Ltd.

term will be used to define the background current which contributes to the total current (the sum of faradaic current from electron transfer and capacitance current). Double-layer capacitance models are available to describe the background current.

Diffusion is governed by the geometry of the electrode, and the diffusivity is quantified by the diffusion coefficient (D). A typical three-electrode cell used in voltammetry is shown in Figure 2.1, where the working, reference and auxiliary electrodes are present and combined with a potentiostat (see Refs [1–3] for details). In principle, if a macrodisk working electrode of the type shown in Figure 2.1 is used, and the reaction of interest at the working electrode occurs as described by Equation (2.1), then mass transfer will occur in accordance with Fick's laws for linear (planar) diffusion. However, there could be a small contribution from radial (edge) diffusion associated with the edge (Figure 2.1, structure A) and perfect sealing between the Teflon, glass or other material holding the electrode in place is needed. If the edge effect is to be removed completely, then the electrode design shown in Figure 2.1 structure B could be employed, although this has rarely been done in recent times. An alternative is to model the diffusion at the edge by invoking Fick's laws for radial diffusion.

A quantitative evaluation of electrode kinetics for even the simplest process is, therefore, a complex matter requiring knowledge of, at least, E^0, k^0, α, D, R_u, and C_{dl}, assuming that the electrode area (A), concentration of electractive species and temperature (T) are known from independent sources. This implies that the modeling of a dynamic electrochemical experiment will require a combination of the Nernst equation, Butler–Volmer kinetics, Ohms law, double-layer theory and Fick's laws. Because of: (i) the number of parameters required in the modeling and their widely varying magnitudes; and (ii) commonly, the lack of analytical mathematical solutions to the theory, the reproducibility of electrode kinetic parameters reported from laboratory to laboratory remains poor, despite the theory and the numerical methods for its solution now being well understood.

In this chapter, a critical overview is provided of quantitative electrode kinetic studies of fast reactions using the ubiquitous macrodisk electrode under conditions of transient

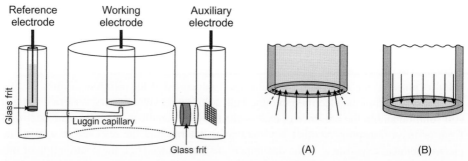

Figure 2.1 *Schematic representation of a three-electrode cell configuration comprising a macrodisk working electrode (constructed from Pt, Au, carbon, etc.) reference electrode (e.g., Ag/AgCl) and an auxiliary electrode (e.g., Pt mesh). In configuration (A), diffusion (planar plus radial) towards a macrodisk electrode is illustrated with a conventional working electrode design. In configuration (B), the macrodisk working electrode is installed inside a tube to remove edge (radial) diffusion.*

voltammetry, with a focus on the analysis of the seemingly simple process given in Equation (2.1). During the past 50 years, the most commonly used voltammetric method for quantifying electrode kinetics has been DC cyclic voltammetry at a stationary macrodisk electrode. However, numerous other waveforms and procedures are now available. The advantages provided by AC waveforms superimposed on the DC potential, as employed in the technique of AC voltammetry, will be considered in this chapter, as will the use of modern *e*-science methods for the comparison of experiment and theory. The use of steady-state conditions achieved with microdisk or rotated electrodes and other methods available for studies of the electrode kinetics will be addressed in other chapters.

Throughout his career, Martin Fleischmann – to whom this book is dedicated – was extremely interested in electrode kinetics. Indeed, his first paper in 1955 [4] was entitled "An investigation of electrochemical kinetics at constant overvoltage – The behaviour of the lead dioxide electrode." Fleischmann commonly used macrodisk electrodes and DC cyclic voltammetry under stationary solution conditions, but he was also very adventurous in developing novel electrochemical cells to overcome problems with resistance or capacitance. He was also active in introducing new methods to access faster kinetic regimes and implementing advanced mathematical approaches, statistics and modeling needed to quantitatively describe the electrode kinetics (see Chapter 1). Thus, to adopt Fleischmann's philosophy the chapter will move beyond the well-established approaches and consider the advantages of introducing sophisticated forms of data analysis, such as semiintegration or applying Fourier transform methods of data analysis when a transient waveform is superimposed onto the ramp used in conventional DC cyclic voltammetry. The overall goal is to design experiments, to develop appropriate mathematical treatments of the theory, and to introduce rigorous theory-experiment comparisons to obtain high-quality and reliable kinetic parameters for experiments at macrodisk electrodes. Indeed, results are now converging in terms of quality and sensitivity towards those obtained with microdisk or hydrodynamic electrodes under near steady-state environments.

2.1 DC Cyclic Voltammetry

2.1.1 Principles

It is convenient to describe the theory of voltammetry by considering the oxidation form of the half-cell reaction given in Equation (2.1). A current–potential-time (*I-E-t*) relationship under these conditions, which constitutes a description of a voltammogram, can be given by the expression:

$$I = F \cdot A \cdot k^0 \cdot \left(C_{x=0}^{Red}(t) \cdot \exp\left((1-\alpha) \cdot \frac{F(E - E^0)}{RT} \right) - C_{x=0}^{Ox}(t) \cdot \exp\left(-\alpha \cdot \frac{F(E - E^0)}{RT} \right) \right)$$

$$(2.2)$$

where $F = 96\ 485$ (C mol^{-1}) is the Faraday constant; A (cm^2) is the surface area of the electrode, k^0 (cm s^{-1}) is the heterogeneous charge transfer rate constant, $C_{x=0}^{Red}(t)$ and $C_{x=0}^{Ox}(t)$ (mol cm^{-3}) are the concentrations of *Red* and *Ox* at the working electrode surface (where x, the distance from the surface, is zero) as a function of time (t(s)), α is the charge transfer coefficient, E (V) is the true potential of the electrode (exclusive of ohmic (IR_u) drop), and E^0 (V) is the formal potential of the *Red/Ox* couple relative to a reference couple [2]. According to the IUPAC convention as adopted in this chapter, oxidation gives rise to positive current, and reduction to negative current [5].

Equation (2.2) describes the electron transfer in terms of Butler–Volmer (BV) parameterization [6], which is based on macroscopic concepts, assuming a linear dependence of the standard free energies of activation for oxidation and reduction on the potential with the proportionality being expressed in terms of the charge transfer coefficient, α. In principle, α may be potential-dependent; however, voltammetric experiments at macroelectrodes associated with fast electrode kinetics and dissolved species are normally analyzed over a small overpotential (η) range (typically, $\eta = E^0 - E \leq 0.2$ V), so the assumption of the constancy of α can be invoked. Ideally, α is expected to be equal to or close to 0.5 for a simple outer sphere process of the type described by Equation (2.1). Significant deviations from 0.5 will result in an asymmetry of the voltammogram, and are often indicative of either an underestimation of the complexity of the actual mechanism or inapplicability of the BV relationship for describing the heterogeneous electrode kinetics. In the latter case, more advanced contemporary theory based predominantly on developments by Marcus [7] and Hush [8] (MH) needs to be employed. However, in practice, the need to invoke the MH formalism, which takes account of the nature and structure of the solvent, reacting species and electrode material on the kinetics, is rarely required in voltammetric studies of fast process at macroelectrodes [9]. In general, BV parameterization and MH theory provide indistinguishable k^0 values and the same dependence of the rate of the electron transfer on potential at low η. Hence, all simulations and theoretical treatments discussed in this chapter are based on the BV relationship.

From an experimental point of view, voltammetry gives rise to an *I-E-t* relationship or voltammogram, when the current caused by an electrochemical transformation is measured as a function of a time-dependent electrode potential. In DC voltammetry, the potential is usually varied by applying a constant scan rate v (V s^{-1}; Figure 2.2) between the potential E_{start} and E_{sw1} or E_{sw2}.

The theory is simplest when a truly linear *E-t* ramp is used; however, the $E(t)$ dependence applied by many contemporary potentiostats is a step function (inset in Figure 2.2) when this type of experiment is usually referred to as "staircase" voltammetry. In the staircase mode, the current is commonly sampled either at the end of each step or is represented as an average or integrated value. Data derived from staircase voltammetry should be interpreted with caution but if the step size is small enough an analysis of data from the staircase potential ramp will produce results consistent with theoretical expectations based on a linear ramp,

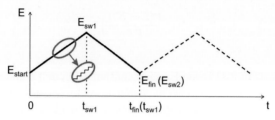

Figure 2.2   *In DC cyclic voltammetry, the potential is swept from E_{st} to switching potential E_{sw1} and to the final (E_{fin}) or second switching potential (E_{sw2}), etc.*

as demonstrated in numerous publications. In this chapter, it will be assumed that either: (i) a potentiostat having the capacity of generating voltammograms corresponding to the truly linear ramp has been used; or (ii) that the magnitude of each potential step in the staircase ramp is sufficiently small that the data obtained are indistinguishable from those obtained with a linear ramp.

Figures 2.3a–c show simulated cyclic voltammograms at a macrodisk electrode with designated k^0 values when initially only the reduced form is present in the bulk solution. For a process with extremely fast electron-transfer kinetics (e.g., $k^0 = 10^4$ cm s^{-1}), usually described as "reversible", the peak-to-peak separation, $\Delta E_p = E_p^a - E_p^c$, approaches the ideal value of $2.218 \cdot RT/F$ (0.057 V at 298 K) when the switching potential is far enough from E^0 [2, 10] (Figure 2.3a). When $k^0 = 10^{-3}$ cm s^{-1}, this gives rise to the so-called "quasi-reversible" case where ΔE_p is larger than for a reversible process (Figure 2.3b). A decrease of k^0 down to 10^{-6} cm s^{-1} gives the so-called "electrochemically irreversible" system, characterized by very large peak-to-peak separations (Figure 2.3c).

An example of a "chemically irreversible" system is also provided (Figure 2.3d), where an electrochemically fast oxidation process is followed by a fast and irreversible chemical transformation of species Ox to a new compound, X. Although there are differences in the profiles of the voltammograms presented in Figure 2.3c and d, which would be detected by an experienced electrochemist, it is easy to confuse "electrochemical" and "chemical"

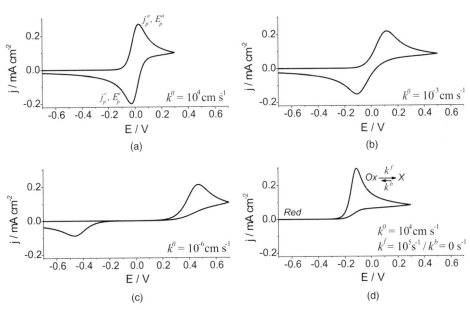

Figure 2.3   *DC cyclic voltammograms (v = 0.10 V s^{-1}) for: (a) reversible (k^0 = 10^4 cm s^{-1}); (b) quasi-reversible (k^0 = 10^{-3} cm s^{-1}); (c) electrochemically irreversible (k^0 = 10^{-6} cm s^{-1}); and (d) chemically irreversible (k^0 = 10^4 cm s^{-1}; k^f = 10^5 s^{-1}; k^b = 0 s^{-1}) one-electron charge transfer processes. Simulations are based on Equation (2.1) and use of the BV formalism for electron transfer with α = 0.50. Other parameters: E^0 = 0.000 V; C$_R^0$ = 1.0 mM; C$_O^0$ = 0.0 mM; all D = 1.0 × 10^{-5} cm^2 s^{-1}; A = 0.07 cm^2; R$_u$ = 0 Ω; C$_{DL}$ = 0 μF cm^{-2}; T = 298 K.*

irreversibility, particularly in a case where the initial potential in a cyclic voltammogram is set to a value that does not allow registration of the reverse peak shown in Figure 2.3c. In this chapter, it is the quasi-reversible process that is of major interest, particularly when the reversible limit is approached – that is, the system has fast electrode kinetics.

2.1.2   Processing DC Cyclic Voltammetric Data

For many years, the dependence of the peak-to-peak separation determined from cyclic voltammograms for a "quasi-reversible" process was almost invariably used to estimate k^0, using data in a "look-up table" provided by Nicholson [11]. This table provides the correlation of ΔE_p and a dimensionless function ψ which is defined as:

$$\psi = \frac{\left(\dfrac{D_{Red}}{D_{Ox}}\right)^{\alpha/2}}{\left(\dfrac{\pi \cdot D_{Red} \cdot n \cdot F}{R \cdot T}\right)^{1/2}} \cdot \frac{k^0}{v^{1/2}} \tag{2.3}$$

The link between ψ and ΔE_p is exemplified in Table 2.1, when $\alpha = 0.50$. A typical strategy employed in the use of this method is to record cyclic voltammograms at increasing values of v, until a significant deviation of ΔE_p from the reversible limit of $2.218 \cdot RT/nF$ is attained. In the case of fast kinetics (k^0 above ca. 0.1 cm s^{-1}) the values of v required for reliable application of the Nicholson method exceed 1 V s^{-1}, which imposes significant limitations. Even a well-designed electrochemical cell and careful alignment of working and reference electrodes (or the use of a Luggin–Huber capillary) does not completely eliminate the influence of uncompensated resistance (R_u), especially when the experiment is performed in an organic solvent containing a supporting electrolyte or in a relatively poorly conducting ionic liquid. The influence of R_u on the cyclic voltammetric response, like slow electrode kinetics, is to increase ΔE_p (Figure 2.4). The voltammograms in Figure 2.4 illustrate that the contribution of R_u and sluggish electrode kinetics are practically indistinguishable, although this ambiguity can be resolved by undertaking measurements

Table 2.1   *Correlation of $\Delta E_p \cdot n$ and parameter ψ defined by Equation (2.3), as derived by Nicholson [11] ($T = 298$ K, $\alpha = 0.5$, $E_{sw1} - E^0 = 0.141/n$).*

ψ	$\Delta E_p \cdot n$ (V)	ψ	$\Delta E_p \cdot n$ (V)
20	0.061	1	0.084
7	0.063	0.75	0.092
6	0.064	0.50	0.105
5	0.065	0.35	0.121
4	0.066	0.25	0.141
3	0.068	0.10	0.212
2	0.072		

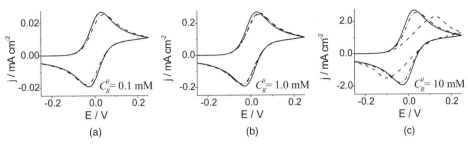

Figure 2.4 *Concentration dependence ($C_R^0 = 0.1$ (a), 1 (b) and 10 mM (c)) of the profile of DC cyclic voltammogram ($v = 0.10$ V s^{-1}) for a reversible process ($k^0 = 10^4$ cm s^{-1}), with $R_u = 0$ (—) and $R_u = 500$ Ω (– – –), and for a quasi-reversible process ($k^0 = 0.012$ cm s^{-1}) with $R_u = 0$ (– · – ·). Other parameters are as in Figure 2.3. Note that (—) and (– – –) examples as well as (– – –) and (– · – ·) are almost indistinguishable in (a) and (b), respectively, unlike in (c).*

at variable concentrations of the electroactive compound (Figure 2.4). With lower current achieved at lower concentrations, the influence of IR_u is smaller (Figure 2.4a) than with higher current resulting from a higher bulk concentration. The latter scenario, with an enhanced IR_u-drop, will cause notably larger ΔE_p values (Figure 2.4c). For a "quasi-reversible" process, ΔE_p should be independent of C_R^0, if the R_u term is negligible.

A second undesirable – but inevitably present – phenomenon, the influence of which can rarely be ignored in a transient voltammetric experiment, is the double-layer capacitance (C_{DL} (F cm^{-2})) associated with the working electrode. This gives rise to the background charging current as the potential is changed. C_{DL} depends on the chemical composition of the electrolyte and nature of the electrode, but is usually less than 50 μF cm^{-2}. Under the fast scan rate conditions needed to measure fast electrode kinetics, the charging current resulting from C_{DL} will make an appreciable contribution to the overall current. This undesirable contribution increases linearly with v, in contrast to the $j_p \propto v^{1/2}$ relationship for the Faradaic current [2] (Figure 2.5). As a result, at high scan rates the DC cyclic voltammograms will be dominated by the capacitive current rather than the response produced by the Faradaic

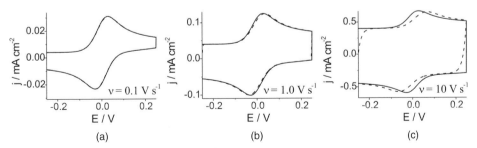

Figure 2.5 *Influence of C_{DL} (40 μF cm^{-2}) on the profile of DC cyclic voltammograms with $C_R^0 = 0.1$ mM as a function of the scan rate. $v = 0.1$ (a), 1 (b) and 10 V s^{-1} (c) for a reversible process ($k^0 = 10^4$ cm s^{-1}) with $R_u = 0$ (—) and 500 Ω (– – –). Other parameters used in this simulation are as in Figure 2.3.*

process of interest. Moreover, this high capacitive current produces an increased IR_u drop and hence an increase in ΔE_p (Figure 2.5c). Finally, the product $R_u{\cdot}C_{DL}$, having the units of seconds, defines the "cell time constant"; this defines the shortest time required for the electrochemical cell to respond correctly to the applied perturbation. In order to achieve high-quality electrode kinetic analysis, on a short timescale (as determined by dE/dt in DC voltammetry and frequency f in AC voltammetry; see below) the $R_u{\cdot}C_{DL}$ time constant should be as small as possible.

Voltammetric data distorted by R_u and C_{DL} (as exemplified in Figures 2.4 and 2.5) are frequently encountered experimentally, and should not to be used for the determination of k^0 via the Nicholson approach (as recognized by Nicholson himself [10]). Unfortunately, complete exclusion of the effect of IR_u-drop is practically impossible in a transient voltammetric experiment, especially in a room-temperature ionic liquid environment [12], where the conductivity is typically lower than in molecular aprotic solvents containing supporting electrolyte such as a tetraalkylammonium salt. In principle, the IR_u-drop could be corrected: (i) instrumentally by the application of positive feedback compensation schemes [13]; (ii) "theoretically," by accounting for the R_u effect when simulating the voltammograms (see below); or (iii) by post treatment of the experimental data to correct for the IR_u-drop. "Immediate" instrumental correction has serious drawbacks, with undercompensation or even overcompensation occurring. As a result, the majority of the voltammetric data acquired using the electronic compensation of resistance are not completely devoid of the IR_u-drop, but are compensated to some (usually unknown) extent. Moreover, instrumental correction does not occur instantaneously but rather with an appreciable time lag that results in phase shifts of the signal. Consequently, the application of positive feedback compensation schemes, followed by quantitative analysis of the voltammetric data, is not recommended. The post IR_u-drop correction of transient voltammetric data requires a complicated numerical mathematical treatment and a precise knowledge of the mechanism, as well as of the kinetic parameters.

The most efficient approach to the analysis of electrode kinetics, including the implementation of a straightforward account of the IR_u-drop and estimation of the contribution of C_{DL}, is theoretical modeling and direct comparison with experimental data. Contemporary versatile electrochemical simulation packages such as DigiSim® [14], DigiElch® [15], COMSOL Multiphysics® [16], KISSA [17], MECSim [18] are available for this purpose. A voltammogram represents the outcome of a sophisticated time-dependent change in concentration profile and an *analytical mathematical solution* of the integral equations, and describing this phenomenon in terms of the electrode kinetics is achievable only for a very limited number of cases. Universal and efficient simulators such as those listed above tackle the theoretical problem *numerically*, using finite difference methods, and describe the system by extending the concepts described in the seminal report of Nicholson and Shain [10]. Arguably, one of the most efficient algorithms for computing voltammetric data developed to date is the fast implicit finite difference algorithm employed by Rudolph (see references in Ref. [14]) and implemented in DigiSim® and DigiElch® software. These packages were, in fact, used to produce the simulated DC voltammograms presented in this chapter (e.g., Figures 2.3–2.5). The discretization of space is inevitably required when the numerical approach is used. In voltammetry, simulations are frequently performed with the aid of an exponentially expanding space grid, which provides high accuracy and stability as well as substantial computational time-saving benefits when compared to evenly discretized space. More complicated advanced adaptive grids have been introduced to provide higher accuracy and efficiency for the computation of fluxes at appreciable distance from the electrode surface, as needed to model homogeneous reactions coupled to electron transfer and when diffusion cannot be described in one spatial dimension. The developers of some electrochemical simulators have now provided the possibility of using MH kinetics as an alternative to the classic BV formalism. Furthermore, possibilities of accounting for diffusion with various electrode geometries, advanced potential ramps, two-dimensional diffusion models, and the influence of adsorption and other features, are offered in the latest versions of universal electrochemical simulators. A 1996 comprehensive review of the simulation methodologies applied to electroanalytical systems by Speiser [19] still provides a useful overview of different approaches for the modeling of voltammetric data. A multitude of useful links to publications devoted to the simulation of electrochemical data can be found in Feldberg's "personal perspective" [20].

An amazing range of possibilities for modelling of voltammetric data has been opened up by access to powerful simulation packages. Nevertheless, simulations do not by themselves remove the physical constraints in the DC voltammetric analysis. Thus, limitations in the determination of very large k^0 values (see Sections 2.3 and 2.4) and distinction between the effects of high R_u and low k^0 (as illustrated in Figure 2.4b) by simulating only a single experimental curve are still doomed to failure.

Another important issue to be addressed with simulation is the determination of the *quality of agreement* between experimental and simulated data. To date, the overwhelming majority of electrochemists compare experiment and theory heuristically, where discrimination between "good" and "bad" fits is done empirically. Obviously, such approaches can produce variation in the interpretation of the same experimental data when analyzed by two scientists having strongly different notions of what "good fit" really means. The introduction of advanced methods of analysis of fitting quality as a function of the parameters used in the simulation, as well as a statistical analysis of the simulation–experiment agreement, seems to be an obvious step. Contemporary approaches for implementing computer-aided fitting procedures for analysis of voltammetric data are addressed below in Section 2.2.

2.1.3   Semiintegration

Semiintegration is a mathematical operation signified by the operator $\frac{d^{-1/2}}{dx^{-1/2}}$, which is applied to some function of the variable x. When applied to transient voltammetry, x is invariably time, while the function to which it is applied may be current, flux density, or concentration [21]. Semiintegration of *I-t* data resulting from the voltammetric analysis of dissolved species produces curves ($M(t)$ ($A\ s^{1/2}$)) that mimic those obtained by steady-state voltammetry. Analysis of the $M(t)$ data can be undertaken using the methods developed predominantly by Savéant *et al.*[1] [22] and Oldham and coworkers [23–26]. The

---

[1] Savéant *et al.* employed the term "convoluted current" $I(t)$ via computation of the convolution integral $I(t) = \dfrac{1}{\pi^{1/2}} \int_0^t \dfrac{i(v)}{(t-v)^{\frac{1}{2}}} dv$, where $i(t)$ is the experimental current curve [22]. Although, the symbolism for $I(t)$ and $M(t)$ is different, both descriptions are equivalent.

semiintegration of discrete experimental data may be performed using the relatively easily programmed formula:

$$M(t) \approx \frac{3}{4} \sqrt{\frac{\delta}{\pi}} \left[I(t) + \sum_{n=1}^{N-1} \left[(n+1)^{3/2} - 2n^{3/2} + (n-1)^{3/2} \right] \cdot I(t - n\delta) \right] \quad (2.4)$$

where $I(t)$ is the current data evenly spaced over the interval from zero to time t, and N is the number of the processed data points, the spacing being $\delta = t/N$.

Figure 2.6a displays the $m(E)$ curves ($m = \frac{M}{A}$ (A $s^{1/2}$ cm^{-2})) obtained by applying Equation 2.4 to voltammetric data simulated for Equation 2.1, using a range of k^0 values (the original transient curves are shown in Figure 2.3). For the reversible case, the backward branch of the semiintegrated cyclic voltammogram exactly retraces the forward branch, while a decrease in k^0 displaces the two parts of the curve and increases the hysteresis. A limiting value of semiintegral attained at large overpotentials (Figure 2.6) is independent of k^0, and equal to:

$$m_{Lim} = F \cdot C_R^0 \cdot \sqrt{D_R} \quad (2.5)$$

and can be easily applied for determination of diffusion coefficients.

Under ideal conditions, when R_u and C_{DL} are negligible, the complete semiintegral data set obtained from a cyclic voltammogram does not depend on v (Figure 2.6b). The influence of R_u again resembles that of sluggish electrode kinetics and produces hysteresis in the $m(E)$ curve, although m_{Lim} remains constant (Figure 2.6b). The parameter that alters the limiting semiintegral value and also displaces forward and backward components of $m(E)$ data is C_{DL} (Figure 2.6c). Accounting for the influence of both R_u and C_{DL} on the semiintegral of the voltammogram is possible by applying appropriate treatment of the data [22, 27]. In the case of C_{DL}, performing blank experiments in the absence of an electroactive compound in the electrolyte, and subtracting from data obtained in its presence, may eliminate the capacitive current and other interfering background signals. However, equality of the charging current response derived from C_{DL} and other background processes (e.g., slow Faradaic processes involving solvent, electrolyte, electrode material or adsorption) in the absence and in

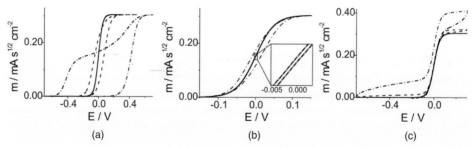

Figure 2.6   *Semiintegral versus potential curves obtained from voltammogram shown in Figure 2.3a (—) and from: (a) voltammograms shown in Figure 2.3b and c ($k^0 = 10^{-3}$ (– – –) and 10^{-6} cm s^{-1} (– · – ·)); (b) voltammograms analogous to that in Figure 2.3a but at v = 10 V s^{-1} (– – –) or with $R_u = 1000 \, \Omega$ (– · – ·); (c) voltammograms analogous to that in Figure 2.3a but with $C_{DL} = 50$ (—) and 300 μF cm^{-2} (– · – ·).*

the presence of the studied compound is always questionable. For this reason, crude subtraction of these two sets of voltammetric data should be applied with caution. Correction for the C_{DL} effect might be avoided by using high concentrations of electroactive compound, which could make the contribution of capacitive currents to the overall signal negligible. However, in this case larger currents will occur and additional care should be taken to minimize R_u.

Apart from allowing the straightforward determination of diffusion coefficients by applying Equation 2.5 to m_{Lim}, the semiintegrated form of the DC cyclic voltammogram can be used for analysis of k^0. In contrast to the classic Nicholson approach employed for the kinetic analysis of DC voltammograms where, in reality, only the two data points giving the separation of the current peak maxima are used for analysis, the method of quantification of kinetics from $m(E)$ data employs all the data available from a single experiment [22,23]. Such an approach provides a useful opportunity to elucidate the potential dependence of the forward and backward rate constants of Equation (2.1) and therefore to make a sober judgment on applicability of BV formalism for the description of heterogeneous charge transfer kinetics. Use of the entire data set for analysis also minimizes the influence of noise. Thus, semiintegration provides an impressive opportunity for all parameters typically sought in voltammetric analysis to be determined from a single experiment – a unique property that led Oldham and coworkers to term this data-processing strategy as "global analysis" [23].

Unfortunately, the application of Equation (2.4) to cyclic voltammetric data and performing data analyses as summarized above has some significant limitations. The most important constraint is that the diffusion of species must be truly planar. As demonstrated in Figure 2.7, even for a typically used macrodisk electrode with diameter of 3 mm, this condition is not strictly met. While voltammograms derived from simulations using diffusion models that take into account or neglect "edge effects" differ only slightly, the corresponding semiintegral curves demonstrate notable noncoincidence (Figure 2.7a), making the data obtained under conditions of two-dimensional diffusion unreliable for calculating D and k^0. The contribution of nonplanar diffusion can be minimized experimentally by using an appropriate working electrode design (Figure 2.1, structure B, which is not always feasible) or by using fast scan rates, which can compromise the quality of the data due to an enhancement of undesired influences associated with R_u and C_{DL}. Apart from being strictly planar,

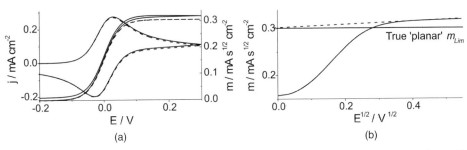

Figure 2.7 *(a) Cyclic voltammograms simulated assuming two-dimensional semi-infinite diffusion to an inlaid disk electrode (—) (electrode A in Figure 2.1) and one-dimensional diffusion to a planar electrode (– – –) (electrode B in Figure 2.1) and the corresponding semiintegral curves; (b) semiintegral curve (—) in linearized coordinates [28]. Parameters are given in Figure 2.3a.*

diffusion must be semi-infinite. The application of Equation (2.4) also requires an equilibrium or a null state to precede the voltammetric experiment, so that the concentrations of Red and Ox are initially uniform.

Advances in mathematical approaches to the semiintegration of electrochemical data allowed Savéant *et al.* [22, 27], as well as Oldham and Mahon, who introduced the so-called "extended semiintegrals" [25], to minimize the above-mentioned constraints, although at the cost of increased complexity in computation methods. The problem of the computation of "edge diffusion effects" with reasonable computational times was elegantly solved by the introduction of "partial sphere approximations" [26, 29], which simplify the two-dimensional diffusion problem into an easily solved one-dimensional one. Estimation of the planar component needed for semiintegral analysis can be performed by "convolutive reshaping," as described by Mahon [29].

Despite the obvious advantages provided by semiintegration (convolutive) analyses of voltammetric data and the ease of implementing the required computational procedures with modern equipment, this approach has not yet achieved the wide recognition it deserves. Rather, most electrochemists still prefer to compare conventional DC cyclic voltammograms with those obtained by simulation. Arguably, the low popularity of semiintegration is a consequence of the absence of versatile commercially available software, which would spare the need for each individual to create their own mathematical code.

2.2   AC Voltammetry

Recognition of the power of DC cyclic voltammetric analysis, along with its limitations, has led to a persistent search to enhance electroanalytical capabilities by introducing new forms of voltammetry. The semiintegration and "global voltammetric analysis" strategies discussed in Section 1.3 are aimed exclusively at refining the data processing stage. A family of voltammetric techniques in which a periodic signal is superimposed onto the DC waveform (Figure 2.8) have been developed to address problems at both instrumental and theoretical levels. Square-wave (SW), AC and pulse voltammetry each represent examples of this type of advanced technique. Each method provides enhancement in the quality and quantity of useful information obtained per experiment, improvement in the separation of faradaic and nonfaradaic components, facilitates deciphering of the electrode mechanism and increases the reliability of quantification of fast heterogeneous and homogeneous kinetics [2].

Initially, each method was assumed to require unique instrumentation, theory and strategy for data analysis. However, recognition of the fact that any periodic signal can be described as a sum of sine and cosine waves represented as a Fourier series means that there is little reason to treat each method in isolation [30]. This understanding opens up the possibility of developing unified data analysis strategies. The *modus operandi* of Fourier-transformed AC voltammetry is exemplified in the case of a sinusoidal AC perturbation in Figure 2.9. The recording of a single AC voltammogram produces a vast array of *I-E-t* data, which when subjected to a Fourier transform (FT) operation to generate the power spectrum, followed by band filtering and an inverse FT (IFT) operation, allows the data to be resolved into smaller blocks of data (harmonics) at a range of frequencies. There is also a DC (aperiodic) component which resembles that found in DC voltammetry. Thus, the FT–band filtering–IFT

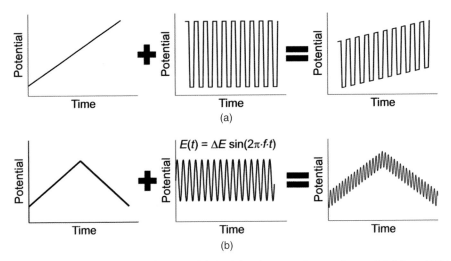

Figure 2.8 *Squarewave (a) and sinusoidal (b) AC voltammetric waveforms. ΔE (V) and f (Hz) are amplitude and frequency of the sinusoidal AC component, respectively.*

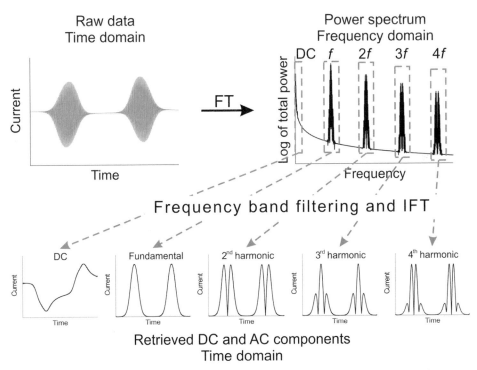

Figure 2.9 *Processing of AC voltammetric data by employing FT–IFT methodology.*

operation sequence provides packages of data consisting of a DC component as well as the fundamental harmonic and higher order harmonics arising from nonlinear elements present in faradaic processes and C_{DL}. Important advantages of FT AC voltammetry are provided by employing large amplitudes of the periodic signal to enrich the harmonic content. The large-amplitude protocol is in contrast to the historically established predilection for use of AC signals of low amplitude in electrochemical impedance spectroscopy (EIS), where linearization of the mathematical models needed to analyze the data has been seen as an advantage.

Each data block acquired by applying the FT–IFT methodology usually contains information on formal potentials, mechanisms, electrode kinetics, mass-transport, R_u and C_{DL}. The virtue of this strategy of data processing of the nonlinear elements is that each recovered component represents a different level of sensitivity to parameters to be determined. Thus, diffusion coefficients might be extracted from aperiodic DC component, in particular, by applying the semiintegration procedure described in Section 2.1.3. Background current derived from double-layer capacitance, which gives a notable contribution in DC voltammetry, also produces a large contribution in the AC fundamental harmonic, but only a very small contribution in the higher-order AC harmonics [31]. Indeed, in fourth- and higher-order harmonics the background current is usually negligible, so highly enhanced faradaic to nonfaradaic current ratios can be obtained for the mechanism in equation 2.1. E^0 values are accessible from all components, although the precise determination of E^0 may not be trivial if the electron transfer step is coupled to a chemical transformation step. Slow heterogeneous electron transfer processes (including completely irreversible background processes derived from the oxidation or reduction of solvent that occur at extreme potentials), sluggish homogeneous chemical reactions and, perhaps most remarkably, radial diffusion mass-transport contribute minimally to the higher-order harmonics, which are predominantly generated by fast electron transfer processes. The fact that higher harmonic components are negligibly influenced by "edge diffusion effects" provides an important advantage of AC voltammetric methods over the DC technique.

Application of the FT–IFT strategy to the analysis of SW voltammetry provides entirely different patterns in the harmonics. Now, all odd harmonics contain both faradaic and nonfaradaic components while even harmonics, being almost purely faradaic, manifest themselves only when departure from reversibility of the electrochemical kinetics occurs or the IR_u-drop is significant [30, 32]. Thus, the electroanalytical advantages furnished by FT AC voltammetry are again moderated by the influence of R_u, the scourge of all voltammetric methods.

The current magnitude of each harmonic component of an AC voltammogram increases with an increase in k^0 (to a limiting value defined by frequency, f (Hz)), while the presence of IR_u-drop suppresses the current. Nevertheless, the accuracy in the determination of k^0 available with FT AC voltammetry is vastly superior to that possible in the DC mode. A careful selection of frequency of the periodic component in FT AC voltammetry provides a straightforward possibility to tune the sensitivity of the method to the kinetics of interest, analogous to varying the scan rate in DC voltammetry. The upper limit of frequency in an experiment is usually determined by the $R_u \cdot C_{DL}$ time constant of the electrochemical cell, and by instrumental limitations.

In contrast to DC voltammetry, where an evaluation of k^0 can be achieved by using "look-up" tables (see Section 2.1.2), the quantitative analysis of FT AC voltammetric data almost always requires simulation–experiment comparisons. Fortunately, if a DC simulation of the faradaic and nonfaradaic components of an electrode process is available, then an exactly analogous computational protocol can be used to simulate any form of AC voltammetry. Commercially available DigiElch® electrochemical simulation software [15] and the freeware MECSim program developed at Monash University [18], allow the possibility of simulating many types of AC voltammetry. Clearly, the application of identical waveforms and mathematical procedures for data treatment (FT–IFT routine, frequency bandwidth filtering, selection of required components from the frequency domain data, etc.) must be used to provide a sensible matching of experimental and simulated data. Although a single AC voltammetric experiment will provide a wealth of information relative to that provided by a DC experiment, and the probability of obtaining a unique solution is markedly higher when FT AC voltammetry is used. Experiment–theory comparisons over a range of frequencies[2] and, most importantly, concentrations of electroactive species should always be considered.

2.2.1 Advanced Methods of Theory–Experiment Comparison

As in DC voltammetry, theory and experiment comparisons of AC voltammetric data have usually been undertaken heuristically, where the experimenter decides empirically whether a satisfactory agreement between the experiment data and the model has been achieved. However, an estimation of the quality of the fit between experimental and simulated data can be gained by examining the residual sum of squares (RSS) or by the mean percentage error (MPE) [34].

Advanced methods of performing numerical simulations and comparing results with experiment are available using *e*-science tools. At Monash University, a family of generic *e*-science tools, called Nimrod, has been developed that allows the results of physical experiments to be integrated with their computational counterparts, and hence to aid the exploration of physical parameters that cannot be directly measured [35]. *E*-science tools exploit features associated with advanced computer infrastructure – that is, huge databases, high-performance computers, and massive memory storage capacity. *E*-science tools [36] provide advantages in data analysis that are applicable to research fields associated with medicine and biology [37, 38], environmental surveys [39], electrochemistry [40] and many others.

Nimrod/G, as used in FT AC voltammetric studies, forms part of the Nimrod tool kit which performs a parameter sweep [35, 40, 41], where each parameter of interest is assigned a set of values. The Nimrod/G tool can generate all possible combinations of these values and executes the simulation outcome corresponding to each combination using the model provided. Normally, the results of the sweep are presented in the form of a contour map for the objective metric where *x* (horizontal axis) and *y* (vertical axis) are used as floating parameters and the others are fixed. The objective metric is defined as the normalized differences between values predicted by a model and the values actually observed. As the experimental and model sinusoids are not necessarily in phase, when the experimental and model outputs are filtered to produce the various harmonics, the data points with the same

---

[2] Variation of frequency may be achieved in a single AC voltammetric experiment by applying a periodic waveform, which contains several carefully selected frequencies [33].

index number for experimental and simulated outputs can mismatch in the time domain. In this situation, the direct point-by-point comparison between model and experiment will provide a spurious measure of the parameter fit. To remedy this problem, a linear spline envelope is computed for the waveform and a metric which computes the dissimilarity between the splines has been developed [30]. The objective function Ψ is calculated as the relative root mean square deviation:

$$\Psi = \sqrt{\frac{\sum_{i=1}^{N}(f^{exp}(x_i) - (f^{sim}(x_i))^2}{\sum_{i=1}^{N}f^{exp}(x_i)^2}}, \tag{2.6}$$

where $f^{exp}(x)$ and $f^{sim}(x)$ are experimental and simulated functions, respectively, and N is the number of data points.

There are many other options available on Nimrod. The inverse problem in FT AC voltammetry requires finding parameters of E^0, k^0, α, R_u and C_{DL} that yield model outputs which agree closely with experimental data. Such a problem may be treated as an optimization problem by minimizing some measure of the difference between the model outputs and experimental data by the Simplex or other algorithm [42]. Nimrod/O [42] provides access to the Simplex method. In the voltammetric context, Simplex optimization algorithms are used to automate the parameter fitting by minimizing the difference between the simulated results and experimental data. Nimrod also may be used for exploring the response of models to their input parameters. For example, Nimrod/E provides a fractional factorial design which allows the values of interest for each parameter to be defined and detected, as well as which interactions should be ignored and which should be estimated.

The Nimrod tools referred to above do not provide statistical analysis, but rather a comparison of simulated data based on a model and experimental data. However, it is planned to introduce to the Nimrod computer grid a statistical form of analysis based on Bayesian probabilities which will be applied to each FT AC experimental data set. In time, major improvements in the reporting of parameters such as E^0, k^0, α, R_u and C_{DL} should emerge when the full power of *e*-science is routinely introduced into the analysis of voltammetric experiments.

2.3   Experimental Studies

The methods of data analysis set out in the previous sections are applied below to three representative systems that differ in their heterogeneous charge transfer kinetics and magnitude of R_u.

2.3.1   Reduction of [Ru(NH$_3$)$_6$]$^{3+}$ in an Aqueous Medium

The hexamineruthenium(III) cation, usually added as the chloride salt [Ru(NH$_3$)$_6$]Cl$_3$, undergoes a one-electron reduction at arguably one of the highest rates of charge transfer measured to date. During the past few decades, the k^0 values reported for the [Ru(NH$_3$)$_6$]$^{3+/2+}$ process have gradually increased from ~0.1 to many cm s^{-1}, probably due to advances in the techniques used in the measurements. The electroreduction of

Table 2.2 *Mid-point potential (E_m versus Ag/AgCl (1 M KCl)), peak-to-peak separation (ΔE_p), reduction peak current (I_p) normalized to $v^{1/2}$ and apparent heterogeneous electron transfer rate constant (k^0_{app}) values estimated on the basis of Nicholson's treatment of DC cyclic voltammetric data obtained with a 3 mm diameter GC electrode for reduction of 2 mM $[Ru(NH_3)_6]^{3+}$ in aqueous 1 M H_2SO_4.*

n (V s^{-1})	E_m (V)	ΔE_p (V)	$10^6 \cdot I_p \, v^{-1/2}$ (A·V$^{-1/2}$·s$^{1/2}$)	k^0_{app} (cm s^{-1})
0.01	−0.253	n.m. / 0.061[a]	n.m. / 88	n.m. / 0.05
0.02	−0.252	0.068 / 0.060	84 / 86	0.01 / 0.07
0.05	−0.251	0.065 / 0.061	83 / 84	0.03 / 0.11
0.10	−0.251	0.065 / 0.062	84 / 84	0.04 / 0.09
0.40	−0.251	0.064 / n.m.	82 / n.m.	0.09 / n.m.
0.60	−0.251	0.066 / n.m.	82 / n.m.	0.08 / n.m.
1.0	−0.251	0.067 / n.m.	82 / n.m.	0.09 / n.m.

[a] ΔE_p values were obtained in instrumental analogue regime with a potential step of 1 mV / staircase regime with potential step of ca. 0.3 mV.
n.m., not measured.

$[Ru(NH_3)_6]^{3+}$ in aqueous electrolytes of sufficiently high concentrations represents a real-life example of an extremely fast charge transfer process that is not necessarily burdened by the influence of R_u. Therefore, the voltammetry of the $[Ru(NH_3)_6]^{3+/2+}$ process should ideally fit the theoretical predictions for an electrochemically reversible process.

Table 2.2 summarizes the data extracted from DC cyclic voltammograms obtained for the reduction of $[Ru(NH_3)_6]^{3+}$ at the macrodisk glassy carbon (GC) electrode, and which are needed for assessing the electrode kinetics when using the classical approach developed by Nicholson [11]. Midpoint potentials derived from averages of the reductive and oxidative peak potentials are independent of the scan rate, and provide a good estimate of E^0. Peak-to-peak separations (ΔE_p) approach the nearly ideal values expected for a reversible process at the medium scan rates, followed by a minor increase at higher values of v. When the potential sweep rate is very low (<0.02 V s^{-1}), mass-transport becomes affected by convection caused by gravity and building vibration, which is inevitably present in the vast majority of laboratories, resulting in a steady-state component and a minor increase in ΔE_p (Table 2.2).

The reduction peak current, normalized to $v^{1/2}$, is almost constant as predicted by the Randles–Ševčik relationship, and the slope of the linear $I_p(v^{1/2})$ dependence can be applied to estimate the diffusion coefficient of $[Ru(NH_3)_6]^{3+}$. By using the D-value estimated in this manner, and by applying the Nicholson method (see Table 2.1) to the ΔE_p data given in Table 2.2, the apparent k^0_{app} values for the $[Ru(NH_3)_6]^{3+/2+}$ process could be reported to be on the order of 0.1 cm s^{-1}. This value is lower than k^0_{app} data reported in some early studies [43–45] and at least two orders of magnitude lower than has been provided in the contemporary reports [46]. This provides a lucid demonstration of one of the limitations of the Nicholson method for quantifying the electrode charge transfer kinetics. In order to reach the required level of accuracy of the Nicholson method, v needs to exceed 10^3 V s^{-1}, but at these scan rates and at millimolar levels of concentration of $[Ru(NH_3)_6]Cl_3$, currents at a macroelectrode are high enough to allow even 3–5 Ω of uncompensated resistance to cause a significant IR_u-drop and hence frustrate the ability to improve the accuracy of determining k^0.

The data in Table 2.2 provide two types of ΔE_p values derived from experiments using different potentiostats, one operating in an analogue regime but with a comparatively high potential step (1 mV), and another measuring voltammograms in the staircase mode at a slightly finer potential step (0.3 mV). The difference in ΔE_p and hence k^0_{app} values derived from using different instrumentation should serve as a warning for experimentalists, particularly when ΔE_p values are near to the reversible limit. A close to ideal reversible ΔE_p value of 0.058–0.059 at $v = 0.112$ V s^{-1} is obtained for the same experiment described in Table 2.2 when using the potentiostat operating in staircase mode but with a potential step as small as 1 μV.

Successful simulations of the experimental DC cyclic voltammograms obtained from the reduction of $[Ru(NH_3)_6]^{3+}$ can in fact be achieved using the k^0_{app} values derived from Nicholson method (Figure 2.10). The remarkably good agreement between the simulated and experimental data exemplified in Figure 2.10 can be partly attributed to using a diffusion model that accounts for the "edge" effect present at a macrodisk electrode. However, it emerged that cyclic voltammograms simulated using any k^0 value above 0.1 cm s^{-1} would differ only slightly from the experimental curves, whereas those acquired under the assumption of lower k^0 values would be in obvious disagreement with the experiment, the difference being more pronounced at higher scan rates (Figure 2.10).

Figure 2.11a shows the effect of the scan rate on the limiting m value of the semiintegral, which does not provide the expected potential-independent "plateau" but rather increases gradually with potential at both $v = 0.1$ and 1.0 V s^{-1}. Subtraction of the background current, which can include electronic noise and a capacitive component, improves the constancy of limiting value of the semiintegral (Figure 2.11b), although the magnitude still exceeds that estimated on the basis of D determined from the Randles–Ševčik relationship and simulation of DC voltammetry (as described above), because of the edge diffusion effect (see Section 1.3). The semiintegral curves shown in Figure 2.11 demonstrate a minimal separation between the forward and backward sweeps in the region of the half-wave potential, in accordance with the reversibility of electrode reaction. k^0_{app} estimated

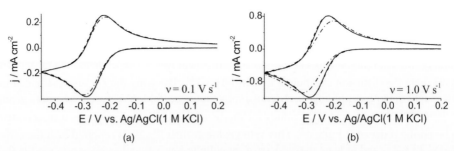

Figure 2.10 *Comparison of experimental background-corrected DC cyclic voltammograms obtained with a 3 mm diameter GC electrode at $v = 0.1$ (a) and 1.0 V s^{-1} (b) for reduction of 2 mM $[Ru(NH_3)_6]^{3+}$ in aqueous 1 M H_2SO_4 (—) with data simulated using $k^0 = 0.1$ (– – –) and 0.01 cm s^{-1} (– · – ·). Other simulation parameters: two-dimensional diffusion model; $E^0 = -0.251$ V; $\alpha = 0.5$; $D = 5 \times 10^{-6}$ cm^2 s^{-1}; $R_u = 5\ \Omega$; $C_{DL} = 0\ \mu F$ cm^{-2}; $T = 298$ K. Experimental and (– – –) simulated data are indiscernible.*

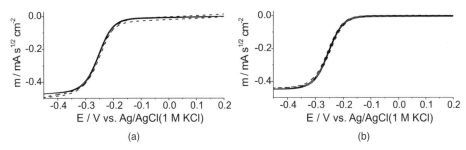

Figure 2.11 *Semiintegrals of (a) raw and (b) background-corrected experimental DC cyclic voltammograms obtained with a 3 mm GC electrode at v = 0.1 (—) and 1.0 V s⁻¹ (– – –) for reduction of 2 mM [Ru(NH₃)₆]³⁺ in aqueous 1 M H₂SO₄.*

from these semiintegral data was 0.05 cm s^{-1}, although by analogy with an analysis of transient voltammetry, this really implies that $k^0_{app} \geq 0.05$ cm s^{-1}.

Experimental FT AC voltammetric data for the reduction of $[Ru(NH_3)_6]^{3+}$ in aqueous 1 M KCl (Figure 2.12) were obtained using a sine wave perturbation having an amplitude of 0.08 V and frequencies of 9 and 219 Hz. Fitting these experimental data with simulated curves using a heuristic approach [34] produces an outcome where any k^0 value above

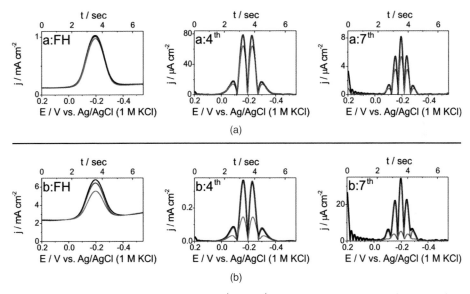

Figure 2.12 *Comparison of fundamental, 4ᵗʰ and 7ᵗʰ harmonic components of FT AC voltammograms obtained with a 3 mm GC electrode at f = 9 (a) and 219 Hz (b) for reduction of 1 mM [Ru(NH₃)₆]³⁺ in aqueous 1 M KCl (black) and data simulated using k⁰ = 5 (a) and 30 cm s⁻¹ (b) (blue) and k⁰ = 0.1 cm s⁻¹ (red). Experimental parameters: ΔE = 0.08 V, v = 0.112 V s⁻¹; Rᵤ = 5 Ω. Simulation parameters: potential-dependent nonlinear C_DL; α = 0.5; E⁰ = −0.194 V; D = 5.9 × 10⁻⁶ cm² s⁻¹. Experimental and blue simulated data are indiscernible in higher harmonics.*

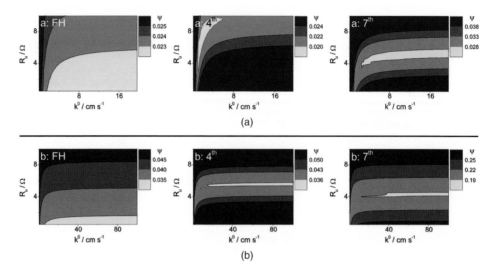

Figure 2.13   *Contour maps for relative difference (Ψ) between the fundamental, 4th and 7th AC harmonic components of the experimental FT AC voltammograms obtained using a 3 mm GC electrode for reduction of 1 mM [Ru(NH$_3$)$_6$]$^{3+}$ in aqueous 1 M KCl at f = 9 (a) and 219 Hz (b) and theoretical data simulated using R_u and k^0 varied over 0.5–10 Ω and 0.1–20 (9 Hz) or 1–100 cm s^{-1} (219 Hz) ranges, respectively. Other parameters are analogous to those listed in the caption for Figure 2.12.*

4 and 30 cm s^{-1} at $f = 9$ and 219 Hz, respectively, provides "good" agreement between experiment and theory. In contrast to a somewhat arbitrary heuristic approach, Nimrod *e*-science analysis provides an unbiased assertion of the quality of fit. This approach is illustrated in Figure 2.13, using a "sweep" experiment that reports the deviation of theory from experiment in terms of relative root mean square values, Ψ (Equation 2.6), for the array of combinations of k^0 and R_u. The lowest Ψ values are indicated in yellow. An increase in either the AC harmonic order or frequency of the AC perturbation – that is, a decrease in the effective time scale of the experiment, shifts the "lower limit" of k^0 for the [Ru(NH$_3$)$_6$]$^{3+/2+}$ process to higher values, which reaches ca. 50–60 cm s^{-1} in the seventh harmonic component of the FT AC voltammogram with $f = 219$ Hz.

Apparently, k^0 for the [Ru(NH$_3$)$_6$]$^{3+/2+}$ process is also too fast to be measured by the FT AC voltammetric method. Consequently, this reaction must be classified as a reversible process under all DC and AC transient voltammetric conditions examined in this chapter. Significantly, it can now be concluded that the results obtained with macro and nanodisk [46] electrodes under transient and steady-state conditions, respectively, are in agreement for the quantification of very fast [Ru(NH$_3$)$_6$]$^{3+/2+}$ electrode kinetics.

2.3.2   Oxidation of Fe^{II}(C$_5$H$_5$)$_2$ in an Aprotic Solvent

Ferrocene is a classical metallocene capable of a fast one-electron oxidation to ferricinium. The "sandwich" structure should ensure a minimal effect of the solvent and support-ing electrolyte on the E^0 value of the Fc$^{0/+}$ process, and IUPAC has suggested this couple as a universal reference potential for studies in nonaqueous media. Analogous to the $[Ru(NH_3)_6]^{3+/2+}$ process, the reported k^0_{app} values for the electro-oxidation of Fc demonstrate a continuous increase with the evolution of instrumentation. However, in contrast to studies on the $[Ru(NH_3)_6]^{3+/2+}$ process in aqueous electrolyte media, the determination of k^0_{app} for the $Fc^{0/+}$ process is usually hampered by a significant IR_u-drop, which is difficult to avoid in nonaqueous solvents, even in the presence of excess electrolyte (usually tetraalkylammonium salts). When using a Luggin capillary to position the (quasi) reference electrode in the vicinity of the surface of the working macrodisk electrode, the uncompensated resistance might be minimized to about 50 Ω, but values below 10 Ω, which are easily achieved in aqueous electrolytes, are hardly accessible.

It is the significant contribution of R_u that makes the ΔE_p values extracted from the DC cyclic voltammetric data obtained with macrodisk electrode for oxidation of Fc notably higher than those determined for the reduction of $[Ru(NH_3)_6]^{3+}$ (*cf.* Tables 2.2 and 2.3). However, k^0_{app} values calculated on the basis of the Nicholson method from ΔE_p for the $Fc^{0/+}$ process emerge as similar to k^0_{app} values estimated for $[Ru(NH_3)_6]^{3+/2+}$ when using the same approach. This is probably due to a notably higher D value for Fc in acetonitrile as compared to that of $[Ru(NH_3)_6]^{3+}$ in aqueous medium [see Equation (2.3)]. Processing the semiintegrals of the background-corrected cyclic voltammograms measured for the $Fc^{0/+}$ process produces k^0_{app}-values as low as about 0.02 cm s^{-1}. In this case, any inaccuracy of the calculations will be aggravated by the influence of appreciable R_u, which was not taken into account in the calculations. In principle, when the exact value of R_u is known, a correction of the experimental data is possible [22].

The ability to account for the influence of R_u, as provided by the majority of contemporary electrochemical simulators, should allow the correct estimation of k^0. However, as in the case of the reduction of $[Ru(NH_3)_6]^{3+}$, the shape of DC voltammograms simulated for the

Table 2.3 *Peak-to-peak separation (ΔE_p) and k^0_{app} values estimated on the basis of Nicholson's treatment [11] using cyclic voltammetric data obtained for oxidation of 1 mM Fc in CH$_3$CN 0.1 M (n-Bu)$_4$NPF$_6$ and reduction of 1 mM [Fe(CN)$_6$]$^{3-}$ in aqueous 1 M KCl with a 3 mm diameter GC electrode.*

	Fc$^{0/+}$		[Fe(CN)$_6$]$^{3-/4-}$	
v (V s^{-1})	ΔE_p (V)	k^0_{app} (cm s^{-1})	ΔE_p (V)	k^0_{app} (cm s^{-1})
0.075[a]	0.064	0.09	n.m.	n.m.
0.089[a]	n.m.	n.m.	0.087	0.008
0.10[b]	0.074 / 0.064	0.03 / 0.10	0.108	0.005
0.40[b]	0.075 / 0.070	0.05 / 0.08	0.138	0.005
0.60[b]	n.m. / 0.070	n.m. / 0.10	0.150	0.005
1.0[b]	0.093 / 0.072	0.04 / 0.11	0.172	0.006

[a] ΔE_p values were obtained in a staircase regime with potential step of ca. 1 μV.
[b] For Fc$^{0/+}$, ΔE_p values were obtained in an analogue instrumental regime with potential step of 1 mV / staircase regime with potential step of ca. 0.3 mV. For [Fe(CN)$_6$]$^{3-/4-}$, ΔE_p values were obtained in an analogue regime with a potential step of 1 mV.
n.m., not measured.

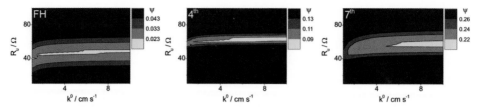

Figure 2.14   *Contour maps for relative difference (Ψ) between the fundamental, 4th and 7th AC harmonic components of the experimental FT AC voltammograms obtained using a 3 mm GC electrode for oxidation of 1 mM Fc in CH_3CN 0.1 M $(n\text{-}Bu)_4NPF_6$ at $f = 72$ Hz and theoretical data simulated using R_u and k^0 varied over the 10–100 Ω and 1–10 cm s^{-1} ranges, respectively. Other parameters: $v = 0.075$ V s^{-1}; potential-dependent nonlinear C_{DL}; $\alpha = 0.5$; $E^0 = 0.0$ V; $D = 2.4 \times 10^{-5}$ cm^2 s^{-1}.*

oxidation of Fc are insensitive to changes in k^0, when its value exceeds about 0.1 cm s^{-1}. Heuristic fitting of the experimental AC voltammetric data obtained at low frequency (9 Hz) with simulated curves predicts that the k^0 of the Fc$^{0/+}$ process would exceed 0.5 cm s^{-1}, while an increase of f to 72 Hz would shift this lower limit to 3 cm s^{-1}. An analysis of the FT AC voltammetric data with the aid of e-science tools (Figure 2.14) again does not provide a precise value of k^0 and, analogously to [Ru(NH$_3$)$_6$]$^{3+/2+}$, predicts that the Fc$^{0/+}$ process is extremely fast ($k^0 \geq 5$ cm s^{-1}) and can be regarded as reversible. At the same time, the contour maps exhibited in Figure 2.14 yield a well-determined value of R_u, which is in a perfect agreement with that measured using EIS.

2.3.3   Reduction of [Fe(CN)$_6$]$^{3-}$ in an Aqueous Electrolyte

The one-electron electroreduction of hexacyanoferrate(III) to hexacyanoferrate(II) (or ferricyanide to ferrocyanide) is a classic example of a quasi-reversible process. Despite being initially regarded as an outer-sphere adiabatic charge-transfer process [44], the electrode kinetics of the [Fe(CN)$_6$]$^{3-/4-}$ transformation exhibits a pronounced dependence on the properties of the electrode, and reproducible results can be achieved only when the working electrode surface is reproducibly pretreated and cleaned, either by polishing or electrochemically [47, 48]. Metal electrodes provide notably faster heterogeneous electron transfer kinetics of [Fe(CN)$_6$]$^{3-/4-}$ as compared to carbon surfaces, possibly due to differences in the density of electronic states (DOS) for these materials, which can impose a notable influence on the rate of adiabatic charge transfer [49].

Application of the Nicholson method to the DC voltammetric data obtained for the reduction of [Fe(CN)$_6$]$^{3-}$ in a highly-conducting 1 M KCl aqueous electrolyte medium produces a k^0-value of about 0.006 cm s^{-1} (Table 2.3) and, not unexpectedly, in this case the k^0-value is in a close agreement with the estimates made by applying semiintegral analysis (0.006 cm s^{-1}) and a heuristic fitting of simulation (0.007 cm s^{-1}) to the same experimental DC data set. A heuristic theory-experiment comparisons of the FTACV data also produced a k^0-value of 0.007 cm s^{-1}. Finally, contour maps displaying relative differences between the harmonic components of the AC voltammograms obtained for the [Fe(CN)$_6$]$^{3-/4-}$ process now provide a definitive answer on the magnitude of k^0 for a given set of experimental data (Figure 2.15).

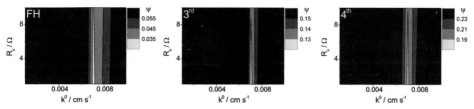

Figure 2.15   *Contour maps for relative difference (Ψ) between the fundamental, 3rd and 4th AC harmonic components of the experimental FT AC voltammograms obtained using a 3 mm GC electrode for reduction of 1 mM [Fe(CN)$_6$]$^{3-}$ in aqueous 1 M KCl electrolyte at f = 9 Hz and theoretical data simulated using R_u and k^0 varied over the 1–10 Ω and 10^{-3} to 10^{-2} cm s^{-1} ranges, respectively. Other parameters: v = 0.089 V s^{-1}; potential-dependent nonlinear C_{DL}; $\alpha = 0.5$; $E^0 = 0.216$ V; D = 7.8 $\times$ 10^{-6} cm^2 s^{-1}.*

2.4   Conclusions and Outlook

By summarizing the apparent heterogeneous charge transfer rate constants determined in Section 2.3 by processing DC and AC voltammetric data obtained for the reduction of [Ru(NH$_3$)$_6$]$^{3+}$ and [Fe(CN)$_6$]$^{3-}$ and oxidation of Fc, the data in Table 2.4 demonstrate the evolution of sensitivity of voltammetric kinetic analysis at macrodisk electrodes. A remarkable refinement of the sensitivity of the voltammetric techniques for quantification of kinetics is clearly observed from the data reported for [Ru(NH$_3$)$_6$]$^{3+/2+}$. Comparison of the lower limits for k^0_{app} by the analysis of AC voltammetry for [Ru(NH$_3$)$_6$]$^{3+/2+}$ and Fc$^{0/+}$ should again remind the reader of the importance of minimizing of R_u in any form of voltammetric analysis. Finally, the [Fe(CN)$_6$]$^{3-/4-}$ case shows that all of the specified voltammetric methods provide a uniformly consistent value for k^0_{app} for a quasi-reversible process that is well removed from the reversible limit and seemingly is not burdened with any interfering factors, other than the impact of electrode surface treatment.

The approach used in the analysis of voltammetric data, and which has evolved over the past 50 years, is briefly surveyed in schematic form in Figure 2.16. At this point, it is deemed necessary to emphasize the importance of varying the concentration in any experiment aimed at studying the kinetics – a simple experimental rule which has often been overlooked in many years. Furthermore, as shown in the final step of the diagram in Figure 2.16, an experimentalist should always seek to reconcile conclusions from voltammetric analysis with the results obtained from independent methods.

Table 2.4 k^0_{app} *(cm s^{-1}) for the reduction of [Ru(NH$_3$)$_6$]$^{3+}$ and [Fe(CN)$_6$]$^{3-}$ ($R_u \leq 10\ \Omega$) and oxidation of Fc ($R_u \approx 60\ \Omega$) estimated by voltammetric analysis at macrodisk electrodes.*

	DC voltammetry			FT AC voltammetry	
Process	Nicholson	Simulation	Semiintegral	Heuristic	e-Science
[Ru(NH$_3$)$_6$]$^{3+/2+}$	>0.1	>0.1	0.05	≥ 30	≥ 60
Fc$^{0/+}$	>0.1	>0.1	0.02	≥ 3	≥ 5
[Fe(CN)$_6$]$^{3-/4-}$	0.006	0.007	0.006	0.007	0.007

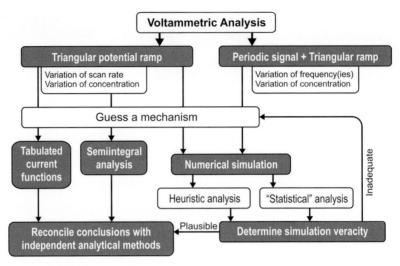

Figure 2.16 *Generalized strategy for voltammetric analysis.*

Foreseeable improvements in the AC voltammetric technique at the instrumental level are probably limited to the enhancement of accessible frequency limits and a wider employment of more sophisticated, specifically designed waveforms. However, it is believed that major advances are available in the data processing stage. The next generation of experimental data analysis strategies could, for example, be focused on a more extensive exploitation of patterns of behavior available in voltammetry and, ultimately, automated pattern recognition. Theoretically, pattern-recognition algorithms could be used to assist in the processing of voltammetric data, especially with FT AC voltammetry using computer-optimized waveforms. In widely employed X-ray structural analysis, experimental X-ray diffraction patterns are compared with those predicted theoretically for thermodynamically feasible structures in order to establish the crystal structure of the analyzed material. A similar strategy could be implemented in the processing of voltammetric data via recognition of the best-fitting electrochemical mechanism from several preset scenarios, for example, immeasurably fast or quasi-reversible electron transfer involving dissolved or surface-confined species, electron transfer coupled to homogeneous chemical process, adsorption, and catalytic schemes. After defining a plausible reaction scheme the program could predict, by means of simulation, the kinetic and thermodynamic parameters of the reactions, as well as parameters such as uncompensated resistance and double-layer capacitance. If the outcome has no obvious physical meaning, then the analysis should be repeated after the introduction of some justified constraints. Decisions on placing constraints on the selected parameters when analyzing voltammetric data are always important in terms of avoiding under- or overparameterization in a model used for the description of the experiment. Clearly, it is preferable to determine as many parameters as possible by independent methods, for example, to measure R_u and estimate C_{DL} by EIS, and to determine diffusion coefficients from near steady-state voltammetry.

The problem of overparameterization can be exemplified by processing the FT AC voltammetric data described above for the reduction of $[Ru(NH_3)]^{3+}$ using an "inappropriate" quasi-reversible rather than reversible model with the aid of the Nimrod/O simplex optimization algorithm. As described in Section 2.2, the Nimrod/O *e*-science tool seeks the "best" values of the unknown parameters in the user-specified range by minimizing the difference between experiment and theory. When the low-frequency ($f = 9$ Hz) and hence not highly kinetically sensitive FT AC voltammetric data are processed with Nimrod/O for the simultaneous determination of E^0, α, k^0 and R_u parameters, the minimal Ψ value (Ψ_{min}) of 0.017 is achieved with obviously far too-low value of k^0 of, for example, 0.4 cm s^{-1}. However, as soon as k^0 is set to the reversible limit, viz. 10^4 cm s^{-1}, the lowest Ψ_{min} value found by Nimrod/O is slightly higher and is now 0.025; that is, the quasi-reversible model is incorrectly predicted to be more favorable than the reversible model. Importantly, simplex optimization of the higher frequency ($f = 219$ Hz) FT ACV data obtained for the $[Ru(NH_3)]^{3+/2+}$ process, where k^0 was either optimized in the range from 1 to 100 cm s^{-1} or was set to 10^4 cm s^{-1}, recovers remarkably close Ψ_{min} values of 0.048 for both models; that is, the reversible model should be used to model the process, with only E^0-values being reported. Analogous conclusions are obtained when applying simplex optimization to the Fc$^{0/+}$ FT ACV data, using both quasi-reversible and reversible models.

Apart from illustrating the importance of the judicious model choice and parameterization used in simulations, the examples provided above highlight the consequences of choosing inappropriate experimental conditions and data analysis combinations. Arguably, a wider implementation of the computer-optimized designed waveforms, the development of "intelligent" pattern-recognition algorithms and the use of advanced statistical methods will help to avoid the pitfalls that are readily encountered in many forms of quantitative voltammetric analysis used today.

References

(1) Brett, C.M.A. and Brett, A.M.O. (1993) *Electrochemistry: Principles, Methods and Applications*, Oxford University Press, New York.
(2) Bard, A.J. and Faulkner, L.R. (2001) *Electrochemical Methods: Fundamentals and Applications*, 2nd edn, John Wiley & Sons, New York.
(3) Zoski, C.G. (2007) *Handbook of Electrochemistry*, Elsevier B.V., Amsterdam.
(4) Fleischmann, M. and Thirsk, H.R. (1955) An investigation of electrochemical kinetics at constant overvoltage. The behaviour of the lead dioxide electrode. Part 5. The formation of lead sulphate and the phase change to lead dioxide. *Transactions of the Faraday Society*, **51**, 71–95.
(5) IUPAC Commission on Electroanalytical Chemistry (1976) Recommendations for sign conventions and plotting of electrochemical data. *Pure and Applied Chemistry*, **45**, 131–134.
(6) (a)Butler, J.A.V. (1924) Studies in heterogeneous equilibria. Part II – The kinetic interpretation of the Nernst theory of electromotive force. *Transactions of the Faraday Society*, **19**, 729–733; (b)Butler, J.A.V. (1924) Studies in heterogeneous equilibria. Part III – A kinetic theory of reversible oxidation potentials at inert electrodes, *Transactions of the Faraday Society*, **19**, 734–739;(c)Erdey-Gruz, T. and Volmer, M. (1930) The theory of hydrogen high tension, *Zeitschrift fur Physikalische Chemie - Leipzig*, **150**, 203–213.
(7) (a)Marcus, R.A. (1964) Chemical and electrochemical electron-transfer theory. *Annual Review of Physical Chemistry*, **15**, 155–196;(b)Marcus, R.A. (1965) On the theory of electron-transfer reactions. VI. Unified treatment for homogeneous and electrode reactions. *Journal of Physical Chemistry*, **43**, 679–701.
(8) (a)Hush, N.S. (1958) Adiabatic rate processes at electrodes. I. Energy-charge relationships. *Journal of Chemical Physics*, **28**, 962–972;(b)Hush, N.S. (1999) Electron transfer in retrospect and prospect 1: Adiabatic electrode processes. *Journal of Electroanalytical Chemistry*, **470**, 170–195.

(9) C.J. Miller, Homogeneous electron transfer at metallic electrodes, in *Physical Electrochemistry, Principles, Methods and Applications* (ed. I. Rubinstein), Monographs in Electroanalytical Chemistry and Electrochemistry, Marcel Dekker, New York, 1995, p. 27.

(10) Nicholson, R.S. and Shain, I. (1964) Theory of stationary electrode polarography single scan and cyclic methods applied to reversible, irreversible, and kinetic systems. *Analytical Chemistry*, **36**, 706–723.

(11) Nicholson, R.S. (1965) Theory and application of cyclic voltammetry for measurement of electrode reaction kinetics. *Analytical Chemistry*, **37**, 1351–1355.

(12) Barrosse-Antle, L.E., Bond, A.M., Compton, R.G. *et al.* (2010) Voltammetry in room temperature ionic liquids: comparisons and contrasts with conventional electrochemical solvents. *Chemistry – An Asian Journal*, **5**, 202–230.

(13) Roe, D.K. (1996) Overcoming solution resistance with stability and grace in potentiostatic circuits, in *Laboratory Techniques in Electroanalytical Chemistry* (eds P.T. Kissinger and W.R. Heineman), Marcel Dekker, New York.

(14) Rudolph, M., Reddy, D.P. and Feldberg, S.W. (1994) A simulator for cyclic voltammetric responses. *Analytical Chemistry*, **66**, 589A–600A.

(15) http://www.gamry.com/products/digielch-electrochemical-simulation-software/. Accessed 25 March 2014.

(16) http://www.comsol.com/products/multiphysics/. Accessed 25 March 2014.

(17) Klymenko, O.V. (2010) Theoretical study of the EE reaction mechanism with comproportionation. *Electrochemistry Communications*, **12**, 1378–1382 and references therein.

(18) http://www.garethkennedy.net/MECSim.html. Accessed 25 March 2014.

(19) Speiser, B. (1996) Numerical simulation of electroanalytical experiments: recent advances in methodology, in *Electroanalytical Chemistry, A Series Of Advances* (eds A.J. Bard and I. Rubinstein), Marcel Dekker, New York, vol. **19**, p. 1.

(20) Feldberg, S.W. (2004) Simulation of electroanalytical systems ~ 1962–2002. A personal perspective: from ad hoc simulations to a generalized simulator, in *Historical Perspectives on the Evolution of Electrochemical Tools* (eds J. Leddy, P. Vanýsek and V. Birss), The Electrochemical Society, New Jersey, USA, p. 191.

(21) Oldham, K.B., Myland, J. and Bond, A.M. (2011) *Electrochemical Science and Technology: Fundamentals and Applications*, John Wiley & Sons and references therein.

(22) Imbeaux, J.C. and Savéant, J.M. (1973) Convolutive potential sweep voltammetry. I. Introduction. *Journal of Electroanalytical Chemistry*, **44**, 169–187 and following papers in this volume of *Journal of Electroanalytical Chemistry*.

(23) Bond, A.M., Henderson, T.L.E. and Oldham, K.B. (1985) A study of electrode kinetics by global analysis of single electrochemical experiment. *Journal of Electroanalytical Chemistry*, **191**, 75–90.

(24) Zoski, C.G., Oldham, K.B. and Mahon, P.J. *et al.* (1991) Global kinetic analysis of cyclic voltammograms at a spherical electrode. *Journal of Electroanalytical Chemistry*, **297**, 1–17.

(25) Mahon, P.J. and Oldham, K.B. (1999) Convolutive modelling of electrochemical processes based on the relationship between the current and the surface concentration. *Journal of Electroanalytical Chemistry*, **464**, 1–13.

(26) Mahon, P.J. (2010) Application of global analysis for the measurement of electrochemical parameters with disk electrodes. *Electrochimica Acta*, **55**, 673–680.

(27) Andrieux, C.P., Garreau, D., Hapiot, P., Pinson, J. and Savéant, J.M. (1988) Fast sweep cyclic voltammetry at ultra-microelectrodes. Evaluation of the method for fast electron-transfer kinetic measurements. *Journal of Electroanalytical Chemistry*, **243**, 321–335.

(28) Oldham, K.B. (1981) Edge effects in semiinfinite diffusion. *Journal of Electroanalytical Chemistry*, **122**, 1–17.

(29) Mahon, P.J. (2011) Convolutive reshaping as a way to simulate voltammetry at disk electrodes. *Electrochimica Acta*, **56**, 2190–2200.

(30) Bond, A.M., Duffy, N.W., Guo, S.-X. *et al.* (2005) Changing the look of voltammetry. Can FT revolutionize voltammetric techniques as it did for NMR. *Analytical Chemistry*, **77**, 186A–195A.

(31) Bond, A.M., Duffy, N.W., Elton, D.M. and Flemming, B.D. (2009) Characterization of nonlinear background components in voltammetry by use of large amplitude periodic perturbations and Fourier transform analysis. *Analytical Chemistry*, **81**, 8801–8808.

(32) Sher, A.A., Bond, A.M., Gavaghan, D.J. *et al.* (2005) Fourier transformed large amplitude square-wave voltammetry as an alternative to impedance spectroscopy: evaluation of resistance, capacitance and electrode kinetic effects via an heuristic approach. *Electroanalysis*, **17**, 1450–1462.

(33) Tan, Y., Stevenson, G.P., Baker, R.E. *et al.* (2009) Designer based Fourier transformed voltammetry: A multi-frequency, variable amplitude, sinusoidal waveform. *Journal of Electroanalytical Chemistry*, **634**, 11–21.

(34) Mashkina, E. and Bond, A.M. (2011) Implementation of a statistically supported heuristic approach to alternating current voltammetric harmonic component analysis: re-evaluation of the macrodisk glassy carbon electrode kinetics for oxidation of ferrocene in acetonitrile. *Analytical Chemistry*, **83**, 1791–1799.

(35) D. Abramson, R. Sosic, J. Giddy and B. Hall (1995) Nimrod: A Tool for Performing Parametised Simulations using Distributed Workstations, The 4th IEEE Symposium on High Performance Distributed Computing, Virginia, August 1995., pp. 112–121.

(36) http://messagelab.monash.edu.au. Accessed 5 February 2014.

(37) Chambers, J., Bethwaite, B., Diamond, N. *et al.* (2012) Parametric computation predicts a multiplicative interaction between synaptic strength parameters that controls properties of gamma oscillations. *Frontiers in Computational Neuroscience*, **6**. Article 53.

(38) Faux, N., Beitz, A., Bate, M. *et al.* eResearch Solutions for High Throughput Structural Biology, *Third IEEE International Conference on e-Science and Grid Computing, 10–13 December 2007,* CS Press, Bangalore, India, pp. 221–227.

(39) Lynch, A.H. (2007) Influence of savanna fire on Australian monsoon season precipitation and circulation as simulated using a distributed computing environment. *Geophysical Research Letters*, **34**, L20801.

(40) Mashkina, E., Peachey, T., Lee, C.-Y. *et al.* (2013) Estimation of electrode kinetic and uncompensated resistance parameters and insights into their significance using Fourier transformed ac voltammetry and e-science software tools. *Journal of Electroanalytical Chemistry*, **690**, 104–110.

(41) Abramson, D., Giddy, J. and Kotler, L. High-performance parametric modeling with Nimrod/G: Killer application for the Global Grid? International Parallel and Distributed Processing Symposium (IPDPS), Cancun, Mexico, May 2000, pp. 520–528.

(42) Lewis, A., Abramson, D. and Peachey, T. (2006) RSCS: A parallel simplex algorithm for the Nimrod/O optimization toolset. *Scientific Programming*, **14**, 1–11.

(43) Gennett, T. and Weaver, M.J. (1984) Reliability of standard rate constants for rapid electrochemical reactions. *Analytical Chemistry*, **56**, 1444–1448.

(44) Deakin, M., Stutts, K.J. and Wightman, R.M. (1985) The effect of pH on some outer-sphere electrode reactions. *Journal of Electroanalytical Chemistry*, **182**, 113–122.

(45) Beriet, C. and Pletcher, D. (1994) A further microelectrode study of the influence of electrolyte concentration on the kinetics of redox couples. *Journal of Electroanalytical Chemistry*, **375**, 213–218.

(46) Sun, P. and Mirkin, M.V. (2006) Kinetics of electron-transfer reactions at nanoelectrodes. *Analytical Chemistry*, **78**, 6526–6534.

(47) Heiduschka, P. and Dittrich, J. (1992) Impedance spectroscopy and cyclic voltammetry at bare and polymer coated glassy carbon electrodes. *Electrochimica Acta*, **37**, 2573–2580.

(48) Patel, A.N., Collignon, M.G. and O'Connell, M.A. *et al.* (2012) A new view of electrochemistry at highly oriented pyrolytic graphite. *Journal of the American Chemical Society*, **134**, 20117–20130 and references therein.

(49) Khoshtariya, D.E., Dolidze, T.D., Zusman, L.D., Waldeck, D.H. (2001) Observation of the turnover between the solvent friction (overdamped) and tunneling (nonadiabatic) charge-transfer mechanisms for a $Au/Fe(CN)_6^{3-/4-}$ electrode process and evidence for a freezing out of the Marcus barrier. *Journal of Physical Chemistry A*, **105**, 1818–1829.

3

Electrocrystallization: Modeling and Its Application

Morteza Y. Abyaneh
University of Uppsala, Department of Philosophy, Sweden

Mathematical modeling of the electrocrystallization processes and its applications has long been a subject dear to Martin Fleischmann. Indeed, both his earliest papers (e.g., Refs [1,2]) and his last two publications [3,4] dealt with this topic.

The study of nucleation is fundamental to the understanding of crystallization. "Heterogeneous" and "homogeneous" nucleation, are terms proposed [5] to differentiate nucleation within a receptive and an inert environment. In the context of electrocrystallization, the terms can be applied to phase formation at preferred sites on the electrode surface and phase formation at surfaces without such sites, respectively. Figure 3.1 [6] illustrates heterogeneous nucleation and shows a scanning electron microscopy (SEM) image of the nuclei of nickel formed on a scratched surface, and on indents.

Two models of nucleation are presented in Figure 3.2: a heterogeneous model (nucleation on an indent); and a spherical-cap model representing homogeneous nucleation. The critical free energy for the formation of the nucleus within the indent and on the smooth surface of the electrode is given by (M. Y. Abyaneh, unpublished results):

$$\left(\Delta G_{\text{het}}^*\right)_{2D} = \frac{8\pi M^2 \sigma_{13}^3 \left[\cos\left[\frac{\alpha - 3\phi}{2}\right] + 3\cos\left[\frac{\alpha - \phi}{2}\right]\right] \csc\left(\frac{\alpha}{2}\right) \sin^3\left(\frac{\phi}{2}\right)}{3(\rho N_A z e \eta)^2} \tag{3.1}$$

$$\left(\Delta G_{\text{hom}}^*\right)_{2D} = \frac{4\pi M^2 \sigma_{13}'^3 [2 - 3\cos(\theta') + \cos^3(\theta')]}{3(\rho N_A z e \eta)^2} \tag{3.2}$$

Developments in Electrochemistry: Science Inspired by Martin Fleischmann, First Edition.
Edited by Derek Pletcher, Zhong-Qun Tian and David E. Williams.
© 2014 John Wiley & Sons, Ltd. Published 2014 by John Wiley & Sons, Ltd.

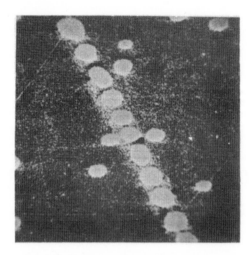

Figure 3.1   *Micrograph of the nuclei of nickel on the surface of vitreous carbon at a magnification of ×5000, showing the preferred nucleation on a particular scratch and on indents; deposition from Watts bath at 50 °C + 10^{-3} M naphthalene-2-sulfonic acid + 10^{-3} M coumarin [6].*

When $\alpha = \pi$, the surface of the electrode is flat and, as expected, Equation (3.1) is reduced to Equation (3.2) for $\phi = \theta'$. If $\alpha = 0$, there is no indent and Equation (3.1) is redundant, as demonstrated by the fact that $\csc(\frac{\alpha}{2}) = 1/0$. The symbols within the above equations are defined either in Figure 3.2 or in Table 3.1. Figure 3.3 shows the manner in which $(\Delta G^*_{\text{het}})_{2D}$ moves towards $(\Delta G^*_{\text{hom}})_{2D}$ as α is gradually increased from 0 to π.

The neatest and most effective way of investigating nucleation and two-dimensional (2D) growth is to record the current as a function of time when a constant potential is applied to a finite (but small) electrode substrate [7]. The prediction of the shape of the current–time transient recorded in a single experiment depends on the location of the nucleus on the electrode surface. However, such a prediction can be made from a repeated number of identical experiments [8, 9].

The first obstacle in modeling the kinetics of electrocrystallization processes is formulating the coverage of an electrode by the depositing phase, Θ. Two geometric representations

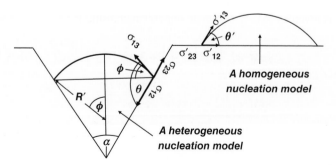

Figure 3.2   *Nucleation on a fault-free surface of an electrode (right-hand side), and nucleation within an indented surface (left-hand side).*

Table 3.1 *Symbols used in text and/or defined in Figure 3.2. Subscripts α and β refer to different phases.*

Symbol	Unit	Definition
A	cm^{-2} s^{-1}	The nucleation rate constant in the hypothetical absence of growth
A'	s^{-1}	The expected frequency of the formation of a nucleus at a given site
A_r		A_β/A_α
c_0	mol cm^{-3}	Bulk concentration
D	cm^2 s^{-1}	Diffusion coefficient of the depositing species
E		The expectation that a given point is covered by growing phase.
F	C mol^{-1}	Faraday constant
h	cm	Height of a monolayer
I		$\dfrac{j}{zFkh}\sqrt{A'/\pi A}$
j	A cm^{-2}	Current density
j_1	A cm^{-2}	Current density at time t_1, resulting from H$_2$ evolution on top of the monolayer
k	mol cm^{-2} s^{-1}	Rate constant for crystal growth in the direction parallel to the electrode surface
k'	mol cm^{-2} s^{-1}	Rate constant for crystal growth in the direction perpendicular to the electrode surface
k_r, k_r'		k_β/k_α, k_β'/k_α'
$l\,(t)$		$j/(zFk_\alpha')$
M	g mol^{-1}	Molar mass of the deposit
N	cm^{-2}	Expected density of nuclei in the hypothetical absence of growth
N_0	cm^{-2}	Density of preferred sites on the electrode surface
N_A	mol^{-1}	Avogadro's number
P_α, P_β		$\dfrac{\pi M_\alpha^2 k_\alpha^2 A_\alpha}{3\rho_\alpha^2}$, $\dfrac{\pi M_\beta^2 k_\beta^2 A_\beta}{3\rho_\beta^2}$
r_c	cm	Critical size of a nucleus
R', α and ϕ		Defined in Figure 3.2.
$R = vt$	cm	Base radius of the growth centers
R_0	cm	Linear dimension of a growth center formed at $t = 0$
S	cm^2	Surface area of the overlapping diffusion zones
t	s	Deposition time
t_1	s	Time prior to the formation of 3D growth centers
T		$A'\gamma t$
$v = \dfrac{Mk}{\rho}$	cm s^{-1}	Speed at which centers grow laterally
z		Number of electrons transferred per ion
ze	C	Charge transferred per ion

(continued)

Table 3.1   (*Continued*)

Symbol	Unit	Definition
γ, κ		$v\sqrt{\pi A/A'}/A',\ \dfrac{k'}{k\tan\theta'}$
η	V	Overpotential
ρ	g cm^{-3}	Density of the deposit
σ_{13}	J cm^{-2}	Interfacial energy (Figure 3.2) for a heterogeneous nucleation
σ'_{13}	J cm^{-2}	Interfacial energy (Figure 3.2) for a homogeneous nucleation
θ and θ'		Contact angles which a nucleus makes with the electrode surface (Figure 3.2)
$\Theta_1, \Theta_2, \ldots, \Theta_n$		Coverage by *only* 1, 2, $\cdots$, or n layers

for the formulation of coverage and overlap on a macroelectrode can be imagined. The first (see Figure 3.4a) illustrates that [9].

$$\Theta = \Theta_1 + \Theta_2 + \Theta_3 + \cdots + \Theta_n + \cdots = \sum_{1}^{\infty} \frac{E^n e^{-E}}{n!} = 1 - \exp(-E) \tag{3.3}$$

where Θ_n is the expected coverage by <u>only</u> n circularly-based growth forms and E is the expectation that any given point on the electrode surface would be covered by the act of growth [10]. In the second geometric representation (Figure 3.4b), the concept is used of $S_{n,ext}$ representing the fractional coverage by <u>at least</u> n growth centers. Thus, [11]:

$$\Theta = S_{1,ext} - S_{2,ext} + S_{3,ext} - \cdots + (-1)^{n+1} S_{n,ext} + \cdots = 1 - \exp(-S_{1,ext}) \tag{3.4}$$

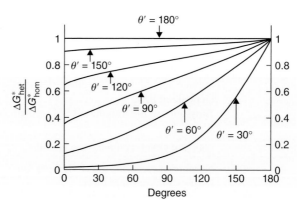

Figure 3.3   *The manner in which $(\Delta G^*_{het})_{2D}$ moves towards $(\Delta G^*_{hom})_{2D}$ for each 30° increase in θ' as α is gradually increased from 0 to π.*

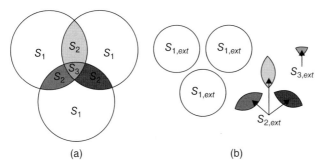

(a)   (b)

Figure 3.4 *Geometric representations of (a) the Evans (a) and (b) the Avrami approach to the statistical formulation of the coverage.*

Thus, for circularly-based growth forms $S_{1,ext}$ is identical with E. However, is this the case for all growth geometries? Figure 3.4 shows that [9] (the limits of the summation for the third equality in Eq. (3.7) of Ref. [9] should be from 1 to ∞):

$$S_{1,ext} = \sum_{1}^{\infty} n\Theta_n = \sum_{1}^{\infty} \frac{nE^n}{n!} e^{-E} = e^{-E} \sum_{0}^{\infty} \frac{(n+1)E^{(n+1)}}{(n+1)!} = Ee^{-E} \sum_{0}^{\infty} \frac{nE^n}{n!} = Ee^{-E}e^{E} = E$$

(3.5)

Since Equation (3.5) was derived independent of any assumptions regarding the shape of the growth centers, it follows that $S_{1,ext} = E$ for all growth forms. However, the use of the Avrami approach requires to write $S_{1,ext} = Ee^{-E}e^{E}$. Thus, calculations involving $S_{1,ext}$ will evidently require a more complex approach than the use of E directly, and this latter route will be the preference when formulating all forms of growth.

3.1 Modeling Electrocrystallization Processes

The kinetics of nucleation are assumed to follow a first-order reaction law. Thus, the number of nuclei in the absence of any subsequent growth is represented by

$$N = \frac{A}{A'}[1 - \exp(-A't)],$$

where $A(\text{cm}^{-2}\text{s}^{-1})$ and $A'(\text{s}^{-1})$ are, respectively, the expected nucleation rate constant and the expected frequency with which a nucleus is to form at a given site. For a homogeneous nucleation $\frac{A}{A'} = \frac{1}{\alpha r_c^2}$, where $\alpha > 4$ and r_c is the critical nucleation size. However, for a heterogeneous formation of nuclei on $N_0(\text{cm}^{-2})$ preferred sites randomly distributed over the surface of an electrode, $\frac{A}{A'} = N_0$.

For the formation of a monolayer of deposit, only the growth of circular discs has been considered [12]. The general nondimensional form of the current–time transient is [13]

$$I = 2\left[T - \gamma + \gamma \exp\left(-\frac{T}{\gamma}\right)\right] \exp\left[-T^2 + 2\gamma T - 2\gamma^2 + 2\gamma^2 \exp\left(-\frac{T}{\gamma}\right)\right]$$

(3.6)

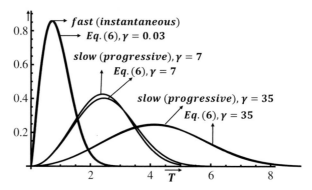

Figure 3.5   *The forms of the transients for 2D growth [Equation (3.6)] for three values of γ.*

where $\gamma = v\sqrt{\pi A/A'}/A'$ and I, v and T are defined in Table 3.1. As $\gamma \to 0$ $(A' \to \infty)$, Equation (3.6) reduces to $I = 2T\exp(-T^2)$, the so-called "instantaneous" nucleation. However, nuclei never form at once when the potential is applied, but always form progressively. Consequently, in order to eliminate any possible confusion, "instantaneous" and "progressive" nucleation will hereafter be referred to as "fast" and "slow" nucleation, respectively. Figure 3.5 reveals that when $0.03 \le \gamma \le 35$, for example, $\gamma = 7$ (the middle transient curve), the current–time transient does not follow the shape derived for the "slow" nucleation $I = T^2 \exp(-T^3/\gamma)/\gamma$.

For this range, the terminology "intermediate" nucleation is introduced. Nucleation is, therefore, "fast" when $\gamma \ge 35$, that is, when $A' \ge 60v\sqrt{A/A'}$, and "slow" when $\gamma \le 0.03$, that is, when $A' \le v\sqrt{A/A'}/20$. However, $A'(\mathrm{s}^{-1})$ and $v\sqrt{A/A'}(\mathrm{s}^{-1})$ are the rates at which sites are either ingested by the act of nucleation or by the act of growth. So, whenever the likelihood of the coverage of a given site by the act of nucleation is 60-fold more or 20-fold less than that of growth, nucleation is termed as "fast" or "slow." The same limits have been shown [14] to hold for the 3D growth models.

The 3D growth models are divided into two categories [15]: (i) shape-preserving topography, which for the sake of brevity is hereafter referred to as Type I; and (ii) shape-changing topography (Type II). Within the Type I category only the growth of right-circular cones [16] has been fully investigated [14, 17]. Hemi-spheroids [14], paraboloids [18], hyperboloids [15] and spherical-caps [19] are examples of Type II category. The current–time transients resulting from nucleation of both, Type I and Type II growth forms, can be derived from [15].

$$
\left.
\begin{aligned}
I(\tau) &= \int_0^\tau \frac{dE}{d\tau} \exp(-E)\,ds \\[2mm]
E &= w^2(\tau, s) - 2\int_0^{w(\tau,s)} \exp\left[-\lambda(\tau - \tau_g)\right] w\,dw \\[2mm]
w(\tau, s) &= \left[\tau^2 + (2\kappa - 1)s^2 - 2\kappa s\tau\right]^{1/2} \\[2mm]
\tau_g &= \kappa s + \sqrt{(1 - \kappa)^2 s^2 + w^2}
\end{aligned}
\right\}
\tag{3.7}
$$

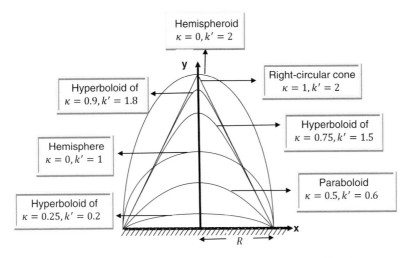

Figure 3.6 *The shapes generated by Equation (3.8) with R = 0.5 and with $1 \geq \kappa \geq 0$, after revolving around the y-axis. Reprinted from Ref. [6] with permission from Elsevier.*

where, E [10] for formulating current–time transients resulting from 3D growth models has been redefined [6]. The current density $j(\mathrm{Acm^{-2}}) = zFk'I(\tau)$, where, τ, κ and λ are defined in the list of symbols.

An equation representing conic sections of revolution is derived by combining Equations (3.14–3.16) and the definition of κ in Equation (3.18) within Ref [15]. Thus,

$$y = \frac{k'}{2\kappa - 1}\{R\kappa - [x^2(2\kappa - 1) + R^2(1 - \kappa)^2]^{1/2}\} \tag{3.8}$$

where $R(\mathrm{cm})$ is the base radius of growth centers. Figure 3.6 represents the shapes generated by Equation (3.8) with $R = 0.5$ and with $1 \geq \kappa \geq 0$. The current–time transients for the growth of right-circular cones [14], hemispheroids [14] and paraboloids [18] are, thus, derived by inserting in Equation (3.7) $\kappa = 1$, $\kappa = 0$ and $\kappa = 0.5$.

The popular approach to modeling current–time transients resulting from nucleation and diffusion-controlled growth of centers mostly follows that described by Davison and Harrison [20]. Most of the later work is based on the equation [21]

$$j = \frac{zFc_0D^{1/2}}{\pi^{1/2}t^{1/2}}\left\{1 - \exp\left[-2\sqrt{2}\pi^{3/2}\left(\frac{Mc_0}{\rho}\right)^{1/2}DN_0t\right]\right\} \tag{3.9}$$

However, Fleischmann's desire to discourage researchers from basing their approach on Equation (3.9) is evident from one of his publications [22], where he points out that it is essentially an empirical equation. So, what does Equation (3.9) represent? For the case of a "fast" nucleation of N_0 right-circular cone growth forms

$$j = \frac{zF\rho}{M}\frac{dV}{dR_0}\cdot\frac{dR_0}{dt} = \frac{zF\rho}{M}\Theta\tan(\theta')\frac{dR_0}{dt} = \frac{zF\rho}{M}\left[1 - \exp\left(-\pi R_0^2 N_0\right)\right]\tan(\delta)\frac{dR_0}{dt} \tag{3.10}$$

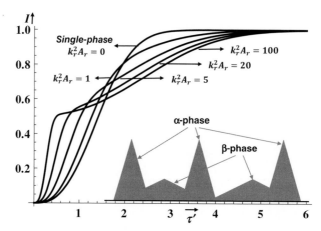

Figure 3.7　*The form of the current versus time transients for a two-phase growth model. Each phase is represented by a Type I growth center of a distinct contact angle; $k'_r = 0.5$.*

Assuming a very low contact angle $\theta' = \tan^{-1}(\frac{2^{1/3}Mc_0}{\pi\rho})^{3/4}$ and a growth rate defined by $\frac{dR_0}{dt} = (\frac{\pi Mc_0}{2\rho})^{1/4}\frac{D^{1/2}}{t^{1/2}}$, we retrieve Equation (3.9) [23]. Thus, Equation (3.9) simply represents right-circular cones growing under a planar diffusion, which implies that the empirically proposed model [21] cannot account for the overlap of diffusion zones.

In many electrocrystallization processes, two or more phases are growing simultaneously. Provided that $k'_\alpha > k'_\beta$, the current–time transients resulting from all forms of simultaneous growth of two phases can be based on [24, 25].

$$I(\tau) = \int_0^{k'_r\tau'} \frac{d(E_\alpha + E_\beta)}{d\tau'} \exp(-E_\alpha - E_\beta)ds + \int_{k'_r\tau'}^{\tau'} \frac{dE_\alpha}{d\tau'} \exp(-E_\alpha)ds \qquad (3.11)$$

The current–time transient patterns of behavior for Type I and Type II categories of two-phase growth are shown in Figure 3.7 [26] and Figure 3.8 [24]. However, when the simultaneous growth of Type I and Type II categories are considered, two distinct combinations (insets in Figures 3.9 and 3.10) are recognized [25]. The current–time transient patterns of behavior for these dissimilar forms of growth are also shown in Figures 3.9 and 3.10. The wide variety of features, which are derived for the two-phase growth models (see Figures 3.7–3.10) help in identifying from the shapes of the recorded transients the manner, rates and topographies with which the individual phases are growing relative to one another.

3.2　Applications of Models

The above models are to be used for analyzing those transients that are recorded under special conditions [27]. In particular, the transients are recorded by applying a two-step

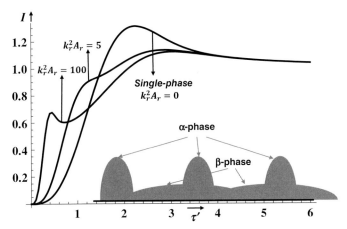

Figure 3.8 *The form of the current versus time transients for a two-phase growth model. Each phase is represented by hemispheroids (Type II) of a distinct ellipticity; $k'_r = 0.5$.*

potential profile to the working electrode. The first profile is to a potential as close as possible to that where nucleation occurs but where no rising current is observed; the second profile is to the potential at which the current–time transients are to be recorded. This procedure ensures the reduction of the initial falling background/charging current, so that the magnitude of this initial current cannot mask the very early stages of electrocrystallization.

Figure 3.11 is a current–time transient recorded during the electrocrystallization of cobalt. It is because of the application of the two-step potential that the formation of a mono-layer of deposit is clearly observed in the time range $0 < t < 20$ s. This figure also shows the fit of the recorded current–time transients to the transients derived for the "slow" nucleation of the three forms of growth center (right-circular cones, hemispheroids, and paraboloids). It can be seen that the rising portion of the transient is insensitive to the growth geometry,

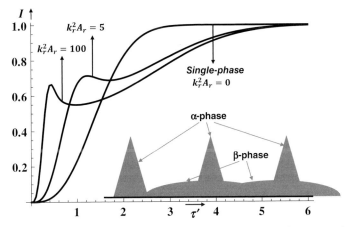

Figure 3.9 *The form of the current versus time transients for a two-phase growth model. One phase is represented by a Type I growth form, and the other by a Type II growth form; $k'_r = 0.5$.*

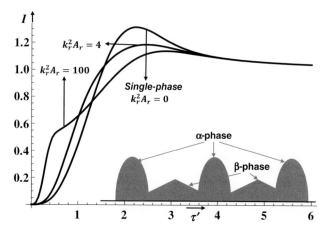

Figure 3.10   *As for Figure 3.9, except that in this case the rate of the outward growth of Type II centers is greater than that of Type I; $k'_r = 0.5$.*

and this allows the kinetic parameters to be determined by analyzing the rising portion of the recorded transients according to the model of growth of right-circular cones. The pronounced current maximum can be accounted by the simultaneous hydrogen evolution on the tops of Type II growth geometries (see Section 3.2.2).

3.2.1   The Deposition of Lead Dioxide

The technique of studying the early stages of the rising transients was pioneered by Fleischmann and Thirsk [1], and first applied to the study of the electrocrystallization of PbO_2 [2]. The early rising portion of a current–time transient is considered to be "slow" or "fast", if, respectively, $(j - j_1)^{1/3}$ or $(j - j_1)^{1/2}$ is proportional to time t [2, 28]. Figure 3.12

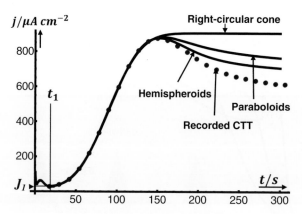

Figure 3.11   *Current versus time transient recorded (. . . .) during the electrocrystallization of cobalt at an applied potential of −0.80 V (SCE) with the best fit (—) to the three models of growth: solution composition 0.85 mol dm^{-3} CoSO$_4$·7H$_2$O, 0.15 mol dm^{-3} CoCl$_2$·6H$_2$O, and 0.58 mol dm^{-3} H$_3$BO$_3$.*

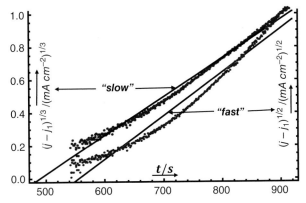

Figure 3.12 *Plots of $(j - j_1)^{1/3}$ and $(j - j_1)^{1/2}$ versus time t for the electrocrystallization of PbO_2 from a solution of 0.1 M $Pb(NO_3)_2$ + 1 M HNO_3 at an applied potential of 1510 mV, temperature 293 K. Reprinted from Ref. [29] with permission from Elsevier.*

shows the plots of $(j - j_1)^{1/3}$ and $(j - j_1)^{1/2}$ against t for the current–time transient recorded during the electrocrystallization of PbO_2 [29]. Not only does the plot of $(j - j_1)^{1/2}$ against t curve upwards, indicating that the current density is proportional to powers of t greater than 2, but the plot of $(j - j_1)^{1/3}$ against t also displays an upward tendency (indicating the impossible scenario that the current density is proportional to powers of time even greater than 3). However, a closer inspection reveals that there are, in fact, two distinct linear portions in the plot of $(j - j_1)^{1/3}$ versus t (Figure 3.13). The existence of two linear portions is an indication of the simultaneous formation of two distinct phases of a deposit [28].

If it is assumed that there exist two phases of a deposit, α and β, each of which has its own specific induction period, t_α and t_β, with $t_\alpha < t_\beta$. The overall current density for $t < t_\alpha$ is zero. However, in the time region $t_\alpha \leq t \leq t_\beta$ only the α-phase is expected to grow, for which $(j - j_1)^{1/3} = (zFk'_\alpha P_\alpha)^{1/3}(t - t_\alpha)$. A value of $(zFk'_\alpha P_\alpha)^{1/3} \approx 0.00215(\text{mA cm}^{-2})^{1/3}\text{s}^{-1}$ is obtained from the slope of the early linear segment of Figure 3.13, with an estimation of

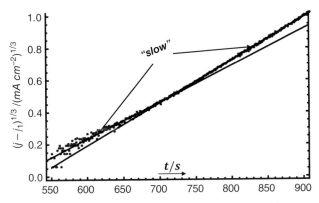

Figure 3.13 *The same data of Figure 3.12, but only plotted for $(j - j_1)^{1/3}$ versus time t, showing two distinct linear portions.*

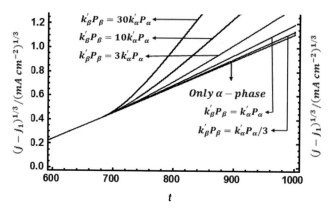

Figure 3.14 Plots of $(j - j_1)^{1/3}$ against time t, as predicted by Equation (3.12) for different values of $k'_\beta P'_\beta/k'_\alpha P'_\alpha$.

$t_\alpha (= 480$ s) from the intercept with the t-axis. The data fit well with the first linear segment up to $t \approx 700$ s, after which it starts to diverge. Thus, for the time range $480 \leq t \leq 700$, $(j - j_1)^{1/3} = 0.00215(t - 488)$. For $t > 700$ s,

$$(j - j_1)^{1/3} = \left[0.0215^3 (t - 488)^3 + zFk'_\beta P_\beta (t - 700)^3 \right]^{1/3} \qquad (3.12)$$

Figure 3.14 shows that, for a two-phase electrocrystallization, the larger the value of $k'_\beta P_\beta$ compared with $k'_\alpha P_\alpha$, the greater is the deviation of the slope between the two portions of the plot. It is therefore not possible to observe from practical measurements a deviation in the slope of the two portions of the plots when $k'_\beta P_\beta < k'_\alpha P_\alpha$. It is thus concluded that, even for cases when a seemingly single linear dependence of $(j - j_1)^{1/3}$ against t is observed, the formation of a second phase cannot be ruled out.

3.2.2   The Electrocrystallization of Cobalt

The formation of a monolayer prior to the onset of the 3D growth of cobalt deposit has clearly been shown in Figure 3.11. The nonlinear regress returns an induction time, t_1 ~25 s, which is the time taken for the monolayer to grow large enough that the 3D growth forms are nucleated on top of the monolayer. It is very likely that the formation of 3D growth centers requires such sites as the junctions of three 2D growth centers, and thus the coverage of the monolayer should pass well beyond the overlap of 2D growth forms in order for the 3D centers to form. The high current density, $j_1 \sim 28$ μA cm^{-2}, during monolayer formation is an indication of the rate of hydrogen evolving on top of the monolayer [28].

None of the theoretically derived models can account for the observed ratio of the current maximum to the steady-state value (see Figure 3.11). An incorrect assumption that the deposition current density is always proportional to the surface area of the growth forms could lead to an incorrect conclusion that the growth of elongated hemispheroids would result in a greater ratio of the current maximum to the steady-state value. It has been shown [30] that the shape of the derived transients is independent of the eccentricity of hemispheroids; however, the current density for the evolution of hydrogen is directly

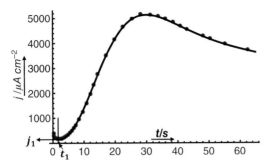

Figure 3.15   *The closest fit (——) of the current–time transient derived for the "fast" nucleation and paraboloidal forms of growth with concurrent evolution of hydrogen (Equation 3.19 of Ref. [18]), to the recorded transient (···) for the electrocrystallization of cobalt at −0.9 V; the solution composition was as in Figure 3.13.*

proportional to the surface area of the deposit [18], and therefore the pronounced maximum current is achieved once the concurrent evolution of hydrogen on tops of elongated Type II growth forms is considered. Figure 3.15 shows a very close fit of the data recorded during the electrocrystallization of cobalt at −0.9 V, to the current–time transient derived for a "fast" nucleation and growth of paraboloids with the concurrent evolution of hydrogen on top of the growth centers.

The higher magnitude of the initial falling background current at −0.9 V, compared with that recorded at −0.8 V (Figure 3.11), masks the transient for formation of the monolayer. However, this masking does not necessarily convey the message that a monolayer, prior to the formation of the 3D growth forms, had *not* been formed. The nonlinear regress (Figure 3.15) reveals that the formation of 3D centers commences at 1.8 s (compared to ~25 s at −0.8 V) after the application of the potential step. This 1.8 s is, indeed, the time taken for enough of the monolayer to form so that the 3D growth forms can nucleate at the junctions where three 2D growth forms meet. If this is the case, then it must be concluded that the rate of formation of the 3D growth forms is dictated by the rate of formation of these junctions, which in turn is determined by the rate of nucleation of the underlying monolayer, a conclusion that was reached in 1981 [27].

The early stages of the rising portion of some (but not all) recorded transients for the deposition of cobalt are characterized by two linear segments when $(j - j_1)^{1/3}$ is plotted against t [28]. This implies that cobalt is also deposited as two distinct phases. In those cases, where only one linear segment is observed (i.e., $k'_\alpha P_\alpha > k'_\beta P_\beta$; Figure 3.14) a very close fit of the recorded transient (Figure 3.15) to that derived for a single-phase growth model is naturally expected. However, when $k'_\alpha P_\alpha < k'_\beta P_\beta$, not only are two distinct linear segments obtained, but the nonlinear fit of the recorded transients would return negative values for the nucleation rates. These findings will be discussed elsewhere.

3.3   Summary and Conclusions

For modeling current–time transients, the first-order nucleation law $N = \frac{A}{A'}[1 - \exp(-A't)]$ has been used as a general form, or $N = N_0[1 - \exp(-A't)]$ as a specific form when the

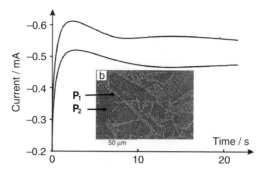

Figure 3.16   *Potential transient performed at −1.1 V and −1.15 V (vs SCE), in a 10 mM [Co(H₂O)₆](NO₃)₂ in 0.1 M MgSO₄ with 0.1 M H₃BO₃ solution (as in Figure 3.3 of Ref. [31]). The inset shows a SEM image of cobalt nucleated on a boron doped diamond surface, after −1.1 V potential has been applied for a 60 s period (as in Figure 3.6b of Ref. [31]). Reprinted from Ref. [31]. Copyright © 2006 Wiley-VCH Verlag GmbH & Co. KGaA, Weinheim.*

preferred number of sites N_0 (cm^{-2}) are randomly spread over the surface of the electrode. It follows that none of the above equations is operative when nuclei are either formed on such substrates (as shown in Figure 3.1) or on boron-doped diamond substrates (see the inset in Figure 3.16) [31]. This is because, in both cases, large areas – such as those on both sides of the scratch in Figure 3.1, or areas surrounding the location P_2 in Figure 3.16 – are not nucleated, which means that the rate of nucleation cannot be supposed to be proportional to the area uncovered, as is the case for the derivation of the above equations; the preferred nucleation on lines such as the one passing through P_1 in Figure 3.16 is obvious.

When using theoretical current–time transients, The following points must be borne in mind:

1. If the recorded transient has either attained a plateau (usually signifying the attainment of a steady-state current for the Type I category of growth models), or has passed beyond a maximum by half of the total deposition time, and the full coverage of the electrode surface is not observed, then the formation of nuclei had not taken place in a random fashion. In such cases, the first-order nucleation law and the subsequent derived transients cannot be applied to the analysis of the recorded transients; see, for example, the recorded transient in Figure 3.16 [31].

2. Transient current refers to the early stages of deposition, which witnesses a changing pattern in current as a direct result of the type of nucleation, the form of growth, and the overlap of growth forms. The flow of current then either attains a plateau (for Type I category of models) or passes through a maximum (Type II category), before attaining a steady-state current density ($= zFk'$). Thus, such a statement with respect to Figure 3.16 as "the current, achieved a steady state after about 10 s" [31] is flawed.

3. It has been noted previously [6, 14] (and also in this chapter) that nucleation is always a progressive process except that, for a "fast" nucleation (the so-called "instantaneous" nucleation), nuclei are forming so fast that it is 60-fold more likely for a given site to be covered by the act of nucleation than by the act of growth. It is, therefore, simply incorrect to relate "fast" nucleation to a scenario in which "*the number of nuclei on the*

surface "immediately" attains the value of N(∞) following the application of a potential of suitable magnitude" [31]. It then follows that in the very early stages the number of nuclei is expected to increase with time, even for a "fast" nucleation, which in one sense can be termed as the transient state of nucleation. In addition, the deposition of cobalt has been shown [28] to follow the formation of an underlying monolayer for which a finite time is necessary, for example, 1.8 s under the condition given in Figure 3.15. Thus, the above explanation can account for the calculated 1 to 2 s delay observed elsewhere [31].

Finally, it is appropriate to refer to the last two publications of Martin Fleischmann [3, 4], which incidentally happened to be on the subject of nucleation. In these papers, he sought to establish an approach to nucleation based on quantum electrodynamics, and built on an earlier conference presentation by the late Preparata [32]. Whilst this is an exciting development, it remains to be seen whether others will seek to build on these investigations.

References

(1) Fleischmann, M. and Thirsk, H.R. (1955) An investigation of electrochemical kinetics at constant overpotential – the behaviour of the lead dioxide electrode. *Transactions of the Faraday Society*, **51**, 71.

(2) Fleischmann, M. and Liler, M. (1958) The anodic oxidation of solution of plumbous salts. *Transactions of the Faraday Society*, **54**, 1370.

(3) Abyaneh, M., Fleischmann, M., Del Gindice, E., and Vitiello, G. (2009) The investigation of nucleation using microelectrodes, Part I the ensemble averages of the times of birth of the first nucleus. *Electrochimica Acta*, **54**, 879.

(4) Abyaneh, M., Fleischmann, M., Del Gindice, E., and Vitiello, G. (2009) The investigation of nucleation using microelectrodes, Part II the second moment of the times of birth of the first nucleus. *Electrochimica Acta*, **54**, 888.

(5) Volmer, M. (1929) Kinetic der Phasenbildung. *Zeitschrift für Elektrochemie*, **35**, 355.

(6) Abyaneh, M.Y. (1982) Calculation of overlap for nucleation and three-dimensional growth of centres. *Electrochimica Acta*, **27**, 1329.

(7) Budevski, E., Staikov, G. and Lorenz, W.J. (2000) Electrocrystallization: nucleation and growth phenomena. *Electrochimica Acta*, **45**, 2559.

(8) Budeski, E., Fleischmann, M., Gabrielli, G., and Labram, M. (1983) Statistical analysis of the 2-D nucleation and electrocrystallisation of silver. *Electrochimica Acta*, **28**, 925.

(9) Abyaneh, M.Y. and Fleischmann, M. (1991) Two-dimensional phase transformation on substrates of infinite and finite size. *Journal of the Electrochemical Society*, **138** (9), 2485.

(10) Evans, U.R. (1945) The laws of expanding circles and spheres in relation to the lateral growth of surface films and the grain-size of metals. *Transactions of the Faraday Society*, **41**, 365.

(11) (a)Avrami, M. (1939) Kinetics of phase change I: general theory. *Journal of Chemical Physics*, **7**, 1103;(b)Avrami, M. (1940) Kinetics of Phase Change II: transformation-time relations for random distribution of nuclei. *Journal of Chemical Physics*, **8**, 212.

(12) Bewick, A., Fleischmann, M. and Thirsk, H.R. (1962) Kinetics of the electrocrystallisation of thin films of calomel. *Transactions of the Faraday Society*, **58**, 2200.

(13) Abyaneh, M.Y. and Fleischmann, M. (1982) The role of nucleation and overlap in electrocrystallization processes. *Electrochimica Acta*, **27**, 1513.

(14) Abyaneh, M.Y. (2002) Extracting nucleation rates from current-time transients Part I: the choice of growth models. *Journal of Electroanalytical Chemistry*, **530**, 82.

(15) Abyaneh, M.Y. (2004) Kinetics of single-phase electrocrystallization processes II: CTTs due to the model of growth of hyperboloids. *Journal of the Electrochemical Society*, **151**, C743.

(16) Armstrong, R.D., Fleischmann, M., and Thirsk, H.R. (1966) Anodic behaviour of mercury in hydroxide ion solutions. *Journal of Electroanalytical Chemistry*, **11**, 208.

(17) Abyaneh, M.Y. and Fleischmann, M. (1981) The electrocrystallisation of nickel Part I. Generalised models of electrocrystallisation. *Journal of Electroanalytical Chemistry*, **119**, 187.

(18) Abyaneh, M.Y. (2004) Kinetics of single-phase electrocrystallization processes I: growth of paraboloids with concurrent evolution of hydrogen. *Journal of the Electrochemical Society*, **151**, C737.

(19) Abyaneh, M.Y. (1986) Formulation of current-time transients due to nucleation and coalescence of spherical-cap growth forms. *Journal of Electroanalytical Chemistry*, **209**, 1.

(20) Davison, W. and Harrison, J.A. (1973) The electrochemical growth of three-dimensional nuclei during a potentiostatic pulse a simulation. *Journal of Electroanalytical Chemistry*, **44**, 213.

(21) Gunawardena, G.A., Hills, G.J., Mortenegro, J., and Scharifker, B.R. (1982) Electrochemical nucleation: part I. general consideration. *Journal of Electroanalytical Chemistry*, **138**, 225.

(22) Abyaneh, M.Y. and Fleischmann, M. (1991) General models for surface nucleation and three-dimensional growth: the effects of concurrent redox reactions and of diffusion. *Journal of the Electrochemical Society*, **138** (9), 2491.

(23) Abyaneh, M.Y. (2006) Modelling diffusion controlled electrocrystallisation processes. *Journal of Electroanalytical Chemistry*, **208**, 196.

(24) Abyaneh, M.Y. (2005) Kinetics of two-phase electrocrystallization processes III: competitive nucleation and growth of hemispheroids. *Journal of the Electrochemical Society*, **152** (11), C776.

(25) Abyaneh, M.Y. (2007) Kinetics of two-phase electrocrystallization processes IV: CTTs associated with the growth of dissimilar geometric shapes. *Journal of the Electrochemical Society*, **154** (1), 5.

(26) Abyaneh, M.Y. (2004) Kinetics of two-phase electrocrystallization processes II: generalized form of transients. *Journal of the Electrochemical Society*, **151** (3), 194.

(27) Abyaneh, M.Y. and Fleischmann, M. (1981) The electrocrystallization of nickel Part II. Comparison of models with the experimental data. *Journal of Electroanalytical Chemistry*, **119**, 197.

(28) Abyaneh, M.Y. and Tajali Pour, A. (1994) The initial nucleation and growth of electrodeposits of cobalt. *Transactions of the Institute of Metal Finishing*, **72** (1), 19.

(29) Abyaneh, M.Y., Saez, V., González-García, J. and Mason, T.J. (2010) Electrocrystallization of lead dioxide: analysis of the early stages of growth, *Electrochimica Acta*, **55**, 3572.

(30) Abyaneh, M.Y. (1995) Generalization of transient equation due to the growth of hemispheroids. *Journal of Electroanalytical Chemistry*, **387**, 29.

(31) Simm, A.O., Ji, X., Banks, C.E., Hyde, M.E. and Compton, R.G. (2006) AFM studies of metal deposition: instantaneous nucleation and the growth of cobalt nanoparticles on boron-doped diamond electrodes. *ChemPhysChem*, **7**, 704.

(32) Preparata, G. (1995) *Q.E.D. Coherence in Matter*, World Scientific, Singapore. ISBN 9810222491 QC173.454.P74.

4

Nucleation and Growth of New Phases on Electrode Surfaces

Benjamin R. Scharifker[1,2] and Jorge Mostany[1]
[1] Universidad Simón Bolívar, Departamento de Química, Venezuela
[2] Universidad Metropolitana, Rectorado, Venezuela

Nucleation, as the onset of a first-order phase transition, is common ground between several disciplines. The formation of short-range ordered aggregates in a melt or solution in polymer science [1], molecular assemblies in biological systems, the condensation of vapors to form mist in the atmosphere, or cavitation phenomena in fluid mechanics [2], all take place through the creation of small aggregates of the condensed phase out of the parent phase.

The description of the kinetics and its dependence of the controlling parameters are of utmost importance, and between the 1920s and 1940s the seminal studies of Volmer and Weber [3], Farkas [4], Kaischew and Stranski [5], Becker and Döring [6] and Zeldovich [7], set the theoretical foundations of classical nucleation theory. Electrochemical nucleation owes much of its theoretical foundation to the field of crystal growth, thoughtfully developed by the Bulgarian crystal growth school [8]. Whereas in the latter, supersaturation is imposed via phase composition, in electrocrystallization the use of suitable electrochemical instrumentation to control the electrical state of the system allows the precise establishment of either the energetics or the kinetics of the process.

The classical nucleation theory, based on Gibbs thermodynamics statements, uses the macroscopic properties characteristic of bulk phases, such as free energies and surface tensions, for the description of small clusters. Contradictory results arose in early studies of electrochemical nucleation [9], where the size of a critical mercury nucleus on a platinum substrate amounted to only a few atoms, with properties that could substantially differ

Developments in Electrochemistry: Science Inspired by Martin Fleischmann, First Edition.
Edited by Derek Pletcher, Zhong-Qun Tian and David E. Williams.
© 2014 John Wiley & Sons, Ltd. Published 2014 by John Wiley & Sons, Ltd.

from the bulk phase [10]. Thus, a different approach to interpret the relationship between electrochemical supersaturation and nucleation rate was devised: the atomistic theory of the nucleation rate [11, 12]. This is a microscopic model of the kinetics of nucleation in terms of atomic interactions, attachment and detachment frequencies to clusters composed of a few atoms, as part of a general nucleation theory based on the steady-state nucleation model [13]. Over the years, both approaches have been understood as limiting cases of a general model, where the classical model works well in the case of low supersaturation and sufficiently large critical clusters, whilst the atomistic model is appropriate for high supersaturation and very active substrates when the critical nuclei are very small [8].

It is evident from this brief outline that, long before the advent of nanosciences, it was common amongst those committed to the study of electrochemical nucleation to refer to aggregates of a few atoms (or even a single atom adsorbed on a surface site! [8]) as critical entities within the framework of available nucleation models. In spite of these early studies and the later developments described below, electrochemical nucleation has been largely absent from the most widely used electrochemical textbooks, and has remained a relatively marginal affair within the mainstream of electrochemical studies devoted to corrosion, metal deposition, electrocatalysis, electrosynthesis, and so forth. There are some notable exceptions though, such as Vetter's discussion of the electrocrystallization overpotential in his classical textbook *Electrochemical Kinetics* [14], or the inclusion of transient techniques for the study of nucleation and growth phenomena in the book *Instrumental Methods in Electrochemistry* [15], authored by the Southampton Electrochemistry Group and strongly influenced by the ideas of Martin Fleischmann. Today, researchers focused on the synthesis and characterization of nanometer aggregates find a mature theoretical framework in the earlier electrochemical nucleation literature, where contributors such as Fleischmann established nucleation as a fundamental electrochemical phenomenon. Some of Martin Fleischmann's most notable contributions are briefly outlined in the following section, while in later sections some more recent advances will be described. The chapter concludes with a brief discussion of some topics where further progress may be expected.

4.1 An Overview of Martin Fleischmann's Contributions to Electrochemical Nucleation Studies

The importance of the analysis of the non-steady state in electrochemical processes was pointed out by Fleischmann [16] back in the 1950s. Previously, most studies away from the steady state involved measuring the overpotential under galvanostatic conditions. Since, however, the rates of electrochemical reactions and, in particular, the rates of nucleation and growth of new phases, were a strong function of the electrode potential, Fleischmann pointed out the advantages of measuring the current as a function of the overpotential for the analysis of systems in the non-steady state. Moreover, he made significant contributions to the design and implementation of fully functional high-speed potentiostats and function generators, as well as appropriate electrochemical cells [17]. These allowed the application of novel sequences of controlled potentials (e.g., double potentiostatic step [18]), that enabled researchers to carry out transient studies aimed to separate experimentally the nucleation rate from steps relating to phase growth processes.

Later, during the 1970s and early 1980s, Fleischmann and coworkers at the University of Southampton conceived a novel approach to study fast kinetics. Mass transport could be controlled, varied and substantially enhanced by diminishing the radius of a disc electrode to micrometric dimensions (<20 μm), providing a simple alternative to the rotating disc electrode. Such electrodes also had the advantages arising from the radial material flux to an almost infinitesimal sink, namely a rapid attainment of the steady state, high current densities with small measured currents, a very high ratio between faradaic and capacitive currents, and a substantial decrease of the ohmic potential drop due to solution resistance [19, 20]. For investigations of nucleation, the low area of such electrodes also offers the possibility to study a small number of nuclei.

Studies on the behavior of the lead dioxide electrode [21], the oxidation of silver sulfate to silver oxide [22], the kinetics of electrocrystallization of thin films of calomel [23] and the electrodeposition of nickel [24], as well as the concepts and applications of micrometer-sized electrodes [25], stand as pioneering research concerning nucleation and growth problems. Wisely designed experimental procedures combined with mathematical models recognizing the chemical nature of the problem led to many new insights.

4.2   Electrochemical Nucleation with Diffusion-Controlled Growth

The possibility of controlling supersaturation – the driving force for nucleation – by manipulating the electrode potential make electrochemical methods very convenient, precise and economical ways of studying heterogeneous nucleation. It is now generally agreed that electrodeposition on a foreign substrate occurs by a process of nucleation, through which ions in solution discharge over "active sites" on the surface (steps, kinks, holes, grain boundaries, chemically modified locations), forming stable nuclei of the new phase that grow further by the incorporation of adsorbed atoms, or by direct attachment of ions from solution. The nucleation process is formulated as the stepwise addition and removal of atomic or molecular species to a given cluster of size n, until the eventual formation of a supercritical nucleus of size n^* that grows irreversibly. The heterogeneous nucleation rate is frequently assumed to be first order with respect to the number of active sites on the surface, thus decaying exponentially with time:

$$dN/dt = AN_0 \exp(-At) \tag{4.1}$$

where N (cm^{-2}) is the number density of nuclei at time t, N_0 (cm^{-2}) is the number density of active sites initially present on the surface, and A (s^{-1}) is the nucleation rate constant per site, a potential-dependent frequency. The nucleation process can be broadly classified into either two-dimensional (2D) or three-dimensional (3D), and phase growth may be kinetically or mass transport-controlled. These categories have fundamental differences regarding the rate limiting step on the overall mechanism: during phase growth, mass transport and charge transfer are consecutive processes, with the slowest being rate-determining. When the rate is limited by the incorporation of (ad)atoms into 2D or 3D growth centers, crystallinity and shape are critical factors. On the other hand, when mass transport controls the rate of the electrocrystallization, simple geometric forms (e.g., hemispheres) may be considered as material "sinks" immersed in a mass transport layer. In this chapter, attention is focused on the latter, with systems characterized by high exchange current densities for the electrodeposition of metal ions, where the overall growth rate is determined by diffusional mass transport to the electrode.

This particular area of electrochemical nucleation and growth has been recently reviewed by Hyde and Compton [26]. From the material summarized therein, several important aspects open to discussion can be identified:

- The difficulty in modeling progressive nucleation.
- The concept of active sites, and the form of the nucleation rate law.
- The real nature of the very first stage of the nucleation of a new phase on the surface of an electrode. In most cases, either the classical or atomistic analyses of its kinetics indicate that single adatoms behave as thermodynamically stable entities that are able to grow irreversibly.

These issues remain unresolved, and some attention will be given to each.

4.3   Mathematical Modeling of Nucleation and Growth Processes

"Instantaneous" and "progressive" nucleation are limiting cases corresponding to fast nucleation on a small number of active sites and slow nucleation on a large number of active sites, respectively. Both extremes are unattainable in real situations and have sparked heated discussions in the literature, as well as some misuse of current theories describing 3D nucleation with diffusion-controlled (3DDC) growth processes. It has been suggested that the observed lag in the nucleation rate in response to changes in the electrode potential is due to a non-steady regime during the early stages of nucleation, rendering the instantaneous nucleation definition unphysical [27]. However, the term "instantaneous" actually refers to processes leading to the nucleation of a new phase occurring in a time-scale much faster than the exhaustion or inhibition of nucleation sites on the surface, due to the subsequent growth of the new phase. Currently available 3DDC models rely on phenomenological descriptions of the statistical geometry of transformations, based on the Kolmogorov–Johnson–Mehl–Avrami (KJMA) theory, as first introduced to electrocrystallization studies by Bewick, Fleischmann and Thirsk in 1962 [23]. The diffusion fields around 3D growth centers with locally spherical symmetry are projected onto the plane of the electrode as planar diffusion zones, and the nucleation and growth process is treated in analogy to a 2D phase transformation [26]. Currently available 3DDC models [28, 29], however, frequently encounter inconsistencies at the progressive nucleation limit, due to mathematical difficulties in handling the overlap of planar diffusion zones representing the diffusion fields of nuclei born at various times [30]. The main intricacy arises from the unreality of modeling such situations with diffusion zones of uniform height. It has been shown that to overcome this major limitation requires a hierarchical treatment of the problem of overlap, with earlier diffusion zones prevailing over later ones [31]. Appropriate formalisms for the rigorous treatment of this statistical geometric problem, however, are still unavailable.

Direct observation of individual nuclei is now possible using *in-situ* atomic force microscopy (AFM) techniques [32] that allow the determination of nucleation and growth rates of Pb nuclei on boron-doped diamond electrodes as a function of time and overpotential. *In-situ* tapping-mode AFM proved to be an efficient tool for imaging the electrodeposition process at the micron scale, and showed an agreement between the kinetic parameters of nucleation and growth under diffusion control obtained from transient analysis using the 3DDC model, with the same quantities observed directly using in-situ AFM. Optical methods have been also used. Wu *et al.* studied the earliest stages of growth of Pb deposits on polycrystalline Cu surfaces at overpotentials, in the presence and absence of Cl$^-$ ions in solution, by a combination of oblique incidence reflectivity difference (OI-RD) and in-situ AFM measurements [33]. This is a fast technique with submonolayer sensitivity that can be used to follow the early stages of electrodeposition processes, even those involving fast kinetics in charge transfer and lattice incorporation steps, as is frequently the case. As described elsewhere in this book, Martin Fleischmann and coworkers pioneered the development of *in-situ* spectroscopic studies of the solid–liquid interface, and the combined use of optical and local probe techniques to study dynamic processes at electrochemical interfaces under potentiostatic control is becoming an extremely valuable tool for understanding the various phenomena involved in the formation and growth of new phases on electrodes.

4.4 The Nature of Active Sites

Experimental nucleation studies are extremely sensitive to the state of the surface. Electrochemists working in this area are fully aware of the extreme care that is needed to obtain reproducible and meaningful results. Polishing electrode surfaces with diamond dust or micrometer-sized alumina to a mirror finish is customary, with additional chemical and electrochemical treatments being common. Depending on the metal being electrodeposited, and the nature of the substrate, anodic dissolution of deposits between experiments may suffice, although leftovers from previous tests frequently enhance nucleation in successive experiments, distorting the systematic measurement of the A and N_0 dependence on the overpotential. Thus, the activity of electrode surfaces towards the nucleation of new phases needs proper definitions and rigorous treatment. In particular, as described by Deutscher and Fletcher [34, 35], the energy required to establish supercritical nuclei on active sites on the surface is not a unique quantity. Rather, it is the mean of a wide range of values arising from surface heterogeneities, leading to dispersion of the nucleation rate due to the high sensitivity of active sites to the interfacial free energies. Consequently, the exponential steady-state nucleation rate law expressed in Equation (4.1) does not hold in the general case, and a site energy-dependent and time-varying function $A(\gamma,t)$ for the nucleation rate should be used instead.

One way of dealing with the surface heterogeneities invariably present at solid–liquid interfaces is to try and avoid them altogether. This may be achieved, for example, by isolating a small region of a crystal face and removing from it dislocations and other surface inhomogeneities with a special technique, to obtain defect free, "quasi-ideal" crystal faces, as ingeniously developed by Budevski and coworkers [36]. This methodology led to very valuable advances in the understanding of fundamental aspects of 2D nucleation, providing also empirical evidence confirming the validity of the classical theory of nucleation. For 3D nucleation with diffusion-controlled growth (3DDC), an interesting alternative approach to the problem of surface heterogeneities has been proposed and a suitable model was developed, describing nucleation at a liquid–liquid interface. This has the advantage of eliminating the heterogeneities of a solid surface, but at the expense of difficulties in defining the supersaturation driving the phase transformation in the interfacial region, and the need to develop transport equations at both sides of the interphase [37, 38]. At the other extreme of surface heterogeneity, the enhanced activity of a particular surface topography has been used advantageously to grow specific structures with attractive properties and potential applications. Typical examples are the growth of nanocrystalline α-MnO$_2$ nanowires for high- performance lithium batteries [39] or photoconductive CdS hemicylindrical shell nanowire ensembles as fast optical gating devices [40], by electrodeposition on the steps of highly ordered pyrolytic graphite (HOPG) surfaces, a method that has been termed electrochemical step edge decoration (ESED).

Due to its extreme sensitivity on the state of the surface, nucleation kinetics is crucially affected by concurrent physical or electrochemical processes, such as an underpotential deposition stage during electrocrystallization [41], the adsorption of ions or blocking agents, or the potential-dependent electrochemistry of the electrode surface. For instance, active sites may appear or disappear from the electrode surface simultaneously with the nuclei of the new phase, rendering it impossible to distinguish between the actual nucleation rates and the rates of appearance and disappearance of active sites, especially if the nucleation experiments are carried out by means of the standard single-step potentiostatic technique. These surface electrochemical reactions have been examined by Milchev *et al.* [42] who developed a polarization routine devised to fix the energetic state of the electrode surface.

Densities of active sites on electrode surfaces have been typically found to be on the order of 10^5 to 10^7 cm^{-2}. It is then possible to isolate single growth centers by restricting the area of the electrode, and this may be realized using microdisc electrodes of sufficiently small radius, exposing approximately 10^{-7} cm^2 of surface area to the solution (e.g., with microdisks with a radius of a few microns). The waiting times for nuclei births on these surfaces may be repeatedly measured under various conditions, in order to probe the statistics of nucleation, as further described below. Instead of successive experiments on a single site, the simultaneous growth of multiple centers on an assembly of microelectrodes sufficiently apart as to avoid the interaction between them may be used to probe surface heterogeneities. Deconvolution of the individual contributions to the overall current makes it possible to construct accurate experimental $N(t)$ curves, and this leads to an interpretation based on the existence of a distribution of activities of the active sites [34], an issue which has important implications in several key areas of electrochemistry, including mineral dissolution reactions, pitting corrosion, gas evolution reactions, and the etching of semiconductors.

Since the growth of nuclei inhibits the nucleation rate in their vicinity [43], information on the dispersion of nucleation rates can be also obtained from the spatial distribution of nuclei on electrode surfaces. If the spread of nucleation rate constants across the surface is broad, then only the most active of sites, with a low number density, will have time to develop nuclei before the surface becomes covered with exclusion zones around them, and the distribution of nearest neighbors will correspond to that of Poisson-distributed point particles [44]. Moreover, a number of experimental systems display instantaneous nucleation at low overpotentials, with the number density of nucleation sites frequently found to depend strongly on the overpotential, turning into progressive nucleation at higher overpotentials, as larger numbers of sites become active for nucleation [45]. The nature of the active sites, the dispersion of their activities, and the dependence of nucleation rates on the state of the surface remain matters that have been minimally explored, and further progress can be expected.

4.5   Induction Times and the Onset of Electrochemical Phase Formation Processes

The clustering process preceding the appearance of nuclei has been regarded as a "birth-and-death" Markov process [25, 46]:

$$n - 2 \rightleftarrows n - 1 \rightleftarrows n^* \tag{4.2}$$

where n^* is the smallest stable aggregate or "critical nucleus," while their further growth is considered a "pure birth" irreversible process:

$$n^* \rightarrow n^* + 1 \rightarrow n^* + 2 \rightarrow \cdots \tag{4.3}$$

The introduction of microelectrode techniques in electrochemistry allowed Pons and Fleischmann [19], as well as others, to examine the statistics of these processes. Early studies on microelectrodes [45, 47] analyzed the induction times before the irreversible growth of the first nucleus on the surface, an event detected by the sudden rise of a faradaic current. Although this requires studies to be conducted at very low currents, the sudden rise is easily detected and effectively identifies with high precision the outcome of the stochastic processes preceding the irreversible growth of the new phase. If the forward rates in Equation (4.3) are considered to be sufficiently fast, then the observed retardation on the growth current occurs during the development of a subcritical cluster; that is, Equation (4.2) applies. By observing the distribution of induction times on a statistically significant number of experiments, the initial kinetics of formation of critical nuclei of Hg [45] and PbO_2 [46] has been defined and analyzed in terms of statistics and the Gibbs energy barriers dictated by the nucleation theory.

One frequently observed result remains paradoxical; this is the interpretation of experimental results when both the classical [3–7] and atomistic [11–13] nucleation theories yield critical nuclei of very small size, in many cases containing even zero atoms [8]. This implies that a single atom discharged onto an active site is already supercritical – that is, it is stable and able to grow irreversibly. As Abyaneh *et al.* recently pointed out [48], such a situation can hardly be considered as 3D nucleation, and 2D formalisms would be more appropriate. With large surface densities of active sites, the very initial step of the phase formation process involves a submonolayer of adsorbed atoms. The activation barrier opposing the formation of stable nuclei vanishes at high supersaturations, and the system becomes unstable, undergoing very fast (i.e., instantaneous) spinodal decomposition, with density fluctuations where the wavelengths are determined by the thermodynamic properties of the system, growing exponentially with time (cf. Ref. [49] and references therein). When the volume fractions of the phases in heterogeneous systems are close to 50%, intricate interconnected patterns evolve, and these are distinctively different from the compact island morphologies expected from nucleation and growth processes. Labyrinthine interconnected structures on the nanoscale have indeed been observed using scanning tunneling microscopy upon stepping the potential of a gold surface undergoing phase transformation in the presence of chloride ions [48]; this substantiates spinodal decomposition as a possible mechanism driving electrochemical phase transformations at high supersaturations. Other experimental evidence also points to mechanisms of phase formation that do not involve a nucleation barrier at high overpotential. The Gibbs energy of formation of critical nuclei may be determined by studying the temperature dependence of the nucleation rate, and very low values of the reversible work of formation of critical nuclei have been found during the nucleation of Ag on vitreous carbon [50], approaching zero at high overpotentials, and also suggesting such a spinodal mechanism for the phase transformation. Although many of these findings still require additional verification, what is evident from this very short discussion is that further experimental studies and theoretical advances of the processes preceding irreversible phase growth are due, and that the necessary experimental tools appear to be already available. Recently, Abyaneh *et al.* revisited studies on the induction times for α-PbO_2, Ag and Hg nucleation on microelectrodes, at high overpotentials [48]. These authors also concluded from an interpretation of their results that the subcritical aggregation processes do not involve the balancing of the positive surface and negative bulk terms of the Gibbs energy as the impeding force for phase formation, as postulated by nucleation theories. Instead, they postulated, as discussed above, that the delay times observed for irreversible phase growth stem from the dynamics of development of coherent domains of the new phase from incoherent clusters, a proposal that was entirely consistent with the general nonequilibrium thermodynamic theory of chemical kinetics and charge transfer [51] that had recently been formalized. Such approaches are fit to address the dynamics of nucleation and growth driven by surface reactions in nonhomogeneous media, as is the case in 3DDC processes and other fields such as intercalation batteries [52].

4.6   Conclusion

Studies on the electrochemical formation of new phases on electrode surfaces have advanced a great deal during the past 60 years. Progress has been built upon fundamental concepts of the equilibrium of heterogeneous systems, statistical and irreversible thermodynamics, the equilibrium structure of surfaces, interfacial and atomic bonding energies, crystal growth, chemical kinetics, mass transfer, stochastic processes, and statistical geometry. Understanding phase formation phenomena in electrochemical systems has had diverse motivations throughout the years. Nowadays, research in this field receives thrust from intense needs for more efficient and economic electrocatalysts for sustainable and environmentally sound systems for the conversion and storage of energy. Other drivers include processes for the microelectronic industries, developments in nanotechnologies, health, new materials, and quantum dots. The field has moved forward by taking full advantage of the new experimental techniques that have been introduced, as well as the development of novel tools for the interpretation of data. In such a multidisciplinary and complex environment, intense interplay between ideas, experiments and mathematical models plays a crucial role. This has been fruitful field for the development of the many imaginative and productive contributions of Martin Fleischmann that will continue being an integral part of the field for the foreseeable future.

References

(1) Gooch, J. (2007) *Encyclopedic Dictionary of Polymers*, Springer, New York, p. 661.
(2) Brujan, E.-A. (2011) *Cavitation in Non-Newtonian Fluids*, Springer-Verlag, Berlin, Heidelberg.

(3) Volmer, M. and Weber, A. (1926) Nucleation in supersaturated structures. *Zeitschrift fur Physikalische Chemie*, **119**, 277–301.

(4) Farkas, L. (1927) Nucleation rate in supersaturated vapours. *Zeitschrift fur Physikalische Chemie*, **125**, 236–242.

(5) Kaischew, R. and Stranski, I. (1934) The kinetic description of the nucleation rate. *Zeitschrift fur Physikalische Chemie, Abteilung B*, **26**, 317–326.

(6) Becker, R. and Döring, W. (1935) The kinetic treatment of nuclear formation in supersaturated vapours. *Annals of Physics*, **24**, 719–752.

(7) Zeldovich, B. (1943) On the theory of new phase formation, cavitation. *Acta Physicochimica URSS*, **18**, 1–22.

(8) Milchev, A. (2002) *Electrocrystallisation, Fundamentals of Nucleation and Growth*, Kluwer, New York.

(9) Tohmfor, G. and Volmer, M. (1938) Germ formation under the influence of electrical charging. *Annals of Physics*, **33**, 109–131.

(10) Tolman, R.C. (1949) The effect of droplet size on surface tension. *Journal of Chemical Physics*, **17**, 333–337.

(11) Milchev, A., Stoyanov, S. and Kaischew, R. (1974) Atomistic theory of electrolytic nucleation: I. *Thin Solid Films*, **22**, 255–265.

(12) Milchev, A. and Stoyanov, S. (1976) Classical and atomistic models of electrolytic nucleation: comparison with experimental data. *Journal of Electroanalytical Chemistry*, **72**, 33–43.

(13) Stoyanov, S. (1973) On the atomistic theory of nucleation rate. *Thin Solid Films*, **18**, 91–98.

(14) Vetter, K.J. (1967) *Electrochemical Kinetics: Theoretical and Experimental Aspects*, Academic Press, New York.

(15) Southampton Electrochemistry Group (1990) *Instrumental Methods in Electrochemistry, Ellis Horwood series in Physical Chemistry*, Ellis Horwood, Chichester.

(16) Fleischmann, M. (1952) Studies in Electrodiffusion. PhD thesis, Imperial College London, University of London.

(17) Bewick, A., Fleischmann, M. and Liler, M. (1959) Some factors in potentiostat design. *Electrochimica Acta*, **1**, 83–105.

(18) Fleischmann, M. and Thirsk, H.R. (1959) The potentiostatic study of the growth of deposits on electrodes. *Electrochimica Acta*, **1**, 146–160.

(19) Pons, S. and Fleishmann, M. (1987) The behaviour of microelectrodes. *Analytical Chemistry*, **59**, 1391A–1399.

(20) Heinze, J. (1993) Ultramicroelectrodes in electrochemistry. *Angewandte Chemie, International Edition in English*, **32**, 1268–1288.

(21) Fleischmann, M. and Thirsk, H.R. (1955) An investigation of electrochemical kinetics at constant overvoltage. The behaviour of the lead dioxide electrode, part 5. The formation of lead sulphate and the phase change to lead dioxide. *Transactions of the Faraday Society*, **51**, 71–95.

(22) Dugdale, I., Fleischmann, M. and Wynne-Jones, W.F.K. (1961) The anodic oxidation of silver sulphate to silver oxide at constant potential. *Electrochimica Acta*, **5**, 229–239.

(23) Bewick, A., Fleischmann, M. and Thirsk, H.R. (1962) Kinetics of the electrocrystallisation of thin films of calomel. *Transactions of the Faraday Society*, **58**, 2200–2216.

(24) Abyaneh, M.Y. and Fleischmann, M. (1981) The electrocrystallisation of nickel: part II. Comparison of models with the experimental data. *Journal of Electroanalytical Chemistry*, **119**, 197–208.

(25) Bindra, P., Fleischmann, M., Oldfield, J.W. and Singleton, D. (1973) Nucleation. *Faraday Discussions of the Chemical Society*, **56**, 180–198.

(26) Hyde, M. and Compton, R. (2003) A review of the analysis of multiple nucleation with diffusion controlled growth. *Journal of Electroanalytical Chemistry*, **549**, 1–12.

(27) Deutscher, R.L. and Fletcher, S. (1998) The deconvolution of nucleation and growth rates from electrochemical current-time transients. *Journal of the Chemical Society, Faraday Transactions*, **94**, 3527–3536.

(28) Scharifker, B.R. and Mostany, J. (1984) Three-dimensional nucleation with diffusion controlled growth: part I. Number density of active sites and nucleation rates per site. *Journal of Electroanalytical Chemistry*, **177**, 13–23.

(29) Heerman, L., Matthijs, E. and Langerock, S. (2001) The concept of planar diffusion zones. Theory of the potentiostatic transient for multiple nucleation on active sites with diffusion-controlled growth. *Electrochimica Acta*, **47**, 905–911.

(30) Scharifker, B.R., Mostany, J., Palomar-Pardavé, M. and González, I. (1999) On the theory of the potentiostatic current transient for diffusion-controlled three-dimensional electrocrystallisation processes. *Journal of the Electrochemical Society*, **146**, 1005–1012.

(31) Mazaira, D., Borrás, C., Mostany, J. and Scharifker, B.R. (2009) Three-dimensional nucleation with diffusion-controlled growth: simulation of hierarchical diffusion zones overlap. *Journal of Electroanalytical Chemistry*, **631**, 22–28.

(32) Hyde, M., Jacobs, R. and Compton, R.G. (2002) In situ AFM studies of metal deposition. *Journal of Physical Chemistry B*, **106**, 11075–11080.

(33) Wu, G.Y., Bae, S.E., Gewirth, A.A. *et al.* (2007) Pb electrodeposition on polycrystalline Cu in the presence and absence of Cl^-: A combined oblique incidence reflectivity difference and in situ AFM study. *Surface Science*, **601**, 1886–1891.

(34) Deutscher, R.L. and Fletcher, S. (1988) Nucleation on active sites: part IV. Invention of an electronic method of counting the number of crystals as a function of time; and the discovery of nucleation rate dispersion. *Journal of Electroanalytical Chemistry*, **239**, 17–54.

(35) Deutscher, R.L. and Fletcher, S. (1990) Nucleation on active sites: part V. The theory of nucleation rate dispersion. *Journal of Electroanalytical Chemistry*, **277**, 1–18.

(36) Budevski, E. (1996) *Electrochemical Phase Formation. An Introduction to the Initial Stages of Metal Deposition*, VCH, Weinheim.

(37) Cheng, Y. and Schiffrin, D.J. (1996) Electrodeposition of metallic gold clusters at the water/1,2-dichloroethane interface. *Journal of the Chemical Society, Faraday Transactions*, **92**, 3865–3871.

(38) Johans, C., Lahtinen, R., Kontturi, K. and Schiffrin, D.J. (2000) Nucleation at liquid | liquid interfaces: electrodeposition without electrodes. *Journal of Electroanalytical Chemistry*, **488**, 99–109.

(39) Li, Q., Olson, J.B. and Penner, R.M. (2004) Nanocrystalline α-MnO_2 nanowires by electrochemical step-edge decoration. *Chemistry of Materials*, **6**, 3402–3405.

(40) Li, Q. and Penner, R.M. (2005) Photoconductive cadmium sulfide hemicylindrical shell nanowire ensembles. *Nano Letters*, **5**, 1720–1725.

(41) Vaskevich, A., Rosemblum, M. and Gileadi, E. (1996) Underpotential–overpotential transition in a silver overlayer on platinum. Part 2. Reversible 2d–3d rearrangement. *Journal of Electroanalytical Chemistry*, **412**, 117–123.

(42) Milchev, A. (1998) Electrochemical nucleation on active sites—what do we measure in reality? Part I. *Journal of Electroanalytical Chemistry*, **457**, 35–46.

(43) Markov, I., Boynov, A. and Toschev, S. (1973) Screening action and growth kinetics of electrodeposited mercury droplets. *Electrochimica Acta*, **18**, 377–384.

(44) Scharifker, B.R., Serruya, A. and Mostany, J. (1992) On the spatial distribution of nuclei on electrode surfaces. *Electrochimica Acta*, **37**, 2503–2510.

(45) Serruya, A., Mostany, J. and Scharifker, B.R. (1993) Spatial distributions and saturation number densities of lead nuclei deposited on vitreous carbon electrodes. *Journal of the Chemical Society, Faraday Transactions*, **89**, 255–261.

(46) Gunawardena, G., Hills, G. and Scharifker, B.R. (1981) Induction times for the formation of single mercury nuclei on a platinum microelectrode. *Journal of Electroanalytical Chemistry*, **130**, 99–112.

(47) Fleischmann, M., Li, L.J. and Peter, L.M. (1989) Molecular level measurements of the kinetics of nucleation of α-PbO_2 on carbon microelectrodes. *Electrochimica Acta*, **34**, 475–483.

(48) Abyaneh, M.Y., Fleischmann, M., Del Giudice, E. and Vitiello, G. (2009) The investigation of nucleation using microelectrodes: I. The ensemble averages of the times of birth of the first nucleus. *Electrochimica Acta*, **54**, 879–887.

(49) Schuster, R., Thron, D., Binetti, M. *et al.* (2003) Two-dimensional nanoscale self-assembly on a gold surface by spinodal decomposition. *Physical Review Letters*, **91**, 066101.

(50) Mostany, J., Scharifker, B.R., Saavedra, K. and Borrás, C. (2008) Electrochemical nucleation and the classical theory: overpotential and temperature dependence of the nucleation rate. *Russian Journal of Electrochemistry (Translation of Elektrokhimiya)*, **44**, 652–658.

(51) Bazant, M.Z. (2013) Theory of chemical kinetics and charge transfer based on nonequilibrium thermodynamics. *Accounts of Chemical Research*, **46**, 1144–1160.

(52) Cogswell, D.A. and Bazant, M.Z. (2013) Theory of coherent nucleation in phase-separating nanoparticles. *Nano Letters*, **13**, 3036–3041.

5

Organic Electrosynthesis

Derek Pletcher
University of Southampton, Chemistry, UK

Martin Fleischmann was not an organic chemist by experience or interest, but he had a substantial impact on the development of organic electrosynthesis. Early in his career, he had come to recognize that it was the applied potential that controlled the rate and selectivity of all electrode reactions, and he was keen to apply this concept to organic synthesis. Related to this idea, he further recognized that the large potential range available in many systems would allow the oxidation/reduction of rather inert molecules. In principle, many syntheses should therefore be possible by electrolysis. Moreover, electrolysis avoids the creation of equimolar or larger quantities of spent redox reagents, and may also achieve the chemical change without the use of toxic and/or expensive reagents or hazardous conditions. The late 1960s was also the period when electrolysis in aprotic solvents (and nonaqueous solvents, in general) was very popular and there was much emphasis on the study of mechanisms, catalyzed by the widespread introduction of cyclic voltammetry into laboratories. All of these ideas remain core today. Fleischmann also recognized that the design of the electrolysis cell, its operating conditions and the choice of electrode materials, had a marked influence on the yield and selectivity of electrosynthetic processes, whether in the laboratory or on a larger scale. This is, however, a story for another chapter.

In particular, the concepts of potential control and the large driving force for chemical change available at electrodes generated two types of investigation. The first type concerned the realization that the first step in many electrode reactions is a simple, reversible one-electron ($1e^-$) transfer to/from the organic molecules and that "stable" intermediates – for example, anion radicals and cation radicals of aromatic compounds and transition

Developments in Electrochemistry: Science Inspired by Martin Fleischmann, First Edition.
Edited by Derek Pletcher, Zhong-Qun Tian and David E. Williams.
© 2014 John Wiley & Sons, Ltd. Published 2014 by John Wiley & Sons, Ltd.

metal complexes with the metal in unusual oxidation states – are relatively common. Such investigations continue today and the resulting literature is very extensive; here, only two early, illustrative examples will be mentioned. Miller *et al.* [1] employed cyclic voltammetry in methylene chloride at 203 K to investigate the stability of the cation radicals of polycyclic aromatic hydrocarbons. The larger molecules showed the voltammetry for reversible 1e⁻ oxidation and it was clear that their cation radicals had substantial half-lives in these conditions; even the cation radical of anthracene could be observed and shown to have a half-life of ~0.5 s. Pickett *et al.* investigated the voltammetry of metal carbonyls in acetonitrile [2] and trifluoroacetic acid [3] and showed that, even at room temperature, unusual species such as the 17e⁻ cation, $Cr(CO)_6{}^+$, could be formed and that these were stable over many seconds.

The second type of investigation envisaged the electrosynthesis of useful molecules from inert substrates, using the power of electrode reactions to convert inert compounds into reactive intermediates. One system studied was the anodic oxidation of aliphatic hydrocarbons to carbenium ions, a reaction possible in several solvents. The Southampton Group were early contributors to such studies proposing fluorosulfonic acid as the solvent [4, 5]. This solvent is highly conducting and difficult to oxidize, which makes it very suitable as a medium for electrolysis and forcing oxidations. It is also a very strong acid capable of stabilizing cations. In order to allow the formation of synthetically interesting products, acetic acid was used as the electrolyte. In this highly acidic solvent, the acetic acid is a base and ionizes by the reaction:

$$CH_3COOH + H^+ \rightarrow CH_3CO^+ + H_2O$$

A typical electrolysis [4, 5] involves the oxidation of cyclohexane to give methyl 2-methylcyclopent-1-enyl ketone in a single step via a mechanism involving carbenium ion intermediates and the chemistry of the acetyl cation.

The current efficiency and chemical yield of this bifunctional product were both >60%. The introduction during the mid-1960s of a large industrial process for a high-tonnage organic compound by the Monsanto Company [6–9] also had a substantial influence on the objectives of research in the 1960s and 1970s. Moreover, interest in electrosynthesis was further stimulated by the later announcement by Monsanto of the introduction of a second version of the technology for the manufacture of adiponitrile from acrylonitrile with a cheaper and simpler cell design, a lower energy consumption, and a two-phase electrolyte leading to a more straightforward product isolation; overall, this second-generation technology therefore had a much enhanced performance. These processes involved the conversion of acrylonitrile to adiponitrile,

$$2CH_2=CHCN + 2H_2O + 2e^- \longrightarrow \begin{array}{c} CH_2CH_2CN \\ | \\ CH_2CH_2CN \end{array} + 2OH^-$$

which was an intermediate in the manufacture of Nylon 66, a polymer produced on a scale of more than one million tons per year in the USA alone. At its peak, electrolysis was producing more than 200 000 tons per year of adiponitrile, and this led to an emphasis on the production of high-tonnage chemicals on an industrial scale such that performance was judged predominantly by the energy consumption and cost of both the electrolysis cell stacks (hence, the current density that could be used) and the related unit processes.

Through the subsequent years the focus of research has changed significantly. First, it quickly became apparent that the number of organic chemicals produced on a large scale is small, and always the requirement is for very cheap technology. Consequently, interest from chemical companies has shifted towards the synthesis of high-cost, low-tonnage chemicals by "green" technology using the ability of electrolysis to carry out selective chemical changes without creating waste streams or using toxic or hazardous chemicals or high temperatures and/or pressures. Many of the transformations of interest involve chemistry with complex, polyfunctional molecules, and in consequence the most critical performance factor is usually the selectivity of the oxidation/reduction with the electrolysis conditions, though conversion and ease of product recovery are also important. Water and methanol have returned as favored solvents because their solutions have low toxicity, acceptable conductivity and clean counter electrode chemistries (see later). Interest in laboratory synthesis was also increased when the synthesis of sub-1 g quantities may be sufficient for testing activity.

5.1 Indirect Electrolysis

A recurring theme has been indirect electrolysis, where an electrode reaction is employed to maintain a reagent in an appropriate oxidation state, while the known chemistry of the reagent can be used to formulate conditions for selective reactions. Initially, the key reactants were metal ions such as $Cr_2O_7^{2-}$, Co^{3+}, Ag^{2+} and Sn^{2+} and, indeed, dichromate-mediated oxidations of montan wax and anthracene have been used industrially. Perhaps the most elegant example in modern technology is the electrolytic oxidation of naphthalene to naphthaquinone, which was subsequently converted catalytically to anthraquinone by reaction with butadiene [10–12]. Cerium(IV) was first generated in aqueous methanesulfonic acid using a membrane-divided, parallel plate cell with a coated titanium anode, and then reacted with solid naphthalene in a separate reactor.

$$Ce^{3+} - e^- \longrightarrow Ce^{4+}$$

The yield of naphthaquinone was >90%. This process was initially developed in the laboratories of W.R. Grace and Company, but then scaled-up to an industrial unit by HydroQuebec and a Canadian company using commercially available cells, membranes, and electrode materials.

The academic world has investigated more sophisticated mediators with a view to increasing the selectivity and sophistication of the catalyzed chemistry. Typical examples are:

- The macrocycles of nickel and cobalt that stabilize the metal(I) oxidation state allowing the use of M(II)L/M(I)L couples to be used for the catalytic reduction of alkyl halides [13, 14]. With alkyl bromides with alkene and alkyne substituents, the reactions can lead to interesting cyclic molecules by intramolecular coupling [15–17].
- TEMPO, a stable organic radical that oxidizes to a stable cation; subsequently, TEMPO$^•$/TEMPO$^+$ has been used for the successful, mediated conversion of alcohols to aldehydes [18–20].

Many of the recent studies [17, 21] have focused on improving the practicality of the mediated procedures (in particular, a convenient and non-toxic solvent and rate of the synthesis (current density)). However, in practice the utility of mediated reactions with such catalysts is commonly limited by the maximum current density that can be achieved and hence, the maximum rate at which product can be formed. Ideally, in synthesis, the mediator should be present at low concentrations with a large excess of the other reactant, but this is only possible with rapid regeneration of the mediator in its electroactive form. This is not necessarily the same as a rapid reaction between the active form of the mediator and the other reactant because the overall catalytic cycle, forming a product and regenerating the mediator, can be a complex sequence of reactions involving the formation of covalently bonded intermediates and their subsequent breakdown. The cleavage of bonds in the covalent intermediate is frequently the rate-determining step in the overall catalytic cycle. Certainly, this is the case with syntheses involving both M(II)L complexes and TEMPO.

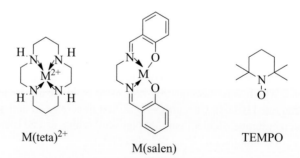

M(teta)$^{2+}$

M(salen)

TEMPO

5.2   Intermediates for Families of Reactions

An attractive concept is to use the electrode reaction to produce a reactive intermediate that can be used for *in-situ* reactions with a variety of substrates, leading to a family of larger molecules. Nematollahi and coworkers [21–23] have described a number of syntheses where an anode is employed to convert catechols to *o*-benzoquinones; the latter are trapped with a nucleophile in a Michael reaction to produce a substituted molecule. As such chemistry is possible with substituted catechols and a wide variety of nucleophiles, the approach has considerable generality. Indeed, it can also be extended to the coupling of *p*-benzoquinones, formed by oxidation of the appropriate hydroquinone, to nucleophiles. Moreover, many of the reactions are possible using buffer solutions in water/solvent mixtures, making them convenient and "green" syntheses. Two examples are:

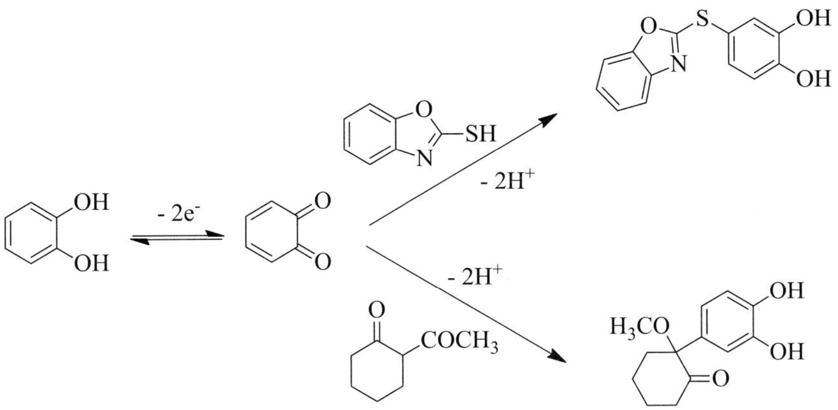

The kinetics of the Michael additions are also conveniently studied by cyclic voltammetry. Utley and colleagues [24, 25] have described a family of reactions based on the cathodic reduction of α,α-dibromo-1,2-dialkylbenzenes to *o*-quinodimethanes and trapping of these metastable intermediates by dienophiles in a Diels–Alder reaction. Here, a typical example is:

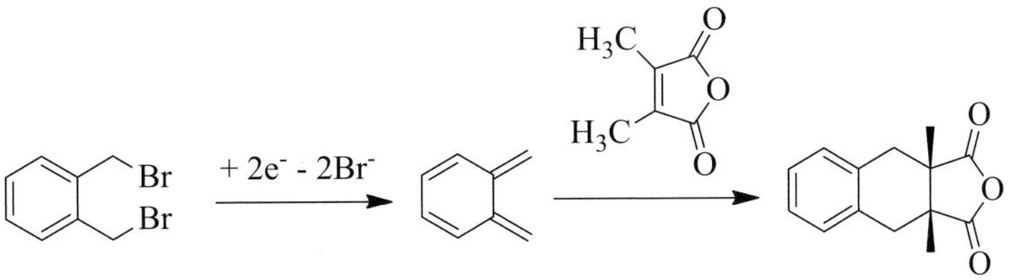

The mechanism of these reactions is not straightforward, however, as the dienophile can act as a redox mediator in the reduction of the α,α-dibromo-1,2-dialkylbenzene to *o*-quinodimethane as well as trapping the reactive intermediate [26].

Yoshida and coworkers [27–29] have shown that it is possible to substantially extend the approach of electrogenerating a metastable intermediate and trapping with a nucleophile or electrophile by employing low temperatures and flow technology combined with the use of non-nucleophilic solvents, and have illustrated the approach with a wealth of examples.

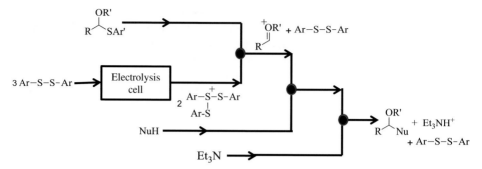

Figure 5.1 *Flow scheme for the indirect generation of alkoxycarbenium ions and their reaction with nucleophiles. Reprinted from Ref. [28]. Copyright © 2011, Wiley-VCH Verlag GmbH & Co. KGaA, Weinheim.*

The early reports showed that a number of cations, normally not stable, could be generated by $2e^-$ oxidation in methylene dichloride at 195 K in a flow cell and then reacted with a number of nucleophiles in a second reactor:

By employing different nucleophiles, a wide range of polyfunctional compounds can be formed in these reactions. The chemistry has been further generalized by introducing a mediator into the chemistry, and this allows the application of even less-stable cations. Figure 5.1 shows the flow scheme for the generation of alkoxycarbenium ions using a disulfide as the mediator.

This is clearly a very flexible procedure that allows the use of a variety of both cationic intermediates and nucleophiles [29]. One illustrative example involves sugar chemistry with the generation of the glycosyl cation:

Anodic methoxylation can be a very general procedure as the first step in introducing functional groups into molecules and the synthesis of polycyclic products. Moreover, it can be a selective synthesis carried out with methanol as the solvent and a high current density using carbon anodes in an undivided cell. The reaction gives particularly good yields with protected nitrogen heterocycles [30–33], where the first step is the oxidation of the N-heterocyclic ring to a cation radical, followed by reaction with the solvent. Two examples of high-yield reactions, carried out on the 100 g scale in flow cells are:

$$- 2e^- - 2H^+, \quad CH_3OH$$

R = H, alkyl, aryl

Further chemistry with the methoxylated products leads to the facile production of polycyclic molecules. Methoxylation of more substituted substrates is well covered in the review by Moeller [34], who has also described a number of reactions where the cation radical generated by the oxidation of an olefin is trapped by an alcohol group in an intramolecular reaction [35–37] to form a cyclic product. Recently, Moeller has applied such reactions to the synthesis of sugar derivatives; an illustrative example is the oxidation of tetramethoxy-furanose derivative:

$$- 2e^- - 2H^+ \quad CH_3OH$$

The reaction is best carried out in a methanol/lithium perchlorate electrolyte when the yield is 85%; the intramolecular coupling reaction is very fast so that the intermolecular methoxylation (i.e., reaction of the cation radical with solvent) does not lead to side products. Moeller *et al.* [38] have also shown that the cation radical intermediates formed in the initial step of oxidation of N-heterocycles can be trapped in an intramolecular reaction, and applied the procedure to form bicyclic products in a single step:

$$- 2e^-$$

Methoxylation has also found application in the industrial production of fine chemicals. Again, the solvent is methanol and undivided, narrow gap cells are suitable. Two reactions carried out in Germany [39] are:

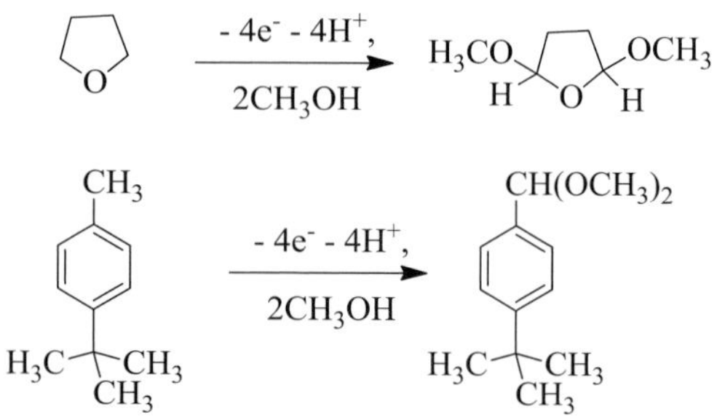

The former reaction is carried out as an indirect electrolysis with bromine as the active intermediate and the product can be converted into the food additive, maltol. The latter reaction is a direct oxidation leading to a product that is readily hydrolyzed to the substituted benzaldehyde. It is also the anode reaction in a paired electrosynthesis where the cathode reaction is the reduction of dimethyl phthalate to phthalide so that the overall cell conversion is:

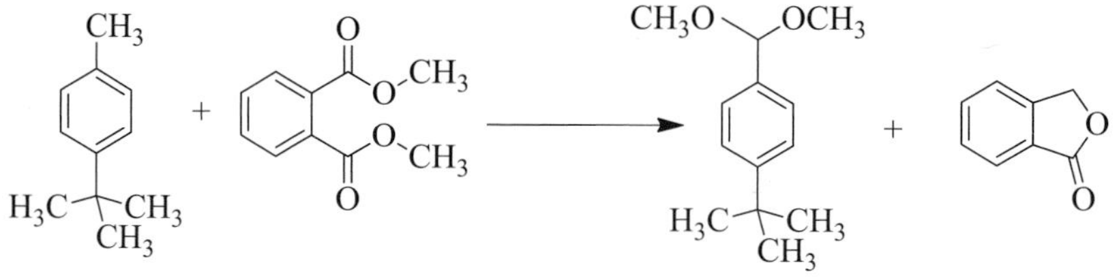

Both products are formed in good yields and are used together in equimolar quantities in downstream chemistry, which makes this electrolysis particularly attractive from an economic viewpoint. This manufacturing process was carried out in a bipolar cell based on a stack of carbon discs with a narrow gap determined by polymer spacers.

5.3   Selective Fluorination

Fluorinated molecules have an important role in the pharmaceutical and fine chemicals industries. For some 60 years, anodic oxidation through the Symons and Phillips processes [40], have been used for the manufacture of perfluorinated small molecules such as alkanes and carboxylic acids. While such processes have met with commercial success, the chemistry is difficult to control and always leads to a mixture of products. The two processes also use very unpleasant and corrosive media, anhydrous hydrogen fluoride, and a HF/KF eutectic at 353 K respectively, that are highly corrosive and difficult to contain or integrate into continuous production. Controlled partial fluorination has proved more of a challenge. Fuchigami and coworkers [41] have, however, developed the application of aprotic solvents such as acetonitrile or 1,2-dimethoxyethane containing either $Et_3N/3HF$ or R_4NF/nHF as the medium for partial fluorination. Such media are less aggressive and can be handled in glass equipment. The same group has also demonstrated the displacement of hydrogen by fluorine in a wide range of molecules with a C–H bond activated by an appropriate substituent. The "parent" reaction was:

$$Y = CF_3, COOEt, CN, COAr, CONH_2, PO(OEt)_2$$

where Y is an electron-withdrawing group. The chemistry can, however, be extended to a wide range of heterocyclic molecules, usually with yields above 80%:

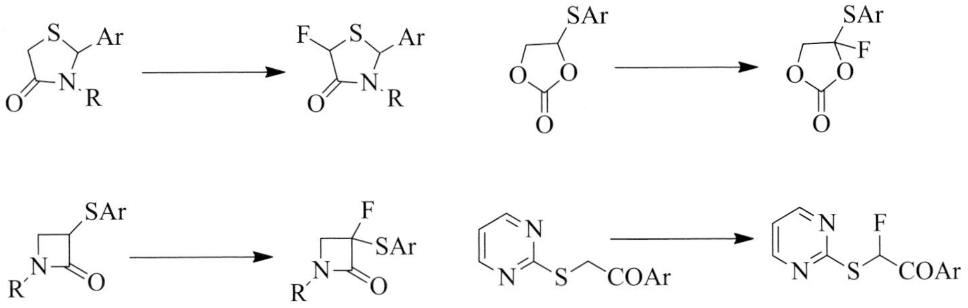

In recent publications, the Fuchigami group [41] have investigated systems without aprotic solvent, so that the medium is an ionic liquid and a similar range of chemistry is possible.

5.4   Two-Phase Electrolysis

Since electrochemistry is generally more suited to aqueous media, while organic chemistry is usually favored by the use of an organic solvent, the concept of two-phase electrolysis can be attractive. The best-known example is the second version of the Monsanto process for adiponitrile [8, 9], in which an emulsion of aqueous electrolyte and acrylonitrile was employed. The roles of the acrylonitrile phase are to maintain the aqueous electrolyte saturated with acrylonitrile, and to extract the adiponitrile back out of the aqueous electrolyte, thereby simplifying product isolation. The organic medium is separated from the aqueous electrolyte, while the reactant and product are separated by distillation. The combination of phase-transfer catalysis with electrolysis introduces new possibilities. Thus, an oxidizing agent may be generated in an aqueous phase, but reacted with an organic molecule in an immiscible organic solvent. For example, benzyl alcohols may be converted selectively to benzaldehydes by the electrolysis of an amyl acetate/aqueous NaBr emulsion also containing tetrabutylammonium ion as the phase-transfer reagent; in this case, the active species, formed at the anode and transferred into the amyl acetate phase, is hypobromite [42]. Another approach envisages the supply of a nucleophile from an aqueous phase to an immiscible organic solvent for an anodic nucleophilic substitution reaction in an organic solvent. Here, an example is the cyanation of naphthalene to 1-cyanonaphthalene that can be achieved in good yield by electrolysis of a methylene dichloride/aqueous NaCN emulsion, again containing tetrabutylammonium ion as the phase-transfer reagent [43]. Other two-phase electrolytic systems are reliant on electron transfer occurring at the interface between the two solvents. An interesting example is the osmium(VIII)-mediated oxidation of olefins to diols (the Sharpless reaction). When the osmium complex contains a chiral ligand, the conversion occurs with a high enantiomeric excess as well as a very high yield. In the electrolytic version of this chemistry, ferrocyanide is oxidized to ferricyanide at an anode, and this converts an Os(VI) complex to the active Os(VIII) species at the interface between the aqueous phase and cyclohexane as the solvent for the organic chemistry [44]. Indeed, the electrosynthesis works well with a low concentration of the osmium reagent. Atobe *et al.* [45] have sought to employ direct electron transfer between an organic molecule and the electrode in an emulsion of an aqueous electrolyte and the organic reactant. In this case, ultrasonication leads to an emulsion where the reactant is present as sub-1 μm droplets in the aqueous phase, and such media have been shown to function well as media for the oxidation of 1-octylamine to 1-cyanoheptane and the hydrogenation of diethyl fumarate. This approach has also been used to generate poly(3,4-ethylenedioxythiophene)(PEDOT) to form layers of electroactive polymer [46].

Okada and Chiba [47] have proposed the application of thermomorphic multiphase systems to electrolysis. On mixing cyclohexane with nitromethane containing lithium perchlorate three phases result, with a cyclohexane-based thermomorphic phase separating the cyclohexane and nitromethane layers. Such systems have been shown to give excellent yields of cyclobutane derivatives formed by the anodic coupling of olefins (see Figure 5.2).

The Fuchigami group have recently proposed a different type of two-phase system with the objective of making electrolysis more convenient and "green". For this, they employed solid silica gel-supported piperidine in a methanol medium for the Kolbe reaction [48]. In the solvent, the piperidine protonates to give the medium some conductivity; the medium then produces almost quantitative yields for the Kolbe dimer, for example:

$$2 \; H_3CO \underset{H_3CO}{\overset{F_3C}{\diagdown}}\!\!-\!\!COO^- \xrightarrow[-\,2CO_2]{-\,2e^-} H_3CO\underset{H_3CO}{\overset{F_3C}{\diagdown}}\!\!\!\diagdown\!\!\!\underset{\underset{OCH_3}{OCH_3}}{\overset{CF_3}{\diagup}}$$

After electrolysis, the solid conducting electrolyte is simply removed by filtration, greatly assisting product isolation. The silica gel-supported piperidine can also be reused for many electrolyses.

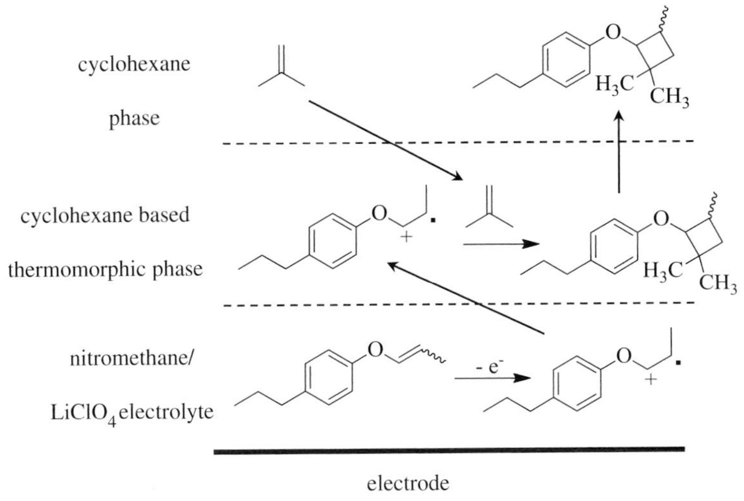

Figure 5.2 *Scheme to illustrate the application of thermomorphic multiphase systems to anodic coupling.*

5.5 Electrode Materials

The selection of electrode materials for electrosynthesis remains a challenge. In principle, different electrode materials can promote different reaction mechanisms [49]. Certainly, there can be no doubt that the choice of electrode material, and even its history and pretreatment, can influence the selectivity and rate of electrosyntheses as well as the importance of competing electrode reactions such as O_2 or H_2 evolution [50]. Hence, in academic and small-scale electrolyses, a wide range of materials has been used. For larger-scale laboratory, pilot-scale and commercial-scale electrosyntheses, the choice is much more restricted after cost and stability within the cell have been taken into account. For oxidations, Pt remains predominant for small-laboratory electrolysis, but attempts to use Pt coatings to mimic bulk Pt have not generally been successful. Likewise, the Ru and Ir oxide coatings on Ti, which have been so successful for the manufacture of large-scale inorganics (e.g., Cl_2, O_2, H_2), have not fulfilled their promise for organic electrosynthesis. In consequence, carbon remains the only choice for many systems although, even then, disintegration can occur unexpectedly and the correct choice of carbon can be important. In aqueous alkali, Ni becomes a viable choice as anode for some electrolyses, although oxidations at Ni do not occur by simple electron transfer but via the mediating couple $Ni(OH)_2/NiO(OH)$ on the surface. Nickel in base is a very effective anode for the conversion of primary alcohols into carboxylic acids and primary amines into nitriles [50, 51], but more recently it has also been used for the oxidation of a wood lignosulfonate into vanillin [52]. For reductions, Hg has provided good performance for small-laboratory electrolyses but its toxicity now makes it an unusual choice. More common are lead (often as an electroplated layer) and carbon, with stainless steel, titanium and nickel as other possibilities. It should also be remembered that some materials are now available in many forms. Reticulated (foam) materials and cloths and meshes (used in multilayer structures) are particularly helpful for increasing the rate of electrosyntheses and hence the space–time yield of cells with no substantial increase in cost.

5.6   Towards Pharmaceutical Products

Currently, the pharmaceutical industry presents many opportunities for electrolytic processes because of the need for new, clean and selective technologies; moreover, the markets for products are for relatively low tonnages. Experience shows, however, that the move from a laboratory scale to production often requires the development of new technology and/or rethinking the relative importance of the criteria determining the process economics.

When the French Company, SNPE, wished to develop new technology for the manufacture of the arylpropionic acid group of anti-inflammatory drugs (see Figure 5.3), they selected the cathodic reduction of chloroethylaromatics in an aprotic medium saturated with carbon dioxide as the preferred route, as it gave good yields on a small laboratory scale:

For scale up [53,54], the reactions presented three main challenges: (i) if the anode reaction involves oxidation of the aprotic solvent, a complex mixture of products would result and the system would become very messy; (ii) the current density for selective conversion is limited by the solubility of carbon dioxide; and (iii) the conductivity of the media is poor. However, these problems were overcome by:

- Using a dissolving metal anode (magnesium); the cell chemistry then leads to magnesium carboxylate aiding product recovery; however, as the anode metal is continuously consumed this type of anode places special restraints on the cell design.
- Employing an elevated pressure.
- Designing a cell with a narrow interelectrode gap.

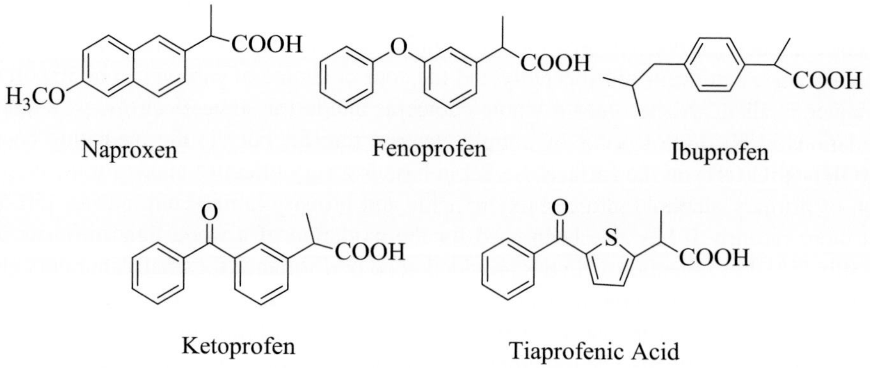

Naproxen   Fenoprofen   Ibuprofen

Ketoprofen   Tiaprofenic Acid

Figure 5.3   *Nonsteroidal, anti-inflammatory drugs with the arylpropionic acid structure.*

Ultimately, this combination of requirements led to the "pencil sharpener" cell, in which the magnesium anode was a rod (diameter 40 cm) with one end shaped as the point of a pencil. The shaped part of the anode sat within a conforming, conical cathode (Pb-plated stainless steel), with the electrodes separated by a thin polymer spacer. The weight of the anode ensured that the anode pressed down on the spacer and that the shape was maintained as metal was lost from the anode. The production unit had two such cells, each with an active electrode area of 0.6 m^2, and was operated with a cell current of 600 A. The operating pressure was 5 bar, and electrolyte was circulated though the cell from a 400 liter reservoir. For the manufacture of fenoprofen the unit was operated in batches, whereby a batch of 60 kg chloroethyldiphenylether was dissolved in 340 kg dimethylformamide that also contained 3 kg tetrabutylammonium bromide. Each batch, when processed for 28 h, led to 50 kg of isolated fenoprofen. The good yield (~80%) showed the electrolytic route to fenoprofen to be very competitive with more traditional syntheses.

The Electrosynthesis Co have described [55] the scale-up of a process for a key step in manufacturing the cephalosporin antibiotic, Ceftibuten (marketed by Schering-Plough):

In this process, the starting material has a very high cost and, hence, the factor that determines the economics of the process is the total conversion of starting material to wanted product, with the highest possible selectivity. Electrochemical reduction is possible in an aqueous phosphate buffer, pH 8, at a temperature of 283–288 K. In this case, the key factor is the choice of cathode material, and tin was found to give the best conversion, though at the cost of a lower current efficiency because of competing hydrogen evolution. The reaction was scaled-up in a commercial parallel plate cell, the FM 21 electrolyzer (electrode areas 0.21 m^2), supplied by ICI that was operated with a Nafion membrane separator and an oxygen-evolving dimensionally stable anode(DSA). When the cell was modified to allow a high-area tin cathode fabricated from several layers of expanded tin mesh to be used, with a current density of 0.1–0.2 A cm^{-2}. The pilot unit was run with batches of 2.84 kg starting material. Obviously, the current efficiency decayed as the starting material was consumed such that, when the batch was driven to almost complete conversion (99%) by using a large charge, the current efficiency fell to an overall value of only 4–6%. This is a clear example of a case where the cost of electricity is of no importance in process economics.

Today, much of the product development conducted in pharmaceutical company laboratories is carried out in microflow systems that allow a total conversion of reactant to product that is rapid, efficient and selective on a scale of 1–100 g for activity testing. The past few years have seen several groups seek to develop electrolysis cell designs that can be used in such equipment. The Southampton group have attempted to develop cells that combine good performance with an appearance similar to other equipment used for routine microflow synthesis. This led to a cell based on a single-patterned microchannel, parallel plate design with an interelectrode gap of 200 μm [56]. When the cell performance was demonstrated using the methoxylation of *N*-formylpyrrolidine [30] as the test reaction, it showed that 1.5 g of product could be formed in less than 1 h, with a conversion >95% in a single pass and with both current efficiency and product selectivity >90%. A cell similar in design is now available commercially [57], and a number of other cells with different degrees of sophistication have been described (see Refs [56, 58–60]). At the same time, an interesting chemistry for microflow cells is being developed. An example of the generation and reaction with nucleophiles of unstable cations was described above [27–29], as were the reactions of anodically generated *o*-quinones with nucleophiles [21–23]. In a recent report [61], these reactions were described in the environment of microflow electrolysis cells, the advantage being that the *o*-quinone could be generated in a microflow cell before being mixed with the nucleophile, thus avoiding the possibility of direct oxidation of the nucleophile at the anode surface. It is difficult to place a separator in most of these microflow cell designs, and this limits their applications to syntheses where the counter electrode chemistry does not interfere or where the overall cell chemistry is fully balanced. An interesting concept has recently been proposed [62] whereby parallel laminar flow within the cell is used to control the separation of the anode and cathode chemistries; this merits further examination as it would lead to a substantial expansion in the possible range of chemistry in microflow systems.

5.7   Future Prospects

So, what are the status and prospects for organic electrosynthesis in 2014? There can be no doubt that electrolysis fits well with the demand for reagentless reactions and "green" technology, as it avoids not only the use of toxic/hazardous reagents but also hazardous conditions in general (most electrolyses are carried out close to ambient temperature and pressure). Certainly, electrolyses do not create waste streams/solids of spent reagents that create environmental hazards. Moreover, there is no shortage of selective and interesting chemistry among reports, often with examples of realistic target molecules. During recent years, there has also been an increasing recognition that electrolysis needs to be presented to synthetic organic chemists in a familiar format. An example is the drive to introduce electrolysis into microflow synthesis using cells that resemble other microflow components and the desirability of conditions that allow the simple recovery of pure products. The need to introduce steps to recycle solvents or electrolyte may also be avoided.

With so many benefits, why has electrolysis not been used more extensively in synthesis, both in the laboratory and in production? The first reason is the predominant use of beaker cells in the chemical literature. Beaker cells inherently have a poor design and commonly have an ill-defined mass transport regime and current distribution, as well as unstated electrode geometry and dimensions, so that the reproduction of electrolysis is impossible. Beaker cells are also seldom capable of generating more than 100 mg of product and so cannot serve as models for scale-up into larger reactors. Although superior and more reproducible flow cell designs have been described, these occur in the electrochemical – or, worse still – in the engineering literature! Second, there is a heavy reliance on Pt and Hg in older reports of organic reactions, which gives organic electrosynthesis a rather quaint image! Third, as chemistry occurs at both the anode and cathode, it is important to consider the overall chemical change in the cell, though this becomes more of an issue as the scale is increased. The target of avoiding byproduct streams must include the counter electrode chemistry, but this again contrasts with reports that usually discuss reactions at only one electrode. It must also be recognized that an electrolyte in the electrolysis medium generally complicates the isolation of a pure product and the recycling of solvents. Lastly, a substantial scale-up requires teams with knowledge of both organic synthesis and electrochemical engineering; in general, the conditions for high-yield chemistry must be maintained while, with increasing scale, issues such as the mass transport regime and current distribution become more important.

During recent years, a significant number of organic electrosyntheses have been scaled-up to pilot or commercial scale, and some of these have been described at conferences and in reports or book chapters. It is important that research teams starting the development of a new electrosynthetic process learn from the labors of others, and do not repeat their efforts and mistakes; to slowly "reinvent the wheel" is not an advance!

References

(1) Byrd, L., Miller, L.L. and Pletcher, D. (1972) The oxidation of aromatic compounds in methylene chloride at −70°C. *Tetrahedron Letters*, **13**, 2411–2415.
(2) Pickett, C.J. and Pletcher, D. (1975) The electrochemical oxidation and reduction of binary metal carbonyls in aprotic solvents. *Journal of the Chemical Society, Dalton Transactions*, 879–886.
(3) Pickett, C.J. and Pletcher, D. (1976) Anodic oxidation of metal carbonyls in trifluoroacetic acid; stabilities of some 17e cations. *Journal of the Chemical Society, Dalton Transactions*, 636–638.
(4) Bertram, J., Fleischmann, M. and Pletcher, D. (1971) The anodic oxidation of alkanes in fluorosulfonic acid; a novel synthesis of $\alpha\beta$-unsaturated ketones. *Tetrahedron Letters*, 349–353.
(5) Bertram, J., Coleman, J.P., Fleischmann, M. and Pletcher, D. (1973) The electrochemical behaviour of alkanes in fluorosulfonic acid. *Journal of the Chemical Society, Perkin Transactions 2*, 374–381.
(6) Baizer, M.M. (1964) Electrolytic reductive coupling, I acrylonitrile. *Journal of the Electrochemical Society*, **111**, 215–222.
(7) Baizer, M.M. and Danley, D.E. (1979) Discovery, development and commercialisation of the electrochemical adiponitrile process, parts I and II. *Chemistry and Industry*, 435–447.
(8) Danley, D.E. (1984) Development and commercialisation of the Monsanto adiponitrile process. *Journal of the Electrochemical Society*, **135**, 435C–442C.
(9) Pletcher, D. and Walsh, F.C. (1990) *Industrial Electrochemistry*, 2nd edn, Kluwer, pp. 298–311.
(10) Kreh, R.P., Spotnitz, R.M. and Lundquist, J.T. (1987) Selective oxidations with ceric methanesulfonate and ceric trifluoromethanesulfonate. *Tetrahedron Letters*, **28**, 1067–1068.
(11) Spotnitz, R.M., Kreh, R.P., Lundquist, J.T. and Press, P.J. (1990) Mediated electrosynthesis with cerium(IV) in methanesulfonic acid. *Journal of Applied Electrochemistry*, **20**, 209–215.
(12) Harrison, S. and Theoret, A. (1999) The electrosynthesis of naphthaquinone and tetrahydroanthraquinone. *Journal of New Materials for Electrochemical Systems*, **2**, 1–9.
(13) (a)Gosden, G., Healy, K.P. and Pletcher, D. (1978) Reaction of electrogenerated square planar nickel(I) complexes with alkyl halides. *Journal of the Chemical Society, Dalton Transactions*, 972–976; (b)Gosden, G. and Pletcher, D. (1980) The catalysis of the electrochemical reduction of alkyl bromides by nickel complexes: the formation of carbon-carbon bonds. *Journal of Organometallic Chemistry*, **186**, 401–409.
(14) Gosden, G., Kerr, J.B., Pletcher, D. and Rosas, R. (1981) The electrochemistry of square planar macrocyclic nickel complexes and the reaction of Ni(I) with alkyl bromides. *Journal of Electroanalytical Chemistry*, **117**, 101–108.

(15) Esteves, A.P., Freitas, A.M., Medeiros, M.J. and Pletcher, D. (2001) Reductive intramolecular cyclisation of unsaturated halides by Ni(II) complexes. *Journal of Electroanalytical Chemistry*, **566**, 39.

(16) Esteves, A.P., Ferreira, E.C. and Medeiros, M.J. (2007) Selective radical cyclisation of propargyl bromoethers to tetrahydrofuran derivatives via electrogenerated nickel(I)(tetramethylcyclam)$^{2+}$. *Tetrahedron*, **63**, 3006–3009.

(17) Medeiros, M.J., Neves, C.S.S., Pereira, A.P. and Dunach, E. (2011) Electroreductive intramolecular cyclisation of bromoalkoxylated derivatives by nickel(I) (tetramethylcyclam)$^{2+}$ in 'green' media. *Electrochimica Acta*, **56**, 4498–4503.

(18) Semmelhack, M.F., Chou, C.S. and Cortes, D.A. (1983) Nitroxyl-mediated electrooxidation of alcohols to aldehydes and ketones. *Journal of the American Chemical Society*, **105**, 4492–4493.

(19) Bobbitt, J.M. and Flores, C.L. (1988) Organic nitrosonium salts as oxidants in organic chemistry. *Heterocycles*, **27**, 509–533.

(20) Kuleshova, J., Hill-Cousins, J.T., Birkin, P.R. *et al.* (2012) TEMPO mediated electrooxidation of primary and secondary alcohols in a microfluidic electrolysis cell. *ChemSusChem*, **5**, 326–331.

(21) Nematollahi, D. and Tammari, E. (2005) Electroorganic synthesis of catecholthioethers. *Journal of Organic Chemistry*, **70**, 7769–7772.

(22) Nematollahi, D., Alimoradi, M. and Rafiee, M. (2007) Kinetic study of electrochemically induced Michael reactions with 2-acetylcyclohexanone and 2-acetylcyclopentanone. *Journal of Physical Organic Chemistry*, **20**, 49–54.

(23) Mazloum-Arkakani, M., Khoshroo, A., Nematollahi, D. and Mirjalili, B. (2012) Electrochemical study of catechol derivatives in the presence of β-diketones: Synthesis of benzofuran derivatives. *Journal of the Electrochemical Society*, **159**, H912–H917.

(24) Eru, E., Hawkes, G.E., Utley, J.H.P. and Wyatt, P.B. (1995) Diels–Alder reactions of *o*-quinodimethanes from the cathodic reduction of α,α-dibromo-1,2-dialkylbenzenes. *Tetrahedron*, **51**, 3033–3044.

(25) Oguntoye, E., Szunerits, S., Utley, J.H.P. and Wyatt, P.B. (1996) Diels–Alder trapping of *o*-quinodimethanes by redox mediated cathodic reduction of α,α-dibromo-*o*-xylene in the presence of hindered dienophiles. *Tetrahedron*, **52**, 7771–7778.

(26) Utley, J.H.P., Ramesh, S., Salvatella, X. *et al.* (2001) Quinodimethanes chemistry. Electrogeneration and reactivity of *o*-quinodimethanes. *Journal of the Chemical Society, Perkin Transactions 2*, 153–163.

(27) Yoshida, J. (2005) Flash chemistry using electrochemical method and microsystems. *Chemical Communications*, 4509–4516.

(28) Saito, K., Ueoka, K., Matsumoto, K. *et al.* (2011) Indirect cation-flow method: the generation of alkoxycarbenium ions and studies on the stability of glycosyl cations. *Angewandte Chemie, International Edition*, **50**, 5153–5156.

(29) Ashikari, Y., Nokami, T. and Yoshida, J. (2011) Integrated electrochemical–chemical oxidation mediated by alkoxysulphonium ions. *Journal of the American Chemical Society*, **133**, 11840–11843.

(30) Nyberg, K. and Servin, R. (1976) Large scale electrolysis in organic systems. III synthesis of α-methoxyalkylamides. *Acta Chemica Scandinavica Series B*, **30**, 640–642.

(31) Shono, T., Hamaguchi, H. and Matsumura, Y. (1975) Anodic oxidation of carbamates. *Journal of the American Chemical Society*, **97**, 4264–4268.

(32) Palasz, P.D. and Utley, J.H.P. (1984) Regioselectivity and the stereochemistry of anodic methoxylation of *N*-acylpiperidines and *N*-acylmorpholines. *Journal of the Chemical Society, Perkin Transactions 2*, 807–813.

(33) Palasz, P.D., Utley, J.H.P. and Hardstone, J.D. (1984) An entry into the quinolizidine and benzoquinolizidine ring systems via anodic oxidation. *Acta Chemica Scandinavica Series B*, **38**, 281–292.

(34) Moeller, K.D. (2000) Synthetic applications of anodic electrochemistry. *Tetrahedron*, **56**, 9527–9554.

(35) Moeller, K.D. (2009) Intramolecular anodic coupling reactions; using cation cation intermediates to trigger new Umpolung reactions. *Synlett*, **8**, 1208–1218.

(36) Xu, G. and Moeller, K.D. (2010) Anodic coupling reactions and the synthesis of C-glucosides. *Organic Letters*, **12**, 2590–2593.

(37) Xu, H. and Moeller, K.D. (2010) Intramolecular anodic olefin coupling reactions: use of the reaction rate to control substrate/product selectivity. *Angewandte Chemie, International Edition*, **49**, 8004–8007.

(38) Cornille, F., Fobian, Y.M., Slomczynska, U. *et al.* (1994) Anodic amide oxidations: conformationally restricted peptide building blocks from direct oxidation of dipeptides. *Tetrahedron Letters*, **35**, 6989–6992.

(39) Pütter, H. (1999) Recent Developments in Industrial Organic Electrosynthesis, at Applied Electrochemistry for the New Millennium, Clearwater Beach, Florida, November 1999.

(40) Childs, W.V., Christensen, L., Klink, F.W. and Kolpin, C.F. (1991) Anodic fluorination, Chapter 26 in *Organic Electrochemistry*, 3rd edn (eds M.M. Baizer and H. Lund), Marcel Dekker.

(41) Fuchigami, T. and Inagi, S. (2011) Selective electrochemical fluorination of organic molecules and macromolecules and macromolecules in ionic liquids. *Chemical Communications*, **47**, 10211–10223.

(42) Pletcher, D. and Tomov, N. (1977) The oxidation of alcohols by electrogenerated hypobromite using a phase transfer reagent. *Journal of Applied Electrochemistry*, **7**, 501–504.

(43) Ellis, S., Pletcher, D., Gough, P. and Korn, S.R. (1982) Electrosynthesis in systems of two immiscible liquids and a phase transfer catalyst. I – the anodic cyanation of naphthalene. *Journal of Applied Electrochemistry*, **12**, 687–691.

(44) Amundsen, A.R. and Balko, E.N. (1992) Preparation of chiral diols by the osmium-catalysed indirect oxidation of olefins. *Journal of Applied Electrochemistry*, **22**, 810–816.

(45) Atobe, M., Ikari, S., Nakabayashi, K. *et al.* (2010) Electrochemical reaction of water-insoluble organic droplets in aqueous electrolytes using acoustic emulsification. *Langmuir*, **26**, 9111–9115.

(46) Asami, R., Atobe, M. and Fuchigami, T. (2005) Electropolymerisation of an immiscible monomer in aqueous electrolytes using acoustic emulsification. *Journal of the American Chemical Society*, **127**, 13160–13161.

(47) Okada, Y. and Chiba, K. (2010) Continuous electrochemical synthetic system using a multiphase electrolyte solution. *Electrochimica Acta*, **55**, 4112–4119.

(48) Kurihara, H., Fuchigami, T. and Tajima, T. (2008) Kolbe carbon-carbon coupling electrosynthesis using solid-supported bases. *Journal of Organic Chemistry*, **73**, 6888–6890.

(49) Couper, A.M., Pletcher, D. and Walsh, F.C. (1990) Electrode materials for electrosynthesis. *Chemical Reviews*, **90**, 837–865.

(50) Amjad, M., Pletcher, D. and Smith, C.Z. (1977) The oxidation of benzyl alcohols and phenylethanols at Ni in alkaline *t*-butanol/water mixtures. *Journal of the Electrochemical Society*, **124**, 203–208.

(51) Schäfer, H.J. (1987) Oxidation of organic compounds at the nickel hydroxide electrode. *Topics in Current Chemistry*, **142**, 101–129.

(52) Smith, C.Z., Utley, J.H.P. and Hammond, J.K. (2011) The electro-oxidative conversion at a laboratory scale of a lignosulfonate into vanillin in an FM01 filter press reactor; preparative and mechanistic aspects. *Journal of Applied Electrochemistry*, **41**, 363–375.

(53) Chaussard, J. (1990) Electrochemical synthesis of fenoprofen, in *Electrosynthesis – From Laboratory, to Pilot, to Production* (eds J.D. Genders and D. Pletcher), The Electrosynthesis Co., pp. 165–176.

(54) Chaussard, J., Troupel, M., Robin, Y. *et al.* (1989) Scale-up of electrocarboxylation reactions with a consumable anode. *Journal of Applied Electrochemistry*, **19**, 345–348.

(55) Chai, D., Genders, J.D., Weinberg, N.L. *et al.* (2002) Ceftibuten: development of a commercial process based on cephalosporin C. Part IV. Pilot-plant scale electrochemical reduction of 3-acetoxymethyl-7(*R*)-glutaroylaminoceph-3-em-4-carboxylic acid 1(*S*)-oxide. *Organic Process Research & Development*, **6**, 178–183.

(56) Kuleshova, J., Hill-Cousins, J.T., Birkin, P.R. *et al.* (2012) The methoxylation of *N*-formylpyrrolidine in a microfluidic electrolysis cell for routine synthesis. *Electrochimica Acta*, **69**, 197–202.

(57) http://www.syrris.com/.

(58) Attour, A., Dirrenberger, P., Rode, S. *et al.* (2011) A high pressure single-pass high conversion electrochemical cell for intensification of organic electrosynthesis processes. *Chemical Engineering Science*, **66**, 480–489.

(59) Kupper, M., Hessel, V., Lowe, H. *et al.* (2003) Micro reactor for electroorganic synthesis in the simulated moving bed-reaction and separation environment. *Electrochimica Acta*, **48**, 2889–2896.

(60) Simms, R., Dubinsky, S., Yudin, A. and Kumasheva, E. (2009) A method for fabricating microfluidic electrochemical reactors. *Lab On A Chip*, **9**, 2395–2397.

(61) Kashiwagi, T., Amemiya, F., Fuchigami, T. and Atobe, M. (2012) In situ electrogeneration of *o*-benzoquinone and high yield reaction with benzenethiols in a microflow system. *Chemical Communications*, **48**, 2806–2808.

(62) Amemiya, F., Matsumoto, H., Fuse, K. *et al.* (2011) Product selectivity control induced by using liquid-liquid parallel laminar flow in a microreactor. *Organic and Biomolecular Chemistry*, **9**, 4256–4265.

6

Electrochemical Engineering and Cell Design

Frank C. Walsh[1] and Derek Pletcher[2]
[1] *University of Southampton, Engineering Sciences, UK*
[2] *University of Southampton, Chemistry, UK*

Although electrochemical technology already had a history extending over almost a century, it was not until the late 1960s that "Electrochemical Engineering" sought to become a recognized academic discipline. Along with other pioneers such as MacMullen, Tobias, Newman, Ibl and Goodridge, Fleischmann sought to introduce topics such as the design and scaling of electrolysis cells, fluid flow, mass transport regimes, current and potential distribution, polar and bipolar electrical connection and bypass currents into university research programmes and postgraduate courses, with mathematical modeling as an important strand of the discipline. It was Fleischmann's decision to seek the funding that allowed the establishment of an Electrochemical Engineering Group in Southampton. This group, led by Martin Fleischmann and Bob Jansson, had extended programmes related to fluidized- and packed-bed electrodes, capillary gap cells, bipolar trickle tower cells and pump cells, taking an interest in the in-cell environment and its influence on performance as well as their applications. The Group also had a substantial interest in the application of their ideas to industrial electrolysis. The concepts developed led to a clearer view of the factors that influence the economics of electrolytic technology and recognition that the dominant factors will depend on the objectives of the process and, maybe, even its location. They were also very influential towards the presentations in various books [1–3] and many short courses.

Along with the availability of new materials, the concepts of electrochemical engineering have led to major enhancements in cell performance for both academic and industrial

Developments in Electrochemistry: Science Inspired by Martin Fleischmann, First Edition.
Edited by Derek Pletcher, Zhong-Qun Tian and David E. Williams.
© 2014 John Wiley & Sons, Ltd. Published 2014 by John Wiley & Sons, Ltd.

practice [1, 4], together with many books and monographs (e.g., Refs [5–11]). Moreover, new electrochemical technology continues to be developed. Due to the dominant requirement for a uniform current distribution, almost all modern electrolysis cells utilize a cell geometry where there is a "uniform gap" between the anode and cathode, and can therefore be regarded as a variant of the parallel plate reactor. Determined by the intended application, differences arise from the relative importance of minimizing energy consumption, increasing cell productivity, maintaining high selectivity or perhaps handling dilute or low-conductivity process solutions. The cell designs, however, incorporate, features such as three-dimensional (3D) electrodes that result from the earlier academic studies.

6.1 Principles of Electrochemical Reactor Design

While a specific cell design usually reflects a particular application, some generalizations and their implications can be helpful. The cell should offer:

- Moderate capital costs, requiring low-cost components and process operation at a high current density.
- Acceptable running costs, determined by a low cell potential and a low pressure drop through the cell (including the inlet and outlet electrolyte flow manifolds).
- When applicable, an undivided cell will simplify engineering needs and allow lower capital and running costs.
- Convenience and reliability in operation, particularly secure gasketting to eliminate electrolyte leakage and long component lifetimes. This also implies adequate design for installation, maintenance and monitoring procedures.
- Appropriate reaction engineering includes controlled (uniform) concentration, current density, potential and flow distributions together with adequate mass transport and provision for the supply of reactants and removal of products.
- Appropriate choice of cell components, both separators and electrode materials. Ion-permeable membranes [13–15] now dominate when a separator is necessary; they introduce selectivity in transport that may be used to achieve the process objectives. The past 30 years has seen great strides in the availability of electrode materials, both catalytic coatings for reactions such as H_2, O_2 and Cl_2 evolution [16–19] and bulk materials in both 2D and 3D forms [20,21]. Both, metals and carbon have become available as sheets, cylinders, clothes, felts, and foams.
- Simplicity and versatility can be the most important factors in achieving an elegant and longlasting design, which is attractive to users.
- Provisions for the future includes a modular design which is easy to scale-up, via an increase in the size or a multiplication of unit cells.

For extended coverage of figures of merit that may be used to quantify cell performance, the reader is referred elsewhere [1, 3]. However, at this point emphasis will be placed on two critical parameters, namely cell potential and the rate of chemical change.

6.1.1 Cell Potential

Together with the current efficiency for the chemical changes at the electrode surfaces, cell potential is the one parameter that determines the energy consumption of the cell. Clearly, the aim is always to minimize the cell potential.

The minimum cell potential is determined by the thermodynamics of the overall chemical change in the cell, and this can only be reduced by changing an electrode reaction (e.g., replacing H_2 evolution by O_2 reduction in a chlor-alkali cell). The equilibrium cell potential difference, E^e_{cell}, is related to the Gibbs free energy change for the cell reaction, ΔG_{cell} by

$$\Delta G_{cell} = -zFE^e_{cell} \tag{6.1}$$

where F is the Faraday constant and z is the number of electrons interchanged. When current is drawn, the cell potential difference can be expressed as:

$$E_{cell} = E^e_{cell} - |\eta^c_{act}| - |\eta^a_{act}| - |\eta^c_{conc}| - |\eta^a_{conc}| - \sum_k IR \tag{6.2}$$

where the first term on the right-hand side is set by thermodynamics. η^c_{act} and η^a_{act} are the overpotentials at the cathode and anode arising from the need to drive the kinetics of charge transfer at the two electrodes. These terms dominate at low current densities and can be minimized by selecting appropriate catalysts for the desired reaction and by operating at a higher temperature. The concentration polarizations terms, η^c_{conc} and η^a_{conc}, become important with low reactant concentrations and/or at high current densities that result in depletion of the reactant at the electrode surfaces. They can be minimized with high-surface-area electrodes, high mass transport flow regimes, and/or the introduction of turbulence by movement of the electrode or the use of turbulence promoters in a flowing electrolyte stream. The IR term is the sum of the electronic and ionic resistances across the cell, and comprises all ohmic drops in the system such as those in the electrolyte(s) within the interelectrode gap, membrane, electrodes, current collectors and external electrical contacts. It is important to minimize energy losses in cells by: (i) reducing the overpotentials at the electrodes by suitable choice of catalytic electrode materials; (ii) ensuring efficient mass transport; and/or (iii) minimizing the resistances within electrodes, across the electrolyte(s) and through the membrane.

6.1.2 The Rate of Chemical Change

The overall rate of chemical change at an electrode process depends on the current density, j (the cell current, I, divided by the electrode area, A) for the desired reaction and the selectivity of the chemical change. The latter is usually discussed in terms of the current efficiency, ϕ, for the desired chemical change and is defined as the fraction of the charge passed used in the desired reaction. The space-time yield, Y_{ST} for an electrolytic reactor may be written

$$Y_{ST} = \frac{1}{V_R} \cdot \frac{dn}{dt} = \frac{\phi I}{zFV_R} \tag{6.3}$$

where V_R is the volume of the reactor and n is the number of moles of product. The common units for space-time yield are mol m^{-3} h^{-1}, obtained by multiplying the right-hand side by 3600.

It is important to differentiate between the extreme types of rate control. Under pure charge transfer control, the rate of electron supply/removal at the electrode surface dominates. The rate is very potential-dependent. The partial current for the desired reaction, I, is exponentially related to the overpotential ($\eta = E - E_e$), and can be written as:

$$I = zFAkc \exp\left(\frac{\alpha zF\eta}{RT}\right) \tag{6.4}$$

The overpotential essential to obtain the required current density can be decreased by an appropriate electrocatalyst (increasing the rate constant, k) or increasing the microscopic area of the electrode surface. In Equation (6.4), α is the transfer coefficient.

Under complete mass transport control, the rate of reactant supply or product removal determines a limiting current, I_L, and this is the maximum possible current for a given reaction. The limiting current, I_L, is determined by the mass transport regime close to the electrode surface, characterized by the mass transfer coefficient, k_m, and related to the relative velocity, v, between the electrode and the electrolyte:

$$I_L = zFcAk_m = Kv^x \tag{6.5}$$

It is important to use a high reactant concentration, c, and to seek a high electrode area, A (here the apparent geometric area) or a sufficiently high mass transport coefficient, k_m by using a sufficiently high relative velocity between electrode and electrolyte, v (resulting from flowing the solution or rotating the electrode) or the use of turbulence promoters. Here, K and x are constants that depend on the electrode geometry, the electrolyte composition and temperature. The velocity power index, x, depends on geometry: typical values are 0.33 for developed laminar flow in a channel, >0.5 for turbulent flow in a channel, and 0.7 for a smooth rotating cylinder electrode (RCE) in turbulent flow. Usually, it is also important that the mass transport regime is uniform to all points on the electrode surface; this can be influenced by the cell geometry, and especially by the design on the electrolyte inlet and exit ports.

6.2   Decisions During the Process of Cell Design

6.2.1   Strategic Decisions

In order to rationalize cell design or aid the selection of a particular type of cell geometry, a binary decision tree [12] is offered in Figure 6.1. This simple diagram can be customized and deployed in two different ways. When used retrospectively, it serves to rationalize a diverse field of existing cell designs by considering their major features. Alternatively, it can be used to aid the selection of a particular design or approaches to scale-up. Although, in this chapter, parallel plate reactors will be emphasized, many of the options are still available within this basic design.

6.2.2   Divided and Undivided Cells

Where possible, a single electrolyte compartment in an undivided cell geometry is favored as it considerably simplifies the construction, electrolyte flow circuit and maintenance needs, while avoiding the potential drop and mass balance problems which can be associated with a microporous separator or ion-exchange membrane.

Division of the cell can, however, offer a number of important benefits; it can separate hazardous mixtures of products; prevent unwanted chemical reactions; prevent reactant or product loss at the other electrode; permit the use of dissimilar electrolytes; and protect an anode from corrosion by aggressive species. In addition, ion-permeable membranes can introduce selective ion migration through the membrane, and this can be used in designing the cell chemistry to achieve greater selectivity. When using a membrane it

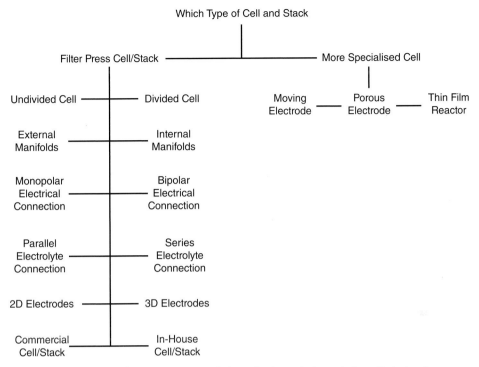

Figure 6.1 *A binary decision tree to aid the selection of electrolytic cell design for a new process. Reproduced from Ref. [12] with permission from IUPAC © 2001.*

is important to establish the ionic (and other) species transporting through it (hence the overall mass balance within the cell, the potential drop across the membrane and the anticipated membrane lifetime/failure modes). The great stability and high current density capability of cation-permeable perfluorocarbon types (such as Nafion®), which have been developed since the early 1970s and are widely used in chlor-alkali, water electrolysis and polymer electrolyte membrane (PEM) fuel cells, leads to their widespread application. In contrast, anion-permeable membranes unfortunately do not have the same stability in many conditions and, in general, have been used only in electrodialysis. Currently, considerable research is being aimed at developing more selective and chemically tailored membranes, or those which have good mechanical stability and ionic conductivity despite being thin. An example of the latter is the use of 25 µm-thick proton-exchange membranes in some fuel cells.

6.2.3   Monopolar and Bipolar Electrical Connections to Electrodes

The electrical connection to electrolysis cells can be either monopolar or bipolar [1, 3]. Monopolar cell stacks require electrical connection to each electrode and expensive control equipment (low voltage, high current), whereas bipolar stacks have electrical connection only to end plates and employ cheaper control equipment (mains supply voltage, lower currents). In addition, a bipolar connection eases the problems of uneven current distribution.

This is especially the case with porous, 3D electrodes such as meshes or foams, when electrical contact only to edges can lead to an uneven current distribution because of the voltage drop through the electrode (due to the passage of current through the electrode structure). The normal strategy is to use stacks with equispaced electrodes of similar size, which will therefore pass the same current and hence develop a similar current density over all of the electrode. Appropriate attention is needed to electrolyte manifolding to achieve uniform velocity profiles across each electrode through the stack. The main drawback with bipolar connection is leakage currents through the electrolyte manifolds internal and external to the cell stack; these must be minimized as they lead to a loss of current efficiency.

6.2.4   Scaling the Cell Current

Maximizing the space-time yield, and thereby minimizing the investment cost in cells, requires the cell current to be maximized. However, different strategies must be used to scale kinetically and mass transport-controlled reactions.

Electrode reactions such as chlorine, oxygen or hydrogen evolution are possible with very high concentrations of reactants, and usually operate under kinetic control; increasing the current can be achieved by developing a better electrocatalyst or by increasing the real surface area of the electrocatalyst on a nanometer–micrometer scale (e.g., by dispersing the catalyst on an inert substrate or by applying the catalyst as a high-area coating).

Many electrolytic processes operate with much lower concentrations of reactants (<1 mM up to a fraction of 1 M) when the maximum current is likely to be determined by mass transport. The approaches are then based on:

- Improving the mass transport regime: In parallel plate cells this is achieved either by increasing the electrolyte flow rate or introducing turbulence promoters into the interelectrode gap.
- Increasing the electrode area on a micrometer–millimeter scale: This can be achieved by roughening the electrode surface or using a 3D structure such as meshes, cloths or foams (these themselves also act as turbulence promoters).

Some idea of the current scaling that can be achieved with dilute reactant solutions is shown in Table 6.1. As the concentration of reactant is increased, the voltage drop through the 3D structure limits the surface area that is used effectively such that only thinner structures are worthwhile. Nonetheless, it is always a strategy that should be considered. Again, it should be noted that the aim should be to have a uniform mass transport regime to all points on the electrode surface, but this will require a careful design of the electrolyte inlets and outlets, as well as the manifolds. Turbulence promoters in the interelectrode gap are also helpful in enhancing the uniformity, as well as increasing mass transport to the electrode.

With gaseous reactants, a high flux of reactant to the site of electron transfer is achieved using a gas diffusion electrode (GDE); originally, such electrodes were developed extensively for fuel cells but are now frequently incorporated into other technologies.

6.2.5   Porous 3D Electrode Structures

As noted above porous, 3D electrodes such as foams, felts, cloths and stacked meshes can lead to a substantial scaling of the cell current, particularly with more dilute solutions. This

Table 6.1 *Enhancement factors (ratio of limiting current for electrode configuration compared to limiting current at a flat plate electrode) for a number of electrode structures measured with dilute reactant solutions. Typical electrolyte flow rate 10 cm s^{-1}; reactant concentration 1 mM.*

Electrode configuration	Enhancement factor	Reference
Flat Plate	1	
Increasing electrolyte flow rate (×10)	2–4	
Flat plate + turbulence promoters	1.5–4	[22]
Roughened surface	3–5	[23]
Four-mesh electrode stack	12–20	[24]
12.5 mm-thick foam	80–170	[25]
12.5 mm-thick felt	3000	[25]

results from both their high surface area (10–500 cm^2 cm^{-3}) and ability to introduce local turbulence. Foams and mesh stacks have a high porosity (20–98% by volume), giving low pressure drops but they may have a mediocre conductivity (which can give rise to a poor potential distribution) and present difficulties in making electrical connections, especially in monopolar stacks.

In the earlier studies, cells were designed specifically for each type of 3D electrode, but more recently 3D electrodes have been incorporated into conventional parallel plate cells and/or stacks. Often, the electrodes are fabricated from cloths or meshes rather than from thicker materials and, as a result, the current scaling is more modest but still significant. Cloth and felt electrodes within parallel plate cells are common in vanadium flow batteries.

6.2.6   Interelectrode Gap

The resistance of the electrolyte phase between the electrodes can be a major contributor to the voltage drop in the cell. Hence, the energy consumption of a process is heavily dependent on the spacing between the anode and cathode. Adopting narrow gap configurations is particularly necessary for processes that operate at a high current density or employ low-conductivity electrolytes.

With undivided, parallel plate cells, narrow gaps are readily achieved by limiting the gap with thin polymers gaskets or spacers. It is also then possible to design cell stacks without resorting to a "filter press design" (see Figure 6.2). Probably the first commercial example of this was the bipolar stack designed for the Monsanto adiponitrile process, using an aqueous/organic emulsion [1, 26, 27]. In this case, a stack of 100 vertical steel plates (each ~1 m × 2 m) and spaced by 2 mm were mounted in a cylindrical pressure vessel. Later, BASF used another bipolar cell stack for several electrolyses with methanol electrolytes [1, 28, 29], in which a series of horizontal carbon discs (diameter ~1 m) were stacked with a separation of 1 mm and the electrolyte was pumped outwards from the center of the discs. Although not a parallel plate design, another way of achieving the same goals is to use a "cylinder in pipe" cell.

With membrane cells, a narrow gap can be achieved by using the "zero gap" concept, where two mesh electrodes are pressed up against the two sides of the membrane so that the membrane fills the interelectrode space. This design has the advantage that the reactant

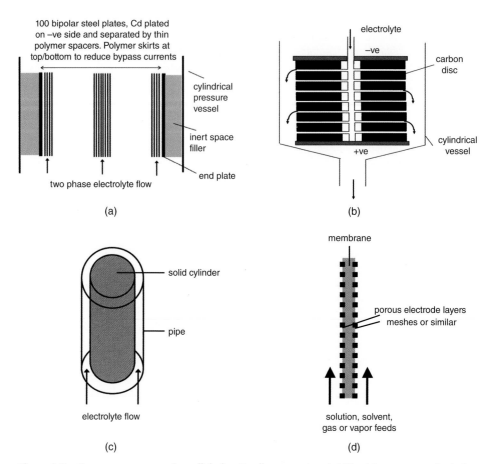

Figure 6.2   *Some narrow gap, "parallel plate" cell geometries. (a) The Monsanto vertical plate stack; (b) The BASF stacked carbon disc cell; (c) A cylindrical cell; (d) A zero-gap membrane cell.*

feeds can be liquid/solution without the electrolyte, or even gases, being present. Moreover, if the electrodes do evolve gas the design forces the gas bubbles out of the interelectrode gap so as to reduce the effect of the gas bubbles in increasing cell resistance.

6.3   The Influence of Electrochemical Engineering on the Chlor-Alkali Industry

The chlor-alkali industry [16, 17] has been one of the great drivers for innovation in electrochemical technology. The reason for this is clear: worldwide, chlorine is manufactured on a scale of some 50 million tons per year at approximately 700 sites, and uses some 15 GW of electrical energy (1–2% of world production). Only a marginal improvement in energy consumption, a more convenient cell operation (less component replacement and/or cell down time), or exit streams closer to the traded forms of the products (particularly 32% NaOH) and/or higher purity would have a major impact on profitability. Hence, new cell designs and – perhaps more importantly – new materials have been developed specifically for the chlor-alkali industry. In contrast, most other applications of electrochemical technology can frequently only test concepts and materials developed for other purposes.

Historically, the conversion of sodium chloride solutions to chlorine and sodium hydroxide was carried out in mercury cells. However, by the 1950s much concern was being expressed about the environmental hazards of the mercury involved, and this led to a demand for a different technology. The first developments were diaphragm cells where Cl_2 and H_2 evolving electrodes were separated by a porous diaphragm, initially asbestos but later polymer alternatives. Unfortunately, these cells suffered from a number of disadvantages. Notably, ion transport through the diaphragm was nonselective, which led to chloride in the sodium hydroxide (due to the transport of chloride from anolyte to catholyte) and the oxygen contamination of the chlorine (due to the transport of hydroxide from catholyte to anolyte); in consequence, the quality and concentration of NaOH that the cell could produce was limited. An additional problem was that the energy consumption could not be improved substantially by upgrading the design of the cell or its components. Together, these problems led to the development of membrane cells in which the electrode reactions were again Cl_2 and H_2 evolution but the separator was a thin membrane that was fabricated from an organic polymer and contained fixed ionic groups designed to be selective to the transport of sodium ions. The overall cell chemistry is shown in Figure 6.3. These revised cells became feasible around 1970 with the commercialization of Nafion^R membranes by Dupont, manufactured from a perfluorinated polymer with side chains leading to sulfonate groups. Dimensionally stable anodes (DSA) with a RuO_2-based coating on titanium and uncatalyzed steel or nickel cathodes were also used.

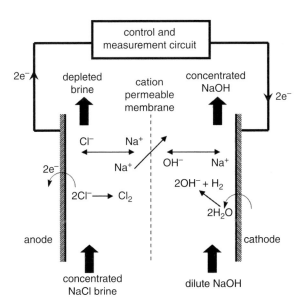

Figure 6.3   *Schematic showing the chemistry of a single chlor-alkali membrane cell.*

Table 6.2   *Advances in membrane cell technology for the chlor-alkali industry. The first cells had a cation-permeable membrane made from a sulfonate polymer resin, catalyst-coated anodes, and steel cathodes.*

Change in technology	Consequences
Improved DSA anode coatings	Lower energy consumption
	Longer time between anode replacement
Catalysed (e.g., high-area NiMo) cathode coatings	Lower energy consumption
Compact stack design with modular cells based on thin frames/gaskets, easily stacked	Ease of cell construction/maintenance
	Lower energy consumption
	Lower process footprint
Bipolar electrical connection	Cheaper control equipment
	Much simplified electrical connection
	More uniform current distribution
Bilayer $-SO_3^-$ polymer/$-COO^-$ polymer membrane	Enables 32% NaOH as cell exit solution
Zero gap cells	Lower energy consumption
	Gas bubbles outside interelectrode gap
	Membrane fully supported
Large area ($\sim$2 m^2), bipolar cells	High space-time yield
	Simplified electrical control
O_2 reduction gas diffusion cathodes	Much lower energy consumption

During the following years a number of improvements were introduced, and these are summarized in Table 6.2. While such advances are listed as single events leading to step improvements, the initial introduction was in many cases followed by continuous R&D, leading to further smaller increments of improvement.

It can be seen that these developments concerned both materials (electrode coatings and membrane) and cell design. The final step in the battle to reduce energy consumption was to replace hydrogen evolution by oxygen reduction as the cathode reaction. Although the cell still produces chlorine and sodium hydroxide, these changes in cell chemistry led to a change in equilibrium cell potential, from $\sim$2.19 V to $\sim$0.96 V. However, in practice the full improvement was not achieved, even when using the most modern catalysts such as rhodium sulfide, because there remained a substantial overpotential associated with oxygen reduction. In addition, the application of oxygen reduction technology requires the introduction of a gas diffusion electrode into the cells, which in turn complicates the design of both cell and stack. Nonetheless, this technology is now available commercially and provides a significant reduction in energy costs.

A consideration of the principles of electrochemical engineering led to the conclusion that the important performance criteria for the Cl_2/NaOH industry are a high current density to minimize the number of cells, a low energy consumption, and long-term, maintenance-free operation, and all of these have now been achieved with membrane cells. Today, the technology is without environmental hazard (although, of course, the sodium hydroxide solution, chlorine and hydrogen must all be handled with care), while cell houses can operate without significant maintenance for over five years. It should be noted, however, that in contrast to many other applications of membrane cells, the fluid flow regime is not critical. The electroactive species are present in high concentrations while the strong gas lift which results from the large flux of gas bubbles rising from the electrodes will alone ensure effective mass transport conditions.

6.4   Parallel Plate Cells

The parallel plate geometry [1, 3] offers uniform current density and potential distribution. The incorporation of this electrode geometry into a plate and frame cell body, particularly in a modular filter-press format, provides a versatile workhorse for many electrochemical reactors. Many developments start with a small, single cell before being scaled-up by increasing the electrode area and then by designing a multiple cell stack in a filter press structure. Parallel plate cells have many advantages:

- The geometry is readily scaled by increasing the electrode area or incorporation into a multiple cell stack, although scaling requires appropriate attention to secure gasketting and the design of electrolyte inlets and outlets to ensure a uniform electrolyte flow.
- A wide range of electrode materials is available, many in suitable or easily modified forms (e.g., plates, meshes, cloths and foams).
- The reaction environment is well understood with regards to fluid flow, mass transport, and concentration distributions.
- Experience of such cells in industrial practice is relatively well documented in the fields of electrosynthesis, electrochemical processing, environmental treatment and flow batteries for energy storage.

For a number of years, several companies have marketed parallel plate cells, including: ICI (the FM01 and FM21 cells [1, 30, 31]); Electrocell (the Microflow, MP, ElectroSyn, and ElectroProd cells [1, 31, 32]); and Electrocatalytic/EA Technology (the Dished Electrode Cell [31, 34]). Each of these companies has offered a range of cells from a laboratory single cell with an electrode area of 10–100 cm^2 to cells with an electrode area of a fraction of 1 m^2 suitable for incorporation into a multicell stack. Unfortunately, this is no longer the case, and developers of electrochemical technology are once again faced with designing and testing their own cells. While this is only inconvenient on a laboratory scale, it is a significant barrier to scale-up as it requires substantial time and the involvement of several disciplines. Nonetheless, much can be learned from the information built up from studies of these commercial cells since, as well as providing typical data, it demonstrates from an experimental viewpoint those factors which are important in determining cell performance.

   For example, the FM21 electrolyser marketed by ICI [1] was originally designed for the chlor-alkali industry but was later applied to other processes. In this case, the unit cell had a membrane and electrodes 1 m wide $\times$ 0.21 m high, and in the chlor-alkali design it was operated in a zero gap configuration with catalyzed lantern blade electrodes (which forced the product gases to leave the interelectrode gap). Up to a 100 cells with a monopolar electrical connection and internal manifolding were mounted between the end plates. As interest in the cell expanded to other applications, a smaller laboratory electrolyser, the FM01-LC (see Figure 6.4) was developed which had an active electrode area of 64 cm^2 but the same ratio of height to width as the FM21, and was intended as an accurate laboratory

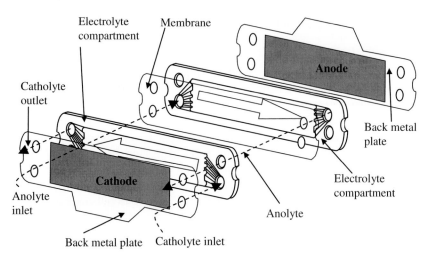

Catholyte outlet

Catholyte inlet

Back metal plate

Figure 6.4 *The FM01-LC electrolyser designed to ape the characteristics of the FM21 electrolyser.*

model for the FM21 electrolyser. Again, a number of cells could be stacked between the end plates, and in order to meet the needs of different applications a variety of electrode designs (e.g., plates, lantern blade, meshes and foams) became available. Studies employing the FM01-LC electrolyser have included:

- The space-averaged mass transport and the mass transport distribution to a flat plate electrode as a function of mean linear flow velocity [22, 35–38].
- Local mass transport studies using segmented electrodes [35].
- Pressure drop as a function of the mean linear flow rate [39].
- The effect of turbulence-promoting meshes on mass transport and current distribution [22, 35, 38]; computational flow modeling [40].
- Mass transport and area enhancement due to 3D electrodes [41].
- Electrosynthesis at carbon and nickel anodes [41–43].
- Flow dispersion in the channel using both experimental residence time distribution [44,45] and computational fluid dynamics mass transport studies, including the effect of manifolds [46,47].

Published experimental studies of the FM21 cell are fewer in number but include comparisons of mass transport in the FM01 and FM21 cells [48,49]. In addition to operation in the chlor-alkali industry [50], the electrolyzer has been used for diverse applications including the removal of nitrate from nuclear waste streams [51], the destruction of contaminated solids via silver(II) [52,53], and the electrosynthesis of a pharmaceutical intermediate [54].

6.5   Redox Flow Batteries

With the current shift towards renewable energy sources, the importance of energy storage is growing rapidly. One approach towards energy storage is based on redox flow batteries [55–57], the most well-developed technology being the all-vanadium flow battery. Such technology for energy storage requires facilities as large or even larger than any existing electrochemical technology, and the batteries have much in common with other flow electrolysis systems.

The Regenesys battery [58] was based on sulfur/polysulfide and bromide/tribromide electrodes, and employed a membrane parallel plate cell. The membrane was Nafion® and the electrodes were composite carbon–polymer plates, sometimes with extended area carbon facings; a polymer mesh (1.8 mm-thick Netlon®) separated each electrode from the ion-exchange membrane. Three sizes of bipolar cell stack were built and tested. Figure 6.5a shows the largest stack, the XL200, which was operated at a current up to 420 A and gave a stack voltage of ~300 V. This bipolar stackstack included 200 cells each with an electrode area of 0.72 m². Each cell frame included gaskets to allow rapid construction and leak-free operation, as well as tortuous inlet and outlet spirals to the manifolds to minimize leakage currents. This modular reactor was conceived with low-cost, production engineering in mind and benefitted from extensive studies of its reaction environment. Figure 6.5b shows a laboratory version of the stack, with only 25 of the XL200 cells; this unit is fully instrumented for fluid flow and mass transport studies and was used for mass transport, fluid dispersion and pressure drop measurements, and also to confirm the conclusions of computer modeling [59,60]. These studies were unusual in that they considered: (i) a stack with full-sized electrodes, albeit with a reduced number (5–25) of bipolar electrodes; and (ii) fluid flow and mass transport in such a small-gap (ca. 2 mm), production-engineered cell.

6.6   Rotating Cylinder Electrode Cells

A rotating cylinder electrode (RCE) with a cylindrical or "pseudo-cylindrical" counterelectrode around it essentially has the geometry of a parallel plate cell. The RCE was developed as a specialty tool for uniform fast electrodeposition in turbulent flow, and for the removal of metal ions from effluents with recovery of the metal in the form of a foil, flake, or powder [61–63]. In the first application, RCE cathodes became the major tool for silver removal from photographic fixer solutions in compact, high-rate units [64], and enabled the recycling of fixer and resale of more than 98 wt% of the silver. Typically, such cells had a stainless steel RCE cathodes of 10–20 cm diameter rotating at speeds up to 1400 rpm, stationary graphite anodes, and were operated at ~50 A.

Later, a membrane-divided RCE cell designed to enable metal deposition in powder form was developed and marketed. The most elaborate of these so-called "Eco-cells" was equipped with scraping devices to facilitate continuous powder removal from the RCE reactor outlet, and was operated potentiostatically, offering a degree of control over metal growth morphology, current efficiency, and selective noble metal deposition [65]. Applications of the unit were focused on the hydrometallurgical extraction of metals and remediation of effluents from chemical industrial process liquors. These devices reached a maximum scale of 10 kA and were capable of copper recovery rates of 40 kg h⁻¹ [61]. A multicompartment, "cascade" version achieved overall conversions of 99% in a single pass [66].

Today, RCE cells find continuing use in controlled metal reclamation; in a recent study, attention was focused on the recovery of precious metals from spent automobile exhaust catalysts [67].

(a)

(b)

Figure 6.5 *(a) The Regenesys XL 200 redox flow battery stack with 200 cells. The unit had an approximate power rating of 120 kW; (b) A laboratory stack with 25 full-sized, bipolar cells, fully instrumented for mass transport, pressure drop, and fluid dispersion studies.*

6.7   Conclusions

Developments in both design and materials have greatly improved the performance of parallel plate cells, including the incorporation of features such as 3D electrodes from academic studies. Unfortunately, "Electrochemical Engineering" has not matured as an academic subject in the way that Martin Fleischmann would have wished, and it remains poorly featured in most electrochemistry courses, research and industrial training programmes. Moreover, from an era when a range of electrolytic reactors could readily be purchased, there has been a return to a time when cells must be designed from first principles by those seeking to upscale electrolytic technology. In consequence, the need to develop a cell stack with adequate performance remains a substantial barrier to new technology. The future would be surely more positive with a more substantial literature on upscaling cells and cell stacks, including the characterization of full-scale industrial cell stacks and computational modeling.

References

(1) Pletcher, D. and Walsh, F.C. (1990) *Industrial Electrochemistry*, 2nd edn, Chapman & Hall, London.
(2) Pletcher, D. (2009) *A First Course in Electrode Processes*, 2nd edn, Royal Society of Chemistry.
(3) Walsh, F.C. (1993) *A First Course in Electrochemical Engineering*, The Electrochemical Consultancy, Romsey.
(4) Marshall, R.J. and Walsh, F.C. (1985) A review of some recent electrolytic cell designs. *Surface Technology*, **24**, 45–77.
(5) Newman, J.S. (1991) *Electrochemical Systems*, 2nd edn, Prentice-Hall.
(6) Kreysa, G. (1986) *Principles of Electrochemical Engineering*, VCH, Weinheim.
(7) Goodridge, F. and Scott, K. (1995) *Electrochemical Process Engineering*, Plenum Press.
(8) Coeuret, F. and Storck, A. (1989) *Eleménts de Génie Electrochimique*, Lavoisier.
(9) Bard, A.J. and Stratmann, M. (eds) (2007) *Electrochemical Engineering, in Encyclopedia of Electrochemistry* (eds D.D. Macdonald and P. Schmuki), vol. **5**, Wiley-VCH.
(10) Pickett, D.J. (1979) *Electrochemical Reactor Design*, 2nd edn, Elsevier.
(11) Wendt, H. and Kreysa, G. (2012) *Electrochemical Engineering: Science and Technology in Chemical and Other Industries*, Springer.
(12) Walsh, F.C. (2001) Electrochemical technology for environmental treatment and clean energy conversion. *Pure and Applied Chemistry*, **73**, 1819–1837.
(13) Davis, T.A., Genders, J.D. and Pletcher, D. (1997) *A First Course in Ion-Permeable Membranes*, The Electrochemical Consultancy.
(14) Strathmann, H. (2004) *Ion-Exchange Membrane Separation Processes in Membrane Science and Technology*, vol. **9**, Elsevier.
(15) Tanaka, Y. (2007) *Ion Exchange Membranes – Fundamentals and Applications*, Elsevier.
(16) Schmittinger, P. (ed.) (2000) *Chlorine: Principles and Industrial Practice*, Wiley-VCH.
(17) O'Brien, T.F., Tilak, B.V. and Hine, F. (eds) (2005) *Handbook of Chlor-Alkali Technology*, vols 1–5, Springer.
(18) Couper, A.M., Pletcher, D. and Walsh, F.C. (1990) Electrode materials for electrosynthesis. *Chemical Reviews*, **90**, 837–865.
(19) Li, X. and Pletcher, D. (2011) Prospects for alkaline zero gap water electrolysers for hydrogen production. *International Journal of Hydrogen Energy*, **36**, 15089–15105.
(20) Pletcher, D. and Walsh, F.C. (1992) Three-dimensional electrodes, in *Electrochemical Technology for a Cleaner Environment* (eds J.D. Genders and N.L. Weinberg), The Electrosynthesis Company Inc.
(21) Walsh, F.C. and Reade, G.W. (1994) Electrochemical techniques for the treatment of dilute metal ion solutions, in *Environmental Oriented Electrochemistry* (ed. C.A.C. Sequeira), Elsevier.
(22) Brown, C.J., Pletcher, D., Walsh, F.C. *et al.* (1993) Studies of space-averaged mass transport in the FM01 electrolyser. *Journal of Applied Electrochemistry*, **23**, 38–43.
(23) Recio, F.J., Herrasti, P., Vasquez, L. *et al.* (2013) Mass transport to a nanostructured nickel electrodeposit of high surface area in a rectangular flow channel. *Electrochimica Acta*, **90**, 507–513.
(24) Lipp, L. and Pletcher, D. (1997) Extended area electrodes based on stacked expanded titanium meshes. *Electrochimica Acta*, **42**, 1101–1111.

(25) Walsh, F.C., Pletcher, D., Whyte, I. and Millington, J.P. (1991) Electrolytic cells with reticulated carbon cathodes for metal ion removal from process streams, part II. Removal of copper (II) from acid sulphate media. *Journal of Applied Electrochemistry*, **21**, 667–671.

(26) Baizer, M.M. and Danley, D.E. (1979) Discovery, development and commercialisation of the electrochemical adiponitrile process, parts I and II. *Chemistry and Industry*, 435–447.

(27) Danley, D.E. (1984) Development and commercialisation of the Monsanto adiponitrile process. *Journal of the Electrochemical Society*, **135**, 435C–442C.

(28) Pütter, H. (1999) Recent Developments in Industrial Organic Electrosynthesis. Applied Electrochemistry for the New Millennium, Clearwater Beach, Florida, November 1999.

(29) Pletcher, D. (2008) Bioelectrosynthesis – electrolysis and electrodialysis, in *Bioelectrochemistry – Fundamentals, Experimental Techniques and Applications* (ed. P.N. Bartlett), John Wiley & Sons, Inc.

(30) Walsh, F.C. and Robinson, D. (1995) Electrochemical synthesis and processing in modern filter-press reactors. *Chemical Technology Europe*, **May/June**, 16–23.

(31) Genders, D.J. (1990) Electrodes, membranes and cell design for electrosynthesis, in *Electrosynthesis – From Laboratory, to Pilot, to Production* (eds J.D. Genders and D. Pletcher), The Electrosynthesis Co.

(32) Carlsson, L., Sandegren, B., Simonsson, D. and Rihosky, M. (1983) Design and performance of a modular, multi-purpose electrochemical reactor. *Journal of The Electrochemical Society*, **130**, 342–346.

(33) Carlsson, L. (1993) Applications of electrocell technology to a cleaner environment, in *Electrochemistry for a Cleaner Environment* (eds N.L. Weinberg and J.D. Genders), The Electrosynthesis Co.

(34) Taama, W.M., Plimley, R.E. and Scott, K. (1996) Mass transport rates in the DEM electrochemical cell. *Electrochimica Acta*, **41**, 543–548.

(35) Brown, C.J., Pletcher, D., Walsh, F.C., Robinson, D. and Hammond, J.K. (1992) Local mass transport effects in the FM01 parallel plate electrolyser. *Journal of Applied Electrochemistry*, **22**, 613–619.

(36) Brown, C.J., Hammond, J.K., Pletcher, D. *et al.* (1992) Mass Transport in Filter Press, Monopolar (FM-type) Electrolysers, Part II – Laboratory Studies in the FM01-LC Reactor. *Dechema Monograph*, **123**, 299–315.

(37) Griffiths, M., Ponce de León, C. and Walsh, F.C. (2005) Mass transport in the rectangular channel of a filter-press electrolyzer (the FM01-LC reactor). *AIChE Journal*, **51**, 682–687.

(38) Brown, C.J., Pletcher, D., Walsh, F.C. *et al.* (1993) Studies of space-averaged mass transport in the FM01-LC laboratory electrolyser. *Journal of Applied Electrochemistry*, **23**, 38–43.

(39) Brown, C.J., Walsh, F.C. and Pletcher, D. (1995) Mass transfer and pressure drop in a laboratory filter-press electrolyser. *Transactions of the Institute of Chemical Engineers*, **73A**, 196–205.

(40) Vázquez, L., Alvarez-Gallegos, A., Sierra, F.Z. *et al.* (2013) CFD evaluation of internal manifold effects on mass transport distribution in a laboratory filter-press flow cell. *Journal of Applied Electrochemistry*, **43**, 453–465.

(41) Brown, C.J., Pletcher, D., Walsh, F.C. *et al.* (1994) Studies of three-dimensional electrodes in the FMO1-LC laboratory electrolyser. *Journal of Applied Electrochemistry*, **24**, 95–106.

(42) Szánto, D.A., Trinidad, P. and Walsh, F.C. (1998) Evaluation of carbon electrodes and electrosynthesis of coumestan and catecholamine derivatives in the FM01-LC electrolyser. *Journal of Applied Electrochemistry*, **28**, 251–258.

(43) Szanto, D.A., Trinidad, P., Whyte, I. and Walsh, F.C. (1996) Electrosynthesis and Mass Transport Measurements in a Laboratory Filter-press Electrolyzer. 4th European Symposium on Electrochemical Engineering, pp. 273–285.

(44) Trinidad, P. and Walsh, F.C. (1996) Hydrodynamic behaviour of the FM01-LC reactor. *Electrochimica Acta*, **41**, 491–502.

(45) Trinidad, P., Ponce-de-León, C. and Walsh, F.C. (2007) The application of flow dispersion models to the FM01-LC laboratory filter-press reactor. *Electrochimica Acta*, **52**, 604–613.

(46) Vázquez, L., Alvarez-Gallegos, A., Sierra, F.Z. *et al.* (2010) Simulation of velocity profiles in a laboratory electrolyser using computational fluid dynamics. *Electrochimica Acta*, **55**, 3437–3445.

(47) Vázquez, L., Alvarez-Gallegos, A., Sierra, F.Z. *et al.* (2010) Prediction of mass transport profiles in a laboratory filter-press electrolyser by computational fluid dynamics modelling. *Electrochimica Acta*, **55**, 3446–3453.

(48) Hammond, J.K., Robinson, D. and Walsh, F.C. (1992) Mass Transport Studies in ICI's Filter-press Monopolar Electrolysers. Part I – Pilot Scale Studies in the FM21-SP Reactor. *Dechema Monograph*, **123**, 279–297.

(49) Robinson, D. (1990) The ICI FM01-LC electrolyzer for electrosynthesis, in *Electrosynthesis – From Laboratory, to Pilot, to Production* (eds J.D. Genders and D. Pletcher), The Electrosynthesis Co.

(50) Girvan, I.J.M., Brereton, C. and Crawford, A.L. (1989) FM21 electrolysers – their application to a range of operations in chlor-alkali plants, in *Modern Chlor-Alkali Technology*, vol. **4** (eds N.M. Prout and J.S. Moorhouse), Elsevier Applied Science.

(51) Steinke, J.L., Hobbs, D.T. and Steeper, T.J. (1997) Pilot Scale Testing of Electrochemical Processes for Nitrate and Nitrite Destruction and Caustic Recovery, 11th International Forum on Electrolysis in the Chemical Industry.

(52) Steele, D.F. (1991) Electrochemistry and waste disposal. *Chemistry in Britain*, **27**, 915–918.

(53) Steele, D.F., Richardson, D., Craig, D.R. *et al.* (1993), Destruction of industrial organic wastes using electrochemical oxidations, in *Electrochemistry for a Cleaner Environment*, (eds N.L. Weinberg and J.D. Genders), The Electrosynthesis Co.

(54) Chai, D., Genders, J.D., Weinberg, N.L. *et al.* (2002) Ceftibuten: development of a commercial process based on cephalosporin C. Part IV. Pilot-plant scale electrochemical reduction of 3-acetoxymethyl-7(R)-glutaroylaminoceph-3-em-4-carboxylic acid 1(S)-oxide. *Organic Process Research & Development*, **6**, 178–183.

(55) Ponce de Leon, C., Fríar-Ferrer, A., Gonsález-Garcia, J. *et al.* (2006) Redox flow cells for energy conversion. *Journal of Power Sources*, **160**, 716–721.

(56) Pletcher, D. (2010) Towards energy storage for renewable generation. *ECS Transactions*, **28** (16), 1–10.

(57) Leung, P.K., Li, X., Ponce de Léon, C. *et al.* (2012) Progress in redox flow batteries, remaining challenges and their applications in energy storage. *RSC Advances*, **2**, 10125–10156.

(58) Price, A., Bartley, S., Male, S. and Cooley, G. (1999) A novel approach to utility scale energy storage. *Power Engineering Journal*, **13**, 122–126.

(59) Ponce-de-León, C., Reade, G.W., Whyte, I. *et al.* (2007) Characterisation of the reaction environment in a filter-press redox flow reactor. *Electrochimica Acta*, **52**, 5815–5821.

(60) Ponce de León, C., Whyte, I., Reade, G.W. *et al.* (2008) Mass transport and flow dispersion in the compartments of a modular, 10 cell filter-press cell stack. *Australian Journal of Chemistry*, **61**, 797–804.

(61) Walsh, F.C. (1992) The role of the rotating cylinder electrode in metal ion removal, in *Electrochemical Technology for a Cleaner Environment* (eds J.D. Genders and N.L. Weinberg), The Electrosynthesis Co. Inc.

(62) Gabe, D.R., Walsh, F.C., Wilcox, G.D. and Gonzalez-Garcia, J. (1998) The rotating cylinder electrode: its continued development and application. *Journal of Applied Electrochemistry*, **28**, 759–780.

(63) Low, C.T., Ponce-de-León, C. and Walsh, F.C. (2005) The rotating cylinder electrode (RCE) and its application to the electrodeposition of metals. *Australian Journal of Chemistry*, **58**, 246–262.

(64) Walsh, F.C. (1986) Cell design considerations for electrolytic removal of silver from photographic processing solution. *Institute of Chemical Engineering Symposium Series*, **98**, 139–149.

(65) Gabe, D.R. and Walsh, F.C. (1990) Recovery of metal from industrial process liquors using a rotating cylinder electrode reactor. *Institute of Chemical Engineering Symposium Series*, **116**, 219–229.

(66) Walsh, F.C., Gardner, N.A. and Gabe, D.R. (1982) The development of eco-cascade cell reactors. *Journal of Applied Electrochemistry*, **12**, 299–309.

(67) Terrazas-Rodríguez, J.E., Gutiérrez-Granados, S., Alatorre-Ordaz, M.A. *et al.* (2011) A comparison of the electrochemical recovery of palladium using a parallel flat plate flow-by reactor and a rotating cylinder electrode reactor. *Electrochimica Acta*, **56**, 9357–9363.

7

Electrochemical Surface-Enhanced Raman Spectroscopy (EC-SERS): Early History, Principles, Methods, and Experiments

Zhong-Qun Tian and Xue-Min Zhang
Xiamen University, State Key Laboratory of Physical Chemistry of Solid Surfaces, China

The field of electrochemical Raman spectroscopy was pioneered by Fleischmann, Hendra and McQuillan of the University of Southampton in 1973. The discovery of surface-enhanced Raman scattering (SERS) was made largely through their efforts and that of Van Duyne from Northwest University during the mid-1970s. With its very high sensitivity and spectral resolution, SERS has been applied to gain meaningful information from an extremely small quantity of species at electrochemical interfaces. With the intrinsically low detection sensitivity of Raman spectroscopy no longer being a disadvantage in its use, electrochemical SERS (EC-SERS) has traveled a tortuous pathway over the past four decades and has developed into a powerful surface diagnostic technique for *in-situ* investigations of adsorption and chemical changes at electrodes. In this chapter, an overview will be provided of the early history of EC-SERS and the principles of SERS outlined to confirm that SERS arises mainly from a nanostructure-based surface plasmon resonance phenomenon. The methodology and measurement procedures – in particular the preparation of various SERS-active substrates – are discussed, and the different approaches employed to utilize the strength and offset the weakness of EC-SERS are highlighted.

Developments in Electrochemistry: Science Inspired by Martin Fleischmann, First Edition.
Edited by Derek Pletcher, Zhong-Qun Tian and David E. Williams.
© 2014 John Wiley & Sons, Ltd. Published 2014 by John Wiley & Sons, Ltd.

Table 7.1 *A comparison of the sensitivity, and energy, time and spatial resolution of electrochemical methods, Raman spectroscopy and scanning tunneling microscopy (STM) in practical electrochemistry study*

	Electrochemistry	EC-Raman spectroscopy	EC-STM
Sensitivity	Submonolayer	Submonolayer[a]	Submonolayer
Energy resolution	10^{-2} V	10^{-4} eV($\sim$1 cm^{-1})[b]	STS is not available[c]
Spatial resolution	0.1–1 μm[d]	1 μm	1 Å
Time resolution	10^{-7}–10^{-6} s	10^{-8} s[e], 10^{-3} s[f]	0.04 s[g]

[a]With surface-enhanced Raman spectroscopy.
[b]With excitation in the visible.
[c]Scanning tunneling spectroscopy; although inelastic electron tunneling spectroscopy provides a higher resolution at about 0.05 eV at very low temperatures (several Kelvin), it is not practical for electrochemistry.
[d]With scanning electrochemical microscopy.
[e]With pulsed laser.
[f]With continuous-wave laser.
[g]0.04 s per STM image.

Today, the structure and dynamics of electrode/solution interfaces play an increasingly important role in electrochemistry, with ramifications of the subject matter extending into areas as diverse as batteries, fuel cells, materials corrosion, clean technologies, semiconductor processing, electrochemical synthesis, the conversion and storage of solar energy, and biological electron-transfer processes [1–3]. Currently, conventional electrochemical methods appear to have approached their limits to meet the growing requirements for characterizing complex and sophisticated systems, especially those related to advanced materials and the life sciences. Since the 1970s, there have been remarkable developments among the *in-situ* nonelectrochemical techniques which provide mechanistic and dynamic information on electrochemical interfaces at the molecular (and/or atomic) level [1, 4–6]. Vibrational spectroscopic methods such as infrared (IR) [7, 8], Raman [9–11], and sum frequency generation (SFG) spectroscopies have been widely used to identify not only interfacial molecules but also the nature of the chemisorption bond [12, 13].

It should be pointed out that each of these techniques has strengths and limitations when studying a particular electrochemical system. Detection sensitivity and resolution are two important criteria when comparing the techniques in terms of their capability to provide high-quality information from interfaces. The resolution can be further classified into energy-resolution, spatial-resolution and time-resolution, and for this purpose a brief comparison of the most frequently used techniques in electrochemistry is provided in Table 7.1.

Detection sensitivity is, undoubtedly, a most important issue as only an extremely small quantity of species will be present at the electrode surface compared to a very much greater amount in bulk solution. In general, existing electrochemical techniques have very high sensitivities that are capable of detecting molecular or atomic changes at the interface in submonolayer quantities; however, unless microelectrodes or nanoelectrodes are used they have a poor time resolution, typically in milliseconds. Whilst nanoelectrodes may be capable of achieving a resolution within microseconds or even nanoseconds, their energy-resolution is poorer at about 10^{-2} eV. In contrast, *in-situ* vibrational spectroscopy has a much higher energy-resolution, with spectral resolutions of approximately 1 cm^{-1} easily being achieved (equivalent to about 10^{-4} eV with visible light excitation). Such a powerful technique is also capable of providing much more detailed structural information with regards to surface bonding, conformation and orientation that, in turn, provides new and powerful insights into various electrochemical interfaces.

Raman spectra can be obtained by using excitations ranging from ultraviolet (UV) to near-IR (NIR) energies. Water, as the most important solvent in electrochemistry, is a very weak Raman scatterer and has almost no absorption in the visible light region. More importantly, Raman spectroscopy can be conveniently applied to *in-situ* measurements of solid–liquid interfaces in both fundamental and practical studies [9–11]. Porous and rough surfaces with high surface areas, which typically are studied with great difficulty when using many surface analysis techniques, can be readily probed with Raman spectroscopy. Furthermore, Raman spectroscopy has a wide spectral window (5 to 4000 cm^{-1}), and requires neither complicated sample preparation nor specialized cell materials. All of these advantages have led to Raman spectroscopy becoming a very convenient and powerful tool for the analysis of electrode/aqueous solution interfaces.

Unfortunately, Raman spectroscopy has encountered major problems due to its intrinsically low detection sensitivity. Notably, as the cross-section of a molecule for the Raman scattering process is about 10^6-fold and 10^{14}-fold smaller than those of IR and fluorescence processes, respectively [11], the Raman-scattered intensity of typically 10^{-8} to 10^{-10} of the incident photon flux is insufficient to detect adsorbed species with (sub)monolayer coverage. Therefore, during the first three decades following its discovery, Raman spectroscopy was not widely used and was not considered for surface analysis. However, following the invention of lasers during the early 1960s, the detection sensitivity of the technique was greatly improved as lasers could provide an ultra-intense light source. Subsequently, as lasers ranging from the UV to the NIR became available, Raman spectroscopy was transformed into a flexible technique that could be used to analyze samples of bulk materials (at the molecular level, termed "normal" Raman measurements) that ranged from lunar rocks to ocean ore, from silicon chips to plastics, and from proteins to drugs [14–18]. However, with regards to interfacial electrochemistry and ultra-trace analysis, for which the number of probe molecules is substantially less than for bulk materials, the detection sensitivity of Raman spectroscopy remained far from satisfactory.

Equation (7.1) gives the Raman intensity expression for a vibrational mode of a molecule following Placzek's polarizability theory with regards to instrumental and surface factors [19]:

$$I_{mn} = \frac{2^7 \pi^5}{3^2 c^4} I_0 \left(v_0 - v_{mn} \right)^4 \sum_{\rho\sigma} \left| \left(\alpha_{\rho\sigma} \right)_{mn} \right|^2 NA\Omega QT_m T_0 \tag{7.1}$$

where N is the number density of the adsorbate (molecules cm^{-2}), A is the surface area illuminated by the laser beam (cm^2), Ω is the solid angle of the collection optics (sr), and $QT_m T_0$ is the product of the detector efficiency, the throughput of the dispersion system and the transmittance of the collection optics, respectively. Even if the probe molecules fully cover a flat surface the number of molecules is only about 10^7 within the laser spot of about 1 μm in diameter. Because, typically, one Raman photon is produced by about 10^8 to 10^{10} incident photons, even with the improvements in $\Omega QT_m T_0$ available in state-of-the-art Raman instruments, it is still not possible to detect electrode surface species with a monolayer coverage. Although the surface Raman scattering intensity scales linearly with the incident laser power (I_0) and the fourth power of energy (v_0), the ultimate sensitivity is limited by the surface damage threshold of the electrode materials. It has, therefore, been highly desirable to develop new methods to remove this limitation.

7.1 Early History of Electrochemical Surface-Enhanced Raman Spectroscopy

The first SERS spectra were obtained from an electrochemical cell during the mid-1970s. In retrospect, it is not surprising that SERS research was initiated from an electrochemical system since, during the late 1960s and early 1970s interest in the study of electrode reactions was shifting from the macroscopic to the microscopic level. Between the late 1960s and early 1980s, spectroelectrochemistry became established with the development of *in-situ* optical spectroscopic methods that provided mechanistic and dynamic information on electrochemical interfaces at the molecular level [4]. Raman spectroscopy was the first vibrational spectroscopy employed to characterize species on solid electrode surfaces by using a light-reflection mode, as aqueous solutions do not lead to notable interferences to the surface signal. The Raman process does have its intrinsic disadvantages, however, owing to its very low detection sensitivity, and therefore electrochemical Raman experiments without some form of signal enhancement proved to be impractical and impossible for many applications.

The first *in-situ* Raman spectroscopic study of an electrochemical system was reported in 1973 by Fleischmann, Hendra and McQuillan, who described Raman spectra for thin films of Hg_2Cl_2, Hg_2Br_2 and HgO, formed on mercury droplets that had been electrodeposited onto platinum electrodes [20]. As these compounds have exceptionally large Raman scattering cross-sections (i.e., they are very good Raman scatterers), a signal from a species composed of only a few monolayers could be detected. These experiments proved, for the first time, the viability of in-situ Raman spectroscopic measurements in electrochemical environments.

In order to overcome the fatal limitation in detection sensitivity for surface species with only monolayer (or less) surface coverage, Fleischmann, Hendra and McQuillan devised a strategy to increase the number of adsorbed molecules and pyridine was chosen as an adsorbate because of its very large Raman cross-section. Pyridine had previously been used in Raman experiments to probe surface sites on oxide catalysts, and in this instance they chose to increase the surface area of a silver electrode in an aqueous electrolyte comprising 0.1 M KCl + 0.05 M pyridine by applying potential-controlled oxidation and reduction cycles. The Raman spectrum obtained from the electrochemically roughened silver electrode was of unexpectedly high quality, and the potential dependence of the signal implied that this was due to the variation in the adsorbed pyridine coverage (as shown in Figure 7.1). All of the major Raman bands were changed markedly in intensity as the potential was changed. The preliminary results were first described in a report in *Faraday Discussions* in 1973 [21, 22], but were then formally published in 1974 [23]. In fact, these were the first SERS measurements to be made, and the roughened silver electrode was the first nanostructure to exhibit SERS activity, though this point was not recognized as such in 1974.

After these preliminary results were published [23], Van Duyne and Jeanmaire tested the hypothesis that such surface roughening would lead to an increased surface area [24] and found – rather surprisingly – that when starting with the degree of roughness used by Fleischmann *et al.* [23] the surface Raman signals were increased as the surface roughness

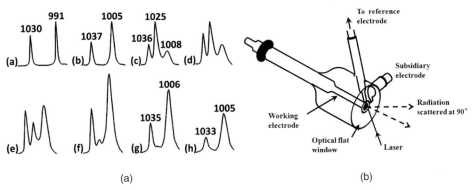

Figure 7.1   (a) SER spectra from pyridine adsorbed onto an electrochemically roughened silver electrode in 0.1 M KCl and 0.05 M pyridine, compared with normal Raman spectra of a: liquid pyridine and b: 0.05 M aqueous pyridine. The applied potentials are: c: 0.0 V; d: −0.2 V; e: −0.4 V; f: −0.6 V; g: −0.8 V; h: −1.0 V; (b) Electrochemistry cells used for the SERS measurements. Reprinted from Ref. [23] with permission from Elsevier.

was decreased. Subsequently, Van Duyne and Jeanmaire devised a procedure to measure the surface enhancement factor, wherein the signal intensity of a specific molecule on the surface could be compared to that of the same molecule in solution. The results of these studies were published in 1977, when the claim was made that " ... some property of the electrode surface or the electrode/solution interface is acting to enhance the effective Raman scattering cross-section", and that this resulted in the Raman intensity from adsorbed pyridine being " ... 5–6 orders of magnitude greater than expected." Independently, in the same year, Albrecht and Creighton reported a similar result from colloidal silver suspensions [25, 26]. The effect was later named surface-enhanced Raman scattering [27].

It is worth noting that, whilst the SERS of pyridine adsorbed at a silver electrode was the first reported case of the technique's ability, it is now a much-studied and classic example. Indeed, after about four decades the system remains one of the best EC-SERS systems in terms of the high quality of spectra obtained and the easy preparation of samples. Certainly, to have chosen this system for subsequent studies was critical to the breakthrough in establishing EC-SERS as a routine process, with great opportunities being offered to design highly sensitive, surface diagnostic techniques that can be applied to not only electrochemical but also to biological and other ambient interfaces. Nowadays, molecular-level investigations using Raman spectroscopy on diverse adsorbates at various transition metal and Group VIII B element electrodes have been realized [28, 29]. Furthermore, the extension of SERS to studies of single-crystal electrodes with truly atomic-smooth surfaces in an electrochemical cell has proved possible during recent years [30, 31]. Clearly, these advances have provided a relatively smooth pathway to the wide application of Raman spectroscopy in electrochemistry.

7.2   Principles and Methods of SERS

It is now well accepted that an enhancement of the electromagnetic field represents a dominant contribution towards the SERS effect [32–34]. Such an enhancement derives from the strong interactions between light (both incident and scattered) and the nanostructured metallic substrates (typically, Au, Ag and Cu). Furthermore, the excitation of surface plasmon resonance (especially the localized version; i.e., localized surface plasmon resonance) contributes most to the enhanced electromagnetic field, and thus is the dominant contribution to a giant enhancement of the SERS effect. Moskovits was the first to introduce the concept of a surface plasmon resonance contribution to SERS when, in 1985, he summarized rather eloquently the electromagnetic enhancement due to surface plasmon resonance as being the dominant contribution to SERS [34]. In some electrochemical systems, chemical enhancement [35] plays an additional role, especially with chemically bonded species, as it is closely associated with the nature of the chemical bond to the metal surface.

Recalling Equation (7.1), which describes the Raman intensity contributed from free molecules, the SERS intensity can expressed in general terms by:

$$I_{SERS} \propto v_{sc}^4 E_{in}^2 E_{sc}^2 \sum_{\rho,\sigma} \left| (\alpha_{\rho\sigma})_{nm} \right|^2 \tag{7.2}$$

where v_{sc} is the frequency of the Raman scattering light. In this case, E_{in} and E_{sc} replace I_0 as the power density of the exciting laser light in Equation (7.1), in order to describe the local electromagnetic (optical electric) fields of the incident and scattering radiation, respectively. The square-terms denote the intensities of the incident and scattering light at the surface, resulting in an electromagnetic enhancement effect [10]. The summation term of $(\alpha_{\rho\sigma})_{mn}$ describes the optical response to intramolecular interactions and the interaction between a molecule and the metal surface, resulting in the chemical enhancement effect.

As the surface-plasmon-enhanced electromagnetic field is the dominant contribution to most cases of SERS, in the next section attention will be focused on the electromagnetic mechanism of SERS.

7.2.1   Electromagnetic Enhancement of SERS

Electromagnetic effects can be divided into several enhancement processes, while the major contribution is from localized surface plasmon resonance. Two requirements are necessary for the presence of localized surface plasmon resonance and great enhancement of SERS: (i) a plasmonic material; and (ii) an appropriate surface roughness or, more exactly, a nanostructured surface. The key materials supporting surface plasmon resonance are always referred to as "plasmonic materials"; these have a real dielectric function with a negative value and an imaginary dielectric function with a positive, but very small, value. The coinage metals (gold, silver, copper) are the most commonly used plasmonic materials. At the plasmon frequency of localized surface plasmon resonance, a large field-induced polarization will occur such that large local fields will be generated at the nanostructured metal surface. (The intensity and frequency of localized surface plasmon resonance depends critically on the size and shape of nanostructure; this will be discussed in Section 7.2.2.) These local fields cause remarkable increases in the Raman scattering intensity of a surface species, which is proportional to the square of the incident radiation field. The nanostructured surface enhances not only the incident laser field but also the Raman scattered field, and therefore the overall enhancement in Raman intensity will scale approximately with the fourth power of the field [as shown in Equation (7.2)].

SERS may also be obtained from the nanostructured transition metals (e.g., Pt, Ru, Rh, Pd, Fe, Co, Ni) in Group VIII-B. However, transition metals do not meet the conditions for good surface plasmon resonance in the visible light region, because the value of the imaginary part of dielectric constants is very large [36, 37]. It should be noted that the surface plasmon is not the only source of enhanced local electromagnetic fields; rather, other types of electromagnetic field, such as the "lightning rod" effect occurring at near curvature points and the dipole image effect, exists on the rough surface. These contributions are much lower than that from the localized surface plasmon resonance, however, and as a result the SERS activity of roughened transition metal surfaces is generally quite low. Experimentally, the average Raman enhancement (from regions of greater field strength, as well as from regions of weaker field strength) may be as high as six or seven orders of magnitude for the nanostructured coinage metals. For transition metals, however, typical surface enhancement factors range from 10 to 10^3, depending on the metal and surface preparations [28, 38].

Overall, SERS is essentially an effect of plasmon-enhanced Raman scattering, and any nanostructures that possess the ability to excite localized surface plasmon resonance in the UV-visible and near-IR regime can also support SERS. The optimization of key factors for localized surface plasmon resonance and Raman scattering will allow an optimization of the conditions for SERS.

7.2.2 Key Factors Influencing SERS

SERS activity, which mainly arises from localized surface plasmon resonance, depends critically on the optical properties, size, shape, and inter-particle space of nanostructures.

7.2.2.1 *Material Dependence*

Besides the coinage metals, alkali metals (Li, Na, K, Rb, Cs) and Al can also function as plasmonic materials, though it is very difficult to prepare stable nanostructured surfaces composed of these chemically active materials. Aluminum is a typical plasmonic material in the UV regime [39–41], while Ga, In, Pt and Rh, as well as their alloys and some complex materials (e.g., TiN), have also been explored as plasmonic materials [42].

7.2.2.2 *Wavelength Dependence*

It should be noted that the SERS effect from coinage metals cannot occur with a too-short wavelength because of the inter-band transitions which transform the incoming optical field into heat instead of enhancing the local electromagnetic field. The typical working wavelength regime for Al is 150–400 nm, for Ag it is 350–1240 nm, for Au 570–1240 nm, and for Cu 600–1240 nm. Therefore, Ag is the best electrode material on which surface SERS can be studied by using various lasers from blue to NIR, while Au and Cu can only be studied using red or NIR lasers.

According to Van Duyne *et al.* [43, 44], in order to obtain SER spectra with enhancement factors as large as possible, the excitation and scattering wavelengths should, if possible, sit at the two sides of a surface plasmon resonance extinction peak [43, 44]. In addition, the excitation wavelength can be selected corresponding to the molecular absorption, and can utilize the resonant Raman effect to further increase the sensitivity, and also avoid interference by the fluorescence background.

7.2.2.3   Distance Dependence

SERS is an interface-sensitive technique, with molecules adsorbed in the first layer at the surface showing the largest enhancements. However, electromagnetic enhancement has a long-range property (i.e., it decays exponentially with increasing distance from the surface, ~1–10 nm) [45–48], which predicts that SERS does not require the adsorbate to be in direct contact with the surface but rather within a certain sensing volume.

Boosted by the long-range effects of electromagnetic enhancement, strategies aimed at "borrowing SERS activity," and thereby extending SERS studies to non-traditional SERS substrates, have been developed [49–63]. Typically, these utilize the enormous electromagnetic field created by high SERS-activity Au or Ag nanostructures. The Raman signals of probed molecules on non-traditional SERS substrates that are spatially separated from the SERS-active substrate can be enhanced to some extent. Figure 7.2 shows, schematically, the electromagnetic field distribution around Au@Pt core–shell nanoparticles under laser irradiation. By effectively borrowing SERS activity from the highly active Au core, weak SERS spectra of adsorbates on the Pt overlayer can be obtained. A detailed discussion of "borrowing SERS activity" in Section 7.2.3.

As shown in Figure 7.2, the electromagnetic field will be attenuated exponentially as the distance between molecules of interest and the surface is increased. As the Raman intensity scales roughly with the fourth power of the electromagnetic field, it can be concluded that the Raman intensity will exhibits an overall r^{-10} distance dependence, where r is

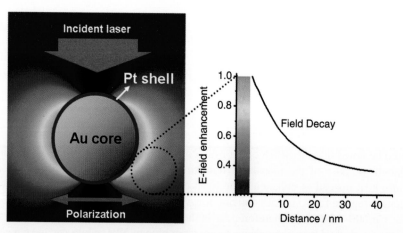

Figure 7.2   *Schematic illustration of the electromagnetic field distribution around Au@Pt core–shell NPs under laser irradiation (left, with 633 nm excitation). The dependence of the electromagnetic field strength (normalized with the strength on the Pt surface) on the distance from the Pt shell is shown in the right-hand plot, indicating that a substantially strong field enhancement can still be obtained on the surface. Reprinted from Ref. [49] with permission from Royal Society of Chemistry.*

the distance between the surface and the molecule [64]. Therefore, the spacer film in the "borrowing SERS activity" has to be ultrathin, normally a few atomic layers.

7.2.2.4 *Size and Shape of the Nanoparticles*

The localized surface plasmon resonance of individual plasmonic nanoparticles depends heavily on the size and shape of each nanoparticle. For instance, the wavelength of the dipolar surface plasmon red shifts with the increase of particle size. However, for much larger nanoparticles new bands for some multipolar modes will appear in the short-wavelength range, while the dipolar band at long-wavelength will be damped. Typically, the size of Au or Ag nanoparticles synthesized for SERS should be less than 150 nm, and larger than 20 nm.

The shape effect in SERS has been recognized as a possible additional source of electromagnetic enhancement. The effect manifests as two concomitant phenomena [65]: (i) there is a shift of the localized surface plasmon resonance; and (ii) there is a modification of the local electromagnetic field on the particle surface, compared to that obtained on a sphere. For instance, the SERS enhancement can indeed be magnified at the tip or corner of nonspherical metallic nanoparticles due to the "lightning rod" effect, when the excitation is polarized along this long axis. However, SERS enhancement factors for most individual particles are too small to be used. In practical terms, Ag or Au dimer, trimer or other aggregates of nanoparticles, are more preferable for SERS.

7.2.2.5 *Electromagnetic Coupling in Nanoparticle Dimer and Aggregates*

In coupled nanoparticle systems, the distance between the nanoparticles plays a key role in the enhancement of the local field and SERS enhancement factor. The smaller the gap size of a gold nanosphere dimer, the stronger the extinction intensity, and the larger the maximum SERS enhancement factor. It is worth noting that, for small gaps, the enhancement factors increase much more quickly than the increase of extinction coefficient. More importantly, when the gap is close to 1–2 nm the enhancement factor in the gap can reach up to 10^8 to 10^{10}, and facilitates single-molecule detection [65].

However, if the gap is reduced to less than 1 nm, then decreases in the local electromagnetic and SERS enhancement are noted [66]. In the sub-nanometer gap regime, the distribution of the electromagnetic field of the dimer nanoparticles cannot be calculated by classical electrodynamics; rather, quantum corrections should be considered, and electron tunneling will cause a reduction in the electromagnetic field strength [67]. If the two nanoparticles touch each other, a lower energy charge-transfer mode will appear in the mid-IR region [68].

7.2.3 "Borrowing SERS Activity" Methods

During the late 1970s and early 1980s, it was recognized that only a few noble metals (mainly Ag, Au and Cu) would provide a large SERS effect if the metal surface roughness or the colloid size were on the scale of several tens of nanometers. However, various research groups have never limited their efforts to extend SERS to the study of other metallic and nonmetallic surfaces. In the 1980s, a strategy based on "borrowing SERS activity" was proposed, either by depositing SERS-active metals onto non-SERS-active

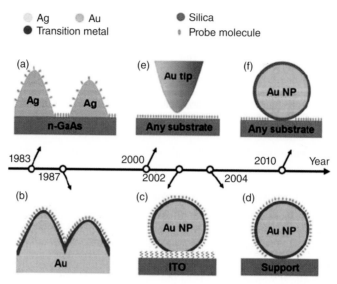

Figure 7.3 *Illustration of development of the "borrowing SERS activity" methods utilizing different nanostructures during the past three decades. For details, see the text.*

substrates (including semiconductors), or by depositing non-SERS-active materials over SERS-active substrates.

The approach of "borrowing SERS activity" was first proposed in 1983 by Van Duyne and coworkers [50], who obtained enhanced Raman signals for molecular species adsorbed onto non-SERS-active materials such as n-GaAs electrodes, by electrodepositing a discontinuous SERS-active overlayer of Ag (Figure 7.3(a)). However, most molecules of interest preferred to adsorb at Ag or Au as the SERS active substrates rather than as non-SERS active materials.

In 1987, the groups of Fleischmann [51, 52] and Weaver [53–55] developed a more effective and feasible method of coating an ultrathin film of a transition metal onto SERS-active Au and Ag electrodes, respectively (Figure 7.3(b)). In this case, SERS-active Ag or Au electrodes were coated with ultrathin films of other metals such as Ni, Co and Fe, Pt, Pd, Rh, and Ru by electrochemical deposition. Subsequently, with the help of long-range electromagnetic enhancement created by the SERS-active substrate underneath, weak SERS spectra could be obtained of adsorbates on the transition metal overlayer [49]. This approach opened the way to SERS studies of Fe corrosion inhibition and passivation [69], whereby Fe monolayers were deposited onto SERS-active Ag from solutions containing 0.1 M $FeSO_4 \cdot (NH_4)_2SO_4$, 0.1 M KCl and pyridine (pH 3.5). This "Fe-on-Ag approach" was used extensively to investigate Fe corrosion inhibitors, as well as to study Fe (and steel) passivity [70]. In 2002, the Weaver group successfully coated an ultrathin (pinhole-free) film of Pt-group metal onto Au nanoparticles immobilized on an indium–tin oxide (ITO) electrode by employing under-potential deposition (UPD) and a chemical redox replacement method (Figure 7.3(c)) [56]. Unfortunately, the whole preparation procedure was very complicated as it involved surface functionalization, electrochemical UPD, and a chemical reaction.

Recently, Tian and colleagues developed a simpler and more straightforward method to prepare transition metal-coated Au nanoparticles, based on a seed-mediated growth method (Figure 7.3(d)) [49, 57, 71]. The procedure commenced with highly monodispersed Au nanoparticles being synthesized according to Frens' method, by reducing $AuCl_4^-$ with sodium citrate. The Au core was then coated with one to ten atomic layers of the desired transition metal through chemical deposition. As discussed in Section 7.2.1, the average enhancement factor can reach 10^7 when the best optimization is made with Au, Ag and Cu (the typical SERS-active substrates), while an enhancement factor of 10^3 can be reached on pure transition metals. By effectively "borrowing" the SERS activity from the highly active core (or substrate) of Au or Ag, the surface enhancement of transition metals can be boosted by two orders of magnitude, while the total enhancement factor (ca. 10^5) can approach the same level as that for conventional coinage SERS substrates (e.g., nonoptimized systems, such as electrochemically roughened electrode surfaces) [49].

7.2.4 Shell-Isolated Nanoparticle-Enhanced Raman Spectroscopy

After carefully examining all conventional SERS systems, some generalities may be recognized, notably that the probed molecules (substance) are adsorbed either at the surface of SERS-active nanoparticles (nanostructures) or the over-coated layer. Moreover, a large number of perturbations in electronic and chemical properties would be induced by electronic contact when a thin transition metal shell is coated onto a SERS-active metal or substrate [72, 73]. The method of tip-enhanced Raman spectroscopy (TERS), which was developed in 2000, utilizes a non-contact working mode (Figure 7.3(e)) that separates the probed substance (molecule or material surface) from a metal tip that acts as the Raman signal amplifier [58–60]. Unfortunately, this technology has been largely limited to molecules with large Raman cross-sections, and has not yet been extended to electrochemical systems. In 2010, Tian *et al.* developed a novel method to circumvent these problems, which involved coating the SERS-active nanoparticles with an ultrathin and inert shell of silica (Figure 7.3(f)). This new technique is referred to as "shell-isolated nanoparticle-enhanced Raman spectroscopy" (SHINERS) [61, 63].

The SHINERS technique is, in one sense, an extension of the core–shell and coatings strategy already described. The reason for the shell, and the way in which the nanoparticles are used, differs significantly in SHINERS, however, and the result has been a considerable step forward in generality. The material of interest is no longer added to the SERS-active coinage metal; rather, an isolating shell is added and the structure is then spread onto a surface of any morphology and any material. So, in another sense, SHINERS is an extension of what was accomplished initially with TERS: the signal amplifier is again separated from the probed surface. With the advent of SHINERS, SERS may be obtained from surfaces with the unlikely morphology of an atomically flat single-crystal and a very unlikely material, an orange skin [30, 31, 74–82]. As the enhancement factors achieved with SHINERS are comparable to those achieved with core–shell nanoparticles and coated (ordered) nanostructures, the probe/analyte molecule generality is also comparable. Both, Tian and colleagues and Wandlowski and coworkers used SHINERS to systematically study the adsorption of pyridine on low-index Au(hkl) and Pt(hkl) single-crystal electrodes [30]. The average enhancement factor reached 10^6 for Au(110) and 10^5 for Pt(110), which was comparable to, or even higher than, the values obtained for bare gold nanoparticles on

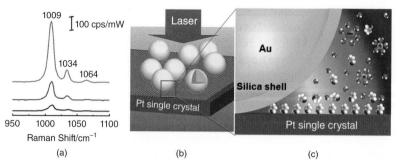

Figure 7.4   *(a) SHINERS spectra of pyridine adsorbed onto Pt(110) (the red line), Pt(100) (the blue line), and Pt(111) (the black line) at 0.00 V [30]. Solution: 0.01 M pyridine + 0.1 M NaClO₄; (b, c) Schematic diagrams of the SHINERS experiment. The electromagnetic field strength is represented by the following color code: red (strong) and blue (weak).*

smooth gold (a widely adopted SERS substrate). Three-dimensional finite-difference time-domain (3D-FDTD) simulations revealed that this large enhancement is due to the transfer of the "hotspots" from nanoparticle–nanoparticle gaps to nanoparticle–surface gaps (as shown in Figure 7.4c). Tian's group also applied SHINERS to examine the Si–H system on smooth Si surfaces and hydrogen adsorption on a Pt (111) surface; the latter is a very important system in surface science, chemistry and industry, including fuel cell applications [61].

7.3   Features of EC-SERS

In EC-SERS systems, both chemical and electromagnetic enhancements can be influenced to some extent by changing the applied electrode potential – that is, the Fermi level of metal and dielectric constant of the interfacial electrolyte. The former in particular can be strongly tuned by potential, leading to drastic changes in interfacial structure and properties. Such properties have led to EC-SERS being one of the most complicated systems in SERS; the features of EC-SERS will be briefly introduced at this point.

7.3.1   Electrochemical Double Layer of EC-SERS Systems

The EC-SERS system generally consists of nanostructured electrodes and an electrolyte. An electrical double layer is formed in the interfacial region such that, when a light illuminates a nanostructured electrode surface to excite a SERS process, a strong optical electric field is established in the electrical double-layer region. As a consequence, there are two types of electric field with distinctively different properties, namely an alternating electromagnetic field and a static electric field (E-field), which coexist in the electrochemical system. The E-field may be quite strong, as the potential drop between the electrode and electrolyte occurs mainly over the compact layer (thickness ~1 nm) and the diffuse layer [6]. This influences the interaction between metal and adsorbate, the orientation of the adsorbate, and the structure of the double layer region, which may in turn cause a redistribution of the surface-localized optical electric field.

By adjusting the electrode potential, the density and polarity of the surface charge will change. Typically, when the electrode potential is more positive than the potential of zero charge (PZC), the surface will be positively charged and the water molecules interact with the surface via the negatively-charged O atom. Meanwhile, other anionic species in the electrolyte may also approach the surface, and molecules of interest may interact with the electrode surface. The pyridine molecule can be used as an example here, as it can repel the surface water and interact with the surface via both the p-orbitals and lone pair orbitals of the N atom in a tilted configuration [83, 84]. When the electrode potential is moved negative of the PZC, the interaction of the electrode with the O atom of water becomes weaker while that with the H atoms become stronger. Cations also become more prevalent than anions adjacent to the surface. Under these conditions, pyridine may interact with the surface via only the lone pair orbital of N atom, which results in a configuration that is normal to the surface. With the further negative movement of the electrode potential, the interaction is weakened further and changes from chemisorbed to physisorbed type, or the pyridine molecule may even be desorbed from the surface.

7.3.2 Electrolyte Solutions and Solvent Dependency

When the electrolyte is changed, it will not only change the double-layer structure of the electrochemical interface but also influence the electrochemical reaction rates, and even the potential window. For example, the double layer will be compressed or expanded with the increase or decrease in the concentration of the electrolyte, respectively. Depending on the electrolyte, nonspecific or specific adsorption may occur on the electrode surface. The specifically adsorbed ions, due to a strong interaction with the metal surface, will possibly induce a shift of surface plasmon resonance. The electrolyte ion may also be coadsorbed with the adsorbates in a competitive or induced way; for example, thiourea is coadsorbed with ClO_4^- and/or SO_4^{2-} on the Ag electrode [38].

A change of solvent will change the optical properties of the electrochemical system. When organic solvents or ionic liquids are used, the electrochemical window will be drastically expanded, and marked changes in the surface physical and chemical properties may be expected [38, 85, 86]. This not only changes the oxidation potential of the metal electrode but also eliminates the evolution of hydrogen via the reduction of water. When the solvent is changed from water to organic or ionic liquid solvents, the surface plasmon resonance frequency of the metal nanostructures will be red-shifted due to an increase in the refractive index of the solvent [87].

From the above points it may be concluded on the one hand that EC-SERS is among the most complicated SERS systems. Yet, on the other hand, it provides more flexibility for investigating the interfacial structure and mechanism of complex systems by changing the experimental conditions, such as the electrode material, electrolyte, solvent, electrode potential and temperature in the fields of both SERS and electrochemistry.

7.4 EC-SERS Experiments

7.4.1 Measurement Procedures for EC-SERS

A layout of the measurement procedure for an electrochemical EC-SERS study is shown schematically in Figure 7.5. Before any study, it is necessary to calibrate both the Raman and electrochemical instruments. As many optical components are very sensitive to their

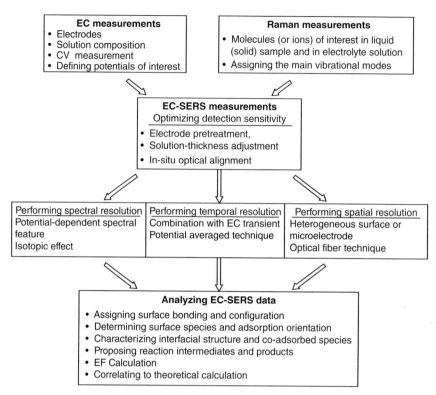

Figure 7.5   *Procedures for performing EC-SERS measurements.*

environment (e.g., temperature and humidity), it is very important to calibrate the sensitivity of the instrument and the accuracy of the frequency. As microprobe Raman instruments are now the dominant instruments on the market, it becomes common practice to use the very sharp and intense peak of a single crystal Si wafer [preferably Si(111)] at 520.6 cm^{-1} to calibrate both the frequency and sensitivity; however, if the excitation line shifts to the UV region it is better to use the 1333.2 cm^{-1} band of diamond as a calibration standard. It is quite common that the collection efficiency of Raman systems will change from day to day. Therefore, in order to compare the intensity of Raman spectra collected on different days it is better to normalize the intensity with that of the 520.6 cm^{-1} peak of a Si wafer.

Before starting a new EC-SERS study, it is important to first obtain the normal Raman spectrum of the species in its original form, such as the pure liquid or solid. In addition, spectra should be collected for the expected product(s) of the electrode reaction to be studied. The Raman spectra of the molecules/materials in the solution to be measured during the *in-situ* study should then be recorded. These good-quality spectra will serve as references to compare with the surface Raman spectra. If the spectrum is too complex, an isotopic study will be very helpful for identifying the vibrational modes. Before the EC-SERS study, it is important to define the electrochemical behavior to obtain the characteristic potential for use in the *in-situ* EC-SERS study.

For some systems, there is no resonance Raman or SERS effect to be utilized, and the sensitivity becomes the main problem. In this case, a potential difference method will be of great help [11]. Here, a spectrum is acquired at potentials where there is no or only a weak surface signal which is subtracted from that at the potential of interest. In addition, a change in the composition of the electrolyte or an isotopic labeling experiment may be considered to identify the surface species and verify its orientation and structure. For temporally resolved studies, electrochemical transient techniques are helpful to understand the surface dynamics and the reconstruction processes of surfaces. For nonuniform surfaces, spatially resolved measurements provide more reliable and complete information on the surface. This is also useful for electrode surfaces that change either chemically or topographically in a microzone upon variation of potential.

7.4.2 Experimental Set-Up for EC-SERS

Figure 7.6 shows the experimental set-up for *in-situ* EC-SERS. This includes a laser to excite the SERS of samples, a Raman spectrometer to disperse and detect the Raman signal, a computer to control the Raman instrument for data acquisition and manipulation, a potentiostat or galvanostat to control the potential of the working electrode, and an EC-SERS cell to accommodate the electrode/electrolyte interface to be studied. It may be

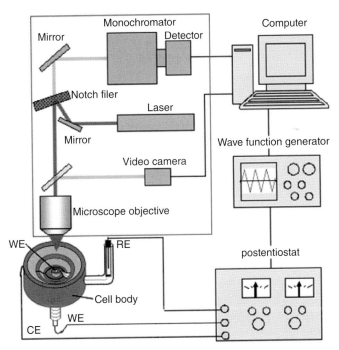

Figure 7.6 *Diagram showing the experimental set-up for EC-SERS, which includes a Raman spectrometer (in the block), potentiostat, computer, wavefunction generator, and an EC-SERS cell. WE, working electrode; CE, counter electrode; RE, reference electrode. Reprinted from Ref. [57] with permission from Royal Society of Chemistry.*

necessary to place a plasma line filter in the incident path for some lasers to obtain a truly monochromatic incident light. The detector of the Raman system can be a single-channel photomultiplier tube (PMT), an avalanche photodiode (APD), or a multichannel charge-coupled device (CCD); currently, the latter has become the dominant configuration. In the case of time-resolved studies, it may be necessary to include a function generator to generate various types of potential/current controls over the electrode, and also to trigger the detector accordingly to acquire the time-resolved SERS signal.

7.4.3   Preparation of SERS Substrates

7.4.3.1   Electrochemical Oxidation and Reduction Cycles

The first SERS spectra were recorded on a roughened silver surface prepared by the method of EC-oxidation/reduction cycles [23]. By application of an oxidation potential to the metal electrode, the electrode was oxidized to soluble ions or an insoluble surface complex; a reduction potential will then reduce these species at the surface, forming surface nanostructures. As silver is one of the most extensively studied SERS substrate materials, it is reasonable to take this as an example to illustrate the oxidation/reduction cycles procedure in detail [88].

Figure 7.7 demonstrates the electrochemical roughening procedures for a Ag electrode in 0.1 M aqueous KCl electrolyte. The morphological evolution of the electrode surface is shown at the bottom of the figure. The electrode potential (the dashed line) is initially

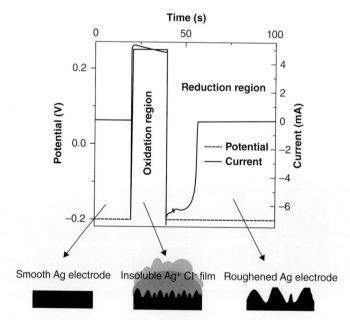

Figure 7.7 *Current–time curve (solid line) and potential–time curve (dashed line) when Ag electrode is roughened in 0.1 M KCl. The corresponding morphological evolution of electrode surface is shown below the graph.*

set negative of the value for oxidation of Ag, after which it is stepped to a more positive potential where the oxidation of Ag occurs. The dwell time of this oxidation process is about 20 s. Due to the presence of Cl⁻ ions, insoluble AgCl is formed on the electrode surface. As the potential is stepped back to the initial potential (in the reduction region), the AgCl is reduced, this gives rise to the roughened nanostructure of silver metal on the electrode surface. It should be noted that the experimental parameters may need to be modified slightly because the crystallinity and purity of Ag electrodes differ.

For electrochemists, EC-oxidation/reduction cycling is a simple and effective method to obtain high-performance SERS-active electrodes. Therefore, this played a significant role during the early stages of development of EC-SERS, and many important experiments have been performed at EC-oxidation/reduction cycled roughened electrodes. For example, Fleischmann *et al.*, for the first time, reported NIR SERS measurements using 1.064 μm excitation at EC-roughened Cu, Ag, and Au electrodes in 1988 [89]. In 1996, the present authors' group were the first to obtain good-quality surface Raman signals from pure transition metals (Pt and Ni) electrode surfaces based on a special EC-oxidation/reduction treatment [90]. In 2003, the SERS excited by UV light (UV-SERS) was first observed unambiguously only from EC-oxidation/reduction roughened transition metals [91].

It should be emphasized that the surfaces resulting from EC-oxidation/reduction cycling are randomly structured (Figure 7.8a), and therefore they lack spot-to-spot and substrate-to-substrate reproducibility in their SERS response. Certainly, this ill-defined geometry is unfavorable for understanding the interfacial structure and maximizing SERS activity.

7.4.3.2   *Ordered Nanostructures*

The first ordered structures for SERS were reported during the early 1980s by Liao *et al.* [95, 96], who obtained regularly ordered SERS substrates by depositing Ag particles over periodic arrays of silica posts that were fabricated by photolithography. Since the 1990s, with the rapid development in the field of nanofabrication, many ordered periodic arrays of nanoparticles for SERS have been made.

In 1995, Natan *et al.* reported that Au and Ag nanoparticles adsorbed onto an organosilane-polymer-modified Si substrate could be used as a SERS-active substrate [97]. Subsequently, in 2001 the same authors [98] proposed a novel approach based on the self-assembled monolayer technique which allows the preparation of regularly arranged monodispersed colloidal Au and Ag nanoparticles on functionalized metal or glass substrates. It is now possible to synthesize or fabricate metal nanostructures of various shapes and sizes with a narrow size distribution. Nanoparticles assembled on an electrically conductive substrate can significantly improve the surface uniformity of the EC-SERS substrate. Therefore, use of nanoparticle sols or assembled nanoparticles as SERS substrates has expanded very rapidly during recent years [92].

Meanwhile, the groups of Van Duyne and Bartlett have made special efforts to broaden the scope of nanosphere lithography to fabricate several new EC-SERS-active structural motifs [94, 99]. This most promising method can be used to produce nanostructured substrates with a precise control over the shape, size and interparticle spacing. In this case, a monolayer or a multilayer of highly ordered nanosphere films is used as the template for vacuum deposition or electrochemical deposition. As a result, three types of structured SERS substrate can be produced: (i) physical vapor deposition on the nanosphere template

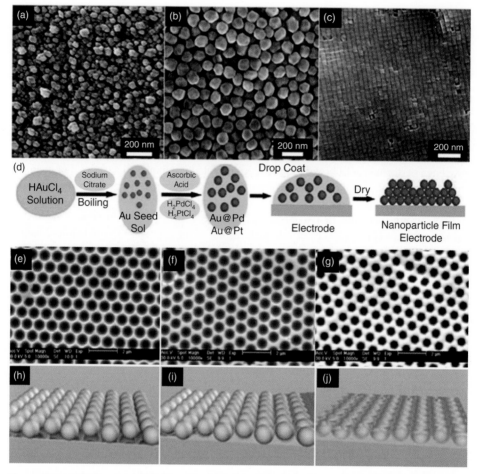

Figure 7.8 *SEM images of three typical SERS-active electrodes for EC-SERS. Reprinted with permission from Ref. [92] Copyright © 2009, Springer-Verlag; (a) An electrochemically rough-ened Pd electrode; (b) Au nanoparticle assembled film electrode and (c) Au-core–Pd-shell (Au@Pd) nanocubes assembled film electrode; (d) Schematic representation for preparing the Au@Pt and Au@Pd nanoparticle film electrodes. Reprinted with permission from Ref. [93] Copy-right © 2008, John Wiley & Sons, Ltd; SEM images of electrodeposited gold nanovoid arrays using 900 nm diameter (D) nanosphere template with film thickness (e) 0.45 D (f) 0.7 D (g) 0.9 D. (h–j) Schematic representations of the preparation (h) assembly of nanospheres (i) Au electrodeposition (j) removal of nanospheres to expose 2D-nanovoid array. Reprinted with permission from Ref. [94] Copyright © 2006, Royal Society of Chemistry.*

leads to the formation of a metal (e.g., Ag) "film over nanosphere" (FON) surface [99]; (ii) the removal of nanospheres by sonicating the entire sample in a solvent results in surface-confined nanoparticles with a triangular footprint [99]; and (iii) electrochemical deposition followed by removal of the spheres leaves a thin nanostructured film containing a regular hexagonal array of uniform sphere voids [94] (Figure 7.8e–j).

All EC-SERS-active systems must possess nanostructures, and the SERS activity is critically dependent on the configuration and composition of nanostructures, as well as the applied electrode potential, by which some new insights into SERS phenomena can be gained. The field of electrochemistry recognizes SERS as a powerful tool to characterize surface molecules and to provide "fingerprint" information of their molecular bonds and molecule–surface bonding. These points will be discussed in Chapter 8.

Acknowledgments

These studies have been made possible by the continuous financial support of the Natural Science Foundation of China (Nos 20433040 and 20673086), and the Ministry of Science and Technology (973 Program Nos 2007CB815303 and 2007CB935603). Whenever work from the authors' group is mentioned in the chapter, it is the great contribution of the self-motivated and hard-working students at Xiamen University that should be acknowledged. The authors sincerely thank Professor James McQuillan at the University of Otago and Professor Patrick Hendra of the University of Southampton for their insightful suggestions and kind help in editing the chapter.

References

(1) Lipkowski, J. and Ross P.N. (eds) (1992) *Adsorption at Electrode Surface*, VCH, New York.
(2) Schmickler, W. and Parsons, R. (1997) *Interfacial Electrochemistry*, Oxford University Press, London.
(3) Bard, A.J. and Faulkner, L.R. (2000) *Electrochemical Methods: Fundamentals and Applications*, 2nd edn, John Wiley & Sons, Inc., New York.
(4) White, R.E., Bockris, J.O'M., Conway, B.E. and Yeager, E. (eds) (1984) *Comprehensive Treatise of Electrochemistry*, vol. **8**, Kluwer, New York.
(5) Abruna, H.D. (ed) (1991) *Electrochemical Interfaces—Modern Techniques for In-situ Interface Characterization*, VCH Verlag Chem., Berlin.
(6) Wieckowski, A. (ed.) (1999) *Interfacial Electrochemistry*, M. Dekker, New York.
(7) Bewick, A. and Pons, S. (1985) *Advances in Infrared and Raman Spectroscopy* (eds R.J.H. Clark and R.E. Hester), John Wiley & Sons, Ltd, UK.
(8) Iwasita, T. and Nart, F.C. (1997) In situ infrared spectroscopy at electrochemical interfaces. *Progress in Surface Science*, **55**, 271–340.
(9) Fleischmann, M. and Hill, I.R. (1984) in *Comprehensive Treatise of Electrochemistry* (eds R.E. White, J.O'M. Bockris, B.E. Conway and E. Yeager), vol. **8**, Kluwer, New York, pp. 373–432.
(10) Birke, R.L., Lu, T. and Lombardi, J.R. (1991) *Techniques for Characterization of Electrodes and Electrochemical Processes* (eds R. Varma and J.R. Selman), John Wiley & Sons, Inc., New York.
(11) Pettinger B. (1992) in *Adsorption at Electrode Surface* (eds J. Lipkowski and P.N. Ross), VCH, New York, pp. 285–345.
(12) Tadjeddine, A. and Le Rille, A. (1999) in *Interfacial Electrochemistry* (ed. A. Wieckowski), M. Dekker, New York, pp. 317–343.
(13) Williams, C.T. and Beattie, D.A. (2002) Probing buried interfaces with non-linear optical spectroscopy. *Surface Science*, **500**, 545–576.
(14) Hochleitner, R., Tarcea, N., Simon, G. *et al.* (2004) Micro-Raman spectroscopy: a valuable tool for the investigation of extraterrestrial material. *Journal of Raman Spectroscopy*, **35**, 515–518.
(15) Choo-Smith, L.P., Edwards, H.G., Endtz, H.P. *et al.* (2002) Medical applications of Raman spectroscopy: from proof of principle to clinical implementation. *Biopolymers*, **67**, 1–9.

(16) Grow, A.E., Wood, L.L., Claycomb, J.L. and Thompson, P.A. (2003) New biochip technology for label-free detection of pathogens and their toxins. *Journal of Microbiological Methods*, **53**, 221–233.

(17) Thomas, G.J. (1999) Raman spectroscopy of protein and nucleic acid assemblies. *Annual Review of Biophysics and Biomolecular Structure*, **28**, 1–27.

(18) McCreery, R.L. (2000) *Raman Spectroscopy for Chemical Analysis*, John Wiley & Sons, Inc., New York.

(19) Galabov, B.S. and Dudev, T. (1996) in *Vibrational Spectra and Structure*, vol. **22** (ed J.R. Durig), Elsevier.

(20) Fleischmann, M., Hendra, P.J. and McQuillan, A.J. (1973) Raman spectra from electrode surfaces. *Journal of the Chemical Society, Chemical Communications*, 80–81.

(21) McQuillan, A.J. (1973) Intermediates in electrochemical reactions. *Faraday Discussions - Royal Society of Chemistry*, **56**, 167–168.

(22) McQuillan, A.J. (2009) The discovery of surface-enhanced Raman scattering. *Notes and Records of the Royal Society*, **63**, 105–109.

(23) Fleischmann, M., Hendra, P.J. and McQuillan, A.J. (1974) Raman spectra of pyridine adsorbed at a silver electrode. *Chemical Physics Letters*, **26**, 163–166.

(24) Jeanmaire, D.L. and Van Duyne, R.P. (1977) Surface Raman spectroelectrochemistry: part I. Heterocyclic, aromatic, and aliphatic amines adsorbed on the anodized silver electrode. *Journal of Electroanalytical Chemistry*, **84**, 1–20.

(25) Albrecht, M.G. and Creighton, J.A. (1977) Anomalously intense Raman spectra of pyridine at a silver electrode. *Journal of the American Chemical Society*, **99**, 5215–5217.

(26) Creighton, J.A. (2010) Contributions to the early development of surface enhanced Raman scattering. *Notes and Records of the Royal Society*, **64**, 175–183.

(27) Van Duyne, R.P. (1979) Laser excitation of Raman scattering from adsorbed molecules on electrode surfaces, in *Chemical and Biochemical Applications of Lasers*, vol. **IV** (ed. C.B. Moore), Academic Press, New York.

(28) Tian, Z.Q., Ren, B. and Wu, D.Y. (2002) Surface-enhanced Raman scattering: From noble to transition metals and from rough surfaces to ordered nanostructures. *The Journal of Physical Chemistry B*, **106**, 9463–9483.

(29) Tian, Z.Q., Yang, Z.L., Ren, B. *et al.* (2006) Surface-enhanced Raman scattering from transition metals with special surface morphology and nanoparticle shape. *Faraday Discussions*, **132**, 159–170.

(30) Li, J.F., Ding, S.Y., Yang, Z.L. *et al.* (2011) Extraordinary enhancement of Raman scattering from pyridine on single crystal Au and Pt electrodes by shell-isolated Au nanoparticles. *Journal of the American Chemical Society*, **133**, 15922–15925.

(31) Honesty, N.R. and Gewirth, A.A. (2012) Shell-isolated nanoparticle enhanced Raman spectroscopy (SHINERS) investigation of benzotriazole film formation on Cu(100), Cu(111), and Cu(poly). *Journal of Raman Spectroscopy*, **43**, 46–50.

(32) Moskovits, M. (2005) Surface-enhanced Raman spectroscopy: a brief retrospective. *Journal of Raman Spectroscopy*, **36**, 485–496.

(33) Moskovits, M. (2012) How the localized surface plasmon became linked with surface-enhanced Raman spectroscopy. *Notes and Records of the Royal Society*, **66**, 195–203.

(34) Moskovits, M. (1985) Surface-enhanced spectroscopy. *Reviews of Modern Physics*, **57**, 783–826.

(35) Otto, A., Mrozek, I., Grabhorn, H. and Akemann, W. (1992) Surface-enhanced Raman scattering. *Journal of Physics: Condensed Matter*, **4**, 1143–1212.

(36) Weaver, J.H. (1975) Optical properties of Rh, Pd, Ir, and Pt. *Physical Reviews B: Solid State*, **11**, 1416–1425.

(37) Ordal, M.A., Bell, R.J., Alexander, R.W. Jr *et al.* (1985) Optical properties of fourteen metals in the infrared and far infrared: Al, Co, Cu, Au, Fe, Pb, Mo, Ni, Pd, Pt, Ag, Ti, V, and W. *Applied Optics*, **24**, 4493–4499.

(38) Tian, Z.Q. and Ren, B. (2004) Adsorption and reaction at electrochemical interfaces as probed by surface-enhanced Raman spectroscopy. *Annual Review of Physical Chemistry*, **55**, 197–229.

(39) Zeman, E.J. and Schatz, G.C. (1987) An accurate electromagnetic theory study of surface enhancement factors for silver, gold, copper, lithium, sodium, aluminum, gallium, indium, zinc, and cadmium. *The Journal of Physical Chemistry*, **91**, 634–643.

(40) Creighton, J.A. and Eadon, D.G. (1991) Ultraviolet-visible absorption spectra of the colloidal metallic elements. *Journal of the Chemical Society, Faraday Transactions*, **87**, 3881–3891.

(41) Sharma, B., Frontiera, R.R., Henry, A.I. *et al.* (2012) SERS: materials, applications, and the future. *Materials Today*, **15**, 16–25.

(42) Boltasseva, A. and Atwater, H.A. (2011) Low-loss plasmonic metamaterials. *Science*, **331**, 290–291.

(43) Stiles, P.L., Dieringer, J.A., Shah, N.C. and Van Duyne, R.P. (2008) Surface-enhanced Raman spectroscopy. *Annual Reviews of Analytical Chemistry*, **1**, 601–626.

(44) Kosuda, K.M., Bingham, J.M., Wustholz, K.L. and Van Duyne, R.P. (2011) Nanostructures and Surface-Enhanced Raman Spectroscopy, in *Comprehensive Nanoscience and Technology* (eds D.L. Andrews, G.S. Scholes and G.P. Wiederrecht), Academic Press, Amsterdam.

(45) Gersten, J. and Nitzan, A. (1980) Electromagnetic theory of enhanced Raman-scattering by molecules adsorbed on rough surfaces. *The Journal of Physical Chemistry*, **73**, 3023–3037.

(46) Nitzan, A. and Brus, L.E. (1981) Theoretical-model for enhanced photochemistry on rough surfaces. *The Journal of Physical Chemistry*, **75**, 2205–2214.

(47) Aravind, P.K., Nitzan, A. and Metiu, H. (1981) The interaction between electromagnetic resonances and its role in spectroscopic studies of molecules adsorbed on colloidal particles or metal spheres. *Surface Science*, **110**, 189–204.

(48) Murray, C.A., Allara, D.L. and Rhinewine, M. (1981) Silver-molecule separation dependence of surface-enhanced Raman-scattering. *Physical Review Letters*, **46**, 57–60.

(49) Tian, Z.Q., Ren, B., Li, J.F. and Yang, Z.L. (2007) Expanding generality of surface-enhanced Raman spectroscopy with borrowing SERS activity strategy. *Chemical Communications*, 3514–3534.

(50) Van Duyne, R.P. and Haushalter, J.P. (1983) Surface-enhanced resonance Raman-spectroscopy of adsorbates on semiconductor electrode surfaces - Tris(bipyridine)Ruthenium(II) adsorbed on silver-modified N-Gaas(100). *The Journal of Physical Chemistry*, **87**, 2999–3003.

(51) Fleischmann, M., Tian, Z.Q. and Li, L.J. (1987) Raman-spectroscopy of adsorbates on thin-film electrodes deposited on silver substrates. *Journal of Electroanalytical Chemistry*, **217**, 397–410.

(52) Fleischmann, M. and Tian, Z.Q. (1987) The induction of SERS on smooth Ag by the deposition of Ni and Co. *Journal of Electroanalytical Chemistry*, **217**, 411–416.

(53) Leung, L.W.H. and Weaver, M.J. (1987) Extending surface-enhanced Raman-spectroscopy to transition-metal surfaces – carbon-monoxide adsorption and electrooxidation on platinum-coated and palladium-coated gold electrodes. *Journal of the American Chemical Society*, **109**, 5113–5119.

(54) Leung, L.W.H. and Weaver, M.J. (1987) Extending the metal interface generality of surface-enhanced Raman-spectroscopy - underpotential deposited layers of mercury, thallium, and lead on gold electrodes. *Journal of Electroanalytical Chemistry*, **217**, 367–384.

(55) Zou, S.Z. and Weaver, M.J. (1998) Surface-enhanced Raman scattering on uniform transition metal films: toward a versatile adsorbate vibrational strategy for solid-nonvacuum interfaces. *Analytical Chemistry*, **70**, 2387–2395.

(56) Park, S., Yang, P.X., Corredor, P. and Weaver, M.J. (2002) Transition metal-coated nanoparticle films: vibrational characterization with surface-enhanced Raman scattering. *Journal of the American Chemical Society*, **124**, 2428–2429.

(57) Wu, D.Y., Li, J.F., Ren, B. and Tian, Z.Q. (2008) Electrochemical surface-enhanced Raman spectroscopy of nanostructures. *Chemical Society Reviews*, **37**, 1025–1041.

(58) Stockle, R.M., Suh, Y.D., Deckert, V. and Zenobi, R. (2000) Nanoscale chemical analysis by tip-enhanced Raman spectroscopy. *Chemical Physics Letters*, **318**, 131–136.

(59) Pettinger, B., Ren, B., Picardi, G. *et al.* (2004) Nanoscale probing of adsorbed species by tip-enhanced Raman spectroscopy. *Physical Review Letters*, **92**, 096101.

(60) Ren, B., Picardi, G., Pettinger, B. *et al.* (2005) Tip-enhanced Raman spectroscopy of benzenethiol adsorbed on Au and Pt single-crystal surfaces. *Angewandte Chemie International Edition*, **44**, 139–142.

(61) Li, J.F., Huang, Y.F., Ding, Y. *et al.* (2010) Shell-isolated nanoparticle-enhanced Raman spectroscopy, *Nature*, **464**, 392–395.

(62) Anema, J.R., Li, J.F., Yang, Z.L. and Ren, B. (2011) Shell-isolated nanoparticle-enhanced Raman spectroscopy: expanding the Versatility of surface-enhanced Raman scattering. *Annual Review of Analytical Chemistry*, **4**, 129–150.

(63) Li, J.F., Tian, X.D., Li, S.B. *et al.* (2013) Surface analysis using shell-isolated nanoparticle-enhanced Raman spectroscopy. *Nature Protocols*, **8**, 52–65.

(64) Kennedy, B.J., Spaeth, S., Dickey, M. and Carron, K.T. (1999) Determination of the distance dependence and experimental effects for modified SERS substrates based on self-assembled monolayers formed using alkanethiols. *The Journal of Physical Chemistry*, **103**, 3640–3646.

(65) Le Ru, E.C. and Etchegoin, P.G. (eds) (2009) *Principles of Surface-Enhanced Raman Spectroscopy and Related Plasmonic Effects*, Elsevier, Amsterdam.

(66) Zuloaga, J., Prodan, E. and Nordlander, P. (2009) Quantum description of the plasmon resonances of a nanoparticle dimer. *Nano Letters*, **9**, 887–891.

(67) Esteban, R., Borisov, A.G., Nordlander, P. and Aizpurua, J. (2012) Bridging quantum and classical plasmonics with a quantum-corrected model. *Nature Communications*, **3**, 825.

(68) Duan, H., Fernández-Domínguez, A.I., Bosman, M. *et al.* (2012) Nanoplasmonics: classical down to the nanometer scale. *Nano Letters*, **12**, 1683–1689.

(69) Mengoli, G., Musiani, M.M., Fleischmann, M. *et al.* (1987) Enhanced Raman scattering from iron electrodes. *Electrochimica Acta*, **32**, 1239–1245.

(70) Rubim, J.C. and Dünnwald, J. (1989) Enhanced Raman scattering from passive films on silver-coated iron electrodes. *Journal of Electroanalytical Chemistry*, **258**, 327–344.

(71) Hu, J.W., Zhang, Y., Li, J.F. *et al.* (2005) Synthesis of Au@Pd core–shell nanoparticles with controllable size. *Chemical Physics Letters*, **408**, 354–359.

(72) Zhang, J., Sasaki, K., Sutter, E. and Adzic, R.R. (2007) Stabilization of platinum oxygen-reduction electrocatalysts using gold clusters. *Science*, **315**, 220–222.

(73) Rodriguez, J.A. and Goodman, D.W. (1992) The nature of the metal-metal bond in bimetallic surfaces. *Science*, **257**, 897–903.

(74) Li, J.F., Anema, J.R., Yu, Y.C. *et al.* (2011) Core-shell nanoparticle based SERS from hydrogen adsorbed on a Rhodium(111) electrode. *Chemical Communications*, **47**, 2023–2025.

(75) Li, J.F., Li, S.B., Anema, J.R. *et al.* (2011) Synthesis and characterization of gold nanoparticles coated with ultrathin and chemically inert dielectric shells for SHINERS applications. *Journal of Applied Spectroscopy*, **65**, 620–626.

(76) Huang, Y.F., Li, C.Y., Broadwell, I. *et al.* (2011) Shell-isolated nanoparticle-enhanced Raman spectroscopy of pyridine on smooth silver electrodes. *Electrochimica Acta*, **56**, 10652–10657.

(77) Li, S.B., Li, L.M., Anema, J.R. *et al.* (2011) Shell-isolated nanoparticle-enhanced Raman spectroscopy based on gold-core silica-shell nanorods. *Zeitschrift für Physikalische Chemie*, **225**, 775–783.

(78) Liu, B., Blaszczyk, A., Mayor, M. and Wandlowski, T. (2011) Redox-switching in a viologen-type adlayer: an electrochemical shell-isolated nanoparticle enhanced Raman spectroscopy study on Au(111)-(1 × 1) single crystal electrodes. *ACS Nano*, **5**, 5662–5672.

(79) Butcher, D.P., Boulos, S.P., Murphy, C.J. *et al.* (2012) Face-dependent shell-isolated nanoparticle enhanced Raman spectroscopy of 2,2′-Bipyridine on Au(100) and Au(111). *The Journal of Physical Chemistry C*, **116**, 5128–5140.

(80) Lin, X.D., Uzayisenga, V., Li, J.F. *et al.* (2012) Synthesis of ultrathin and compact Au@MnO_2 nanoparticles for shell-isolated nanoparticle-enhanced Raman spectroscopy (SHINERS). *Journal of Raman Spectroscopy*, **43**, 40–45.

(81) Qian, K., Liu, H.L., Yang, L.B. and Liu, J.H. (2012) Functionalized shell-isolated nanoparticle-enhanced Raman spectroscopy for selective detection of trinitrotoluene. *Analyst*, **137**, 4644–4646.

(82) Uzayisenga, V., Lin, X.D., Li, L.M. *et al.* (2012) Synthesis, characterization, and 3D-FDTD simulation of Ag@SiO_2 nanoparticles for shell-isolated nanoparticle-enhanced Raman spectroscopy. *Langmuir*, **28**, 9140–9146.

(83) Birke, R.L. and Lombardi, J.R. (1988) Surface-Enhanced Raman Spectroscopy, in *Spectroelectrochemistry – Theory and Practice* (ed. R.J. Gale), Plenum, New York.

(84) Wu, D.Y., Ren, B., Xu, X. *et al.* (2003) Periodic trends in the bonding and vibrational coupling: pyridine interacting with transition metals and noble metals studied by surface-enhanced Raman spectroscopy and density-functional theory. *The Journal of Physical Chemistry*, **119**, 1701–1709.

(85) Ali, A.H. and Foss, C.A. (1999) Electrochemically induced shifts in the plasmon resonance bands of nanoscopic gold particles adsorbed on transparent electrodes. *Journal of The Electrochemical Society*, **146**, 628–636.

(86) Santos, V.O., Alves, M.B., Carvalho, M.S. *et al.* (2006) Surface-enhanced Raman scattering at the silver electrode/ionic liquid (BMIPF$_6$) interface. *The Journal of Physical Chemistry B*, **110**, 20379–20385.

(87) Underwood, S. and Mulvaney, P. (1994) Effect of the solution refractive index on the color of gold colloids. *Langmuir*, **10**, 3427–3430.

(88) Tian, Z.Q. and Ren, B. (2003) Raman spectroscopy of electrode surfaces, in *Encyclopedia of Electrochemistry*, Wiley-VCH Verlag GmbH & Co. KGaA.

(89) Crookell, A. (1988) Surface-enhanced Fourier transform Raman spectroscopy in the near infrared. *Chemical Physics Letters*, **149**, 123–127.

(90) Ren, B., Huang, Q.J., Cai, W.B. *et al.* (1996) Surface Raman spectra of pyridine and hydrogen on bare platinum and nickel electrodes. *Journal of Electroanalytical Chemistry*, **415**, 175–178.

(91) Ren, B., Lin, X.F., Yang, Z.L. *et al.* (2003) Surface-enhanced Raman scattering in the ultraviolet spectral region: UV-SERS on rhodium and ruthenium electrodes. *Journal of the American Chemical Society*, **125**, 9598–9599.

(92) Lin, X.M., Cui, Y., Xu, Y.H. *et al.* (2009) Surface-enhanced Raman spectroscopy: substrate-related issues. *Analytical and Bioanalytical Chemistry*, **394**, 1729–1745.

(93) Fang, P.P., Li, J.F., Yang, Z.L. *et al.* (2008) Optimization of SERS activities of gold nanoparticles and gold-core–palladium-shell nanoparticles by controlling size and shell thickness. *Journal of Raman Spectroscopy*, **39**, 1679–1687.

(94) Mahajan, S., Abdelsalam, M., Suguwara, Y. *et al.* (2007) Tuning plasmons on nano-structured substrates for NIR-SERS. *Physical Chemistry Chemical Physics*, **9**, 104–109.

(95) Liao, P.F., Bergman, J.G., Chemla, D.S. *et al.* (1981) Surface-enhanced Raman scattering from microlithographic silver particle surfaces. *Chemical Physics Letters*, **82**, 355–359.

(96) Liao, P.F. and Stern, M.B. (1982) Surface-enhanced Raman scattering on gold and aluminum particle arrays. *Optics Letters*, **7**, 483–485.

(97) Freeman, R.G., Grabar, K.C., Allison, K.J. *et al.* (1995) Self-assembled metal colloid monolayers: an approach to SERS substrates. *Science*, **267**, 1629–1632.

(98) Nicewarner-Peña, S.R., Freeman, R.G., Reiss, B.D. *et al.* (2001) Submicrometer metallic barcodes. *Science*, **294**, 137–141.

(99) Willets, K.A. and Van Duyne, R.P. (2007) Localized surface plasmon resonance spectroscopy and sensing. *Annual Review of Physical Chemistry*, **58**, 267–297.

8

Applications of Electrochemical Surface-Enhanced Raman Spectroscopy (EC-SERS)

Marco Musiani[1], Jun-Yang Liu[2] and Zhong-Qun Tian[2]
[1]IENI-CNR, Padova, Italy
[2]Xiamen University, State Key Laboratory of Physical Chemistry of Solid Surfaces, China

In this chapter some applications of electrochemical-SERS (EC-SERS) are reviewed, highlighting how potential-dependent surface spectroscopy can provide access to important information on chemical bonding, orientation and the electrochemical reaction of molecules at electrode surfaces. Such studies continue to lead to a much-enhanced understanding of interfacial structure and the mechanism of electrode reactions. The fundamental aspects of SERS are provided in Chapter 7.

Raman spectroscopy, as a vibrational spectroscopy, can record "fingerprint" spectra from electrodes and provide much insight into a variety of surface and interfacial structures and processes at the molecular level; typical examples are the qualitative determination of surface bonding, conformation, and orientation. Raman spectroscopy invariably uses lasers in the ultraviolet (UV) through to the near-infrared (NIR) range. More importantly, the technique can be applied *in-situ* to investigate solid–liquid and solid–solid interfaces of both fundamental and practical importance, and can also be used flexibly to study high-surface-area porous electrode materials, to which many surface techniques are not applicable. Consequently, Raman spectroscopy is among the most promising methods for use in electrochemistry. One major disadvantage of Raman spectroscopy is its intrinsically low detection sensitivity, although by using roughed surfaces the sensitivity of EC-SERS can be enhanced by several orders of magnitude.

Developments in Electrochemistry: Science Inspired by Martin Fleischmann, First Edition.
Edited by Derek Pletcher, Zhong-Qun Tian and David E. Williams.
© 2014 John Wiley & Sons, Ltd. Published 2014 by John Wiley & Sons, Ltd.

Since the discovery of EC-SERS [1–3], Fleischmann and coworkers, as well as other spectroelectrochemitsts, have employed this technique to study the electrode–electrolyte interface, greatly enhancing knowledge of topics that include: the adsorption of organic molecules and their coadsorption with inorganic species [1, 4–14]; the structure of water at the interface [5, 6, 9, 15]; electrodeposition [6, 13, 16]; electropolymerization [17, 18]; corrosion inhibition [19–24]; passivity [25–27]; and surface redox reactions [28–30]. Moreover, during the four decades since the first reports of EC-SERS the possibilities for such studies have been expanded by combining the technique with more recently developed electrode materials, such as (core–shell) nanoparticles and single crystals. For brevity, in the following review attention will be focused on:

- Pyridine adsorption on different metals
- Interfacial water on different metals
- Coadsorption of thiourea with inorganic anions
- Electrodeposition additives
- Corrosion inhibition and passivity
- Lithium batteries
- Intermediates in electrocatalysis

8.1 Pyridine Adsorption on Different Metal Surfaces

Pyridine is adsorbed strongly onto several metals and has a large Raman cross-section. The frequency and relative intensity of pyridine SERS bands are very sensitive to electrode potential and to the surface properties of the metal, and for these reasons pyridine has been the most common probe molecule used in experimental and theoretical investigations aimed at broadening the application of SERS to new electrode materials and at assessing the respective roles of electromagnetic and chemical enhancement mechanisms [31, 32]. Pyridine might also be the best example to illustrate how to perform EC-SERS measurements and theoretical calculations to analyze the adsorption properties of electrochemical systems.

During the past four decades, much research effort has been directed towards expanding the substrate and molecule generality of SERS. Based on highly sensitive, confocal Raman microscopy and the development of new methods for preparing roughened transition metal surfaces [31] and transition metal-coated Au nanoparticles [33] (here, transition metals are group VIII-B elements), Tian's group was able to obtain good-quality surface Raman spectra for pyridine adsorbed at Pt, Pd, Ru, Rh, Fe, Co and Ni electrodes, all of which substrates had previously been considered SERS-inactive. The SERS spectra of pyridine were acquired on coinage metals, Fe group metals and Pt group metals over a wide potential range.

Figure 8.1a shows the spectra of pyridine adsorbed onto roughened Ag, Au, Cu and Pt electrodes at the potentials where their intensities reach a maximum. Only a few bands have been assigned to different vibrational modes of pyridine [34]. The changes between normal Raman (data not shown) and SERS spectra at different potentials and different electrodes were studied by combining experiments and theoretical calculations. It was established that the ring breathing mode, v_1, and the symmetric triangular ring deformation, v_{12}, would

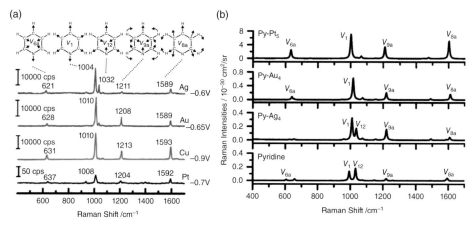

Figure 8.1   *(a) Surface-enhanced Raman spectra of pyridine adsorbed on roughened Ag, Au, Cu and Pt electrode at the potential where the SERS intensity is a maximum (vs SCE), with the bands assignment; (b) Simulated Raman spectra of pyridine interacting with Ag_4, Au_4 and Pt_5 clusters in the off-resonance-Raman scattering process.*

blue-shift from the liquid/solution phase to adsorption state at open circuit potential (OCP). The relative intensity of the v_{12} mode to the v_1 mode was significantly decreased upon interaction with transition metals (e.g., Pt), but not with Ag. When a negative potential is applied to achieve maximum intensity, the peak frequencies are red-shifted, while the relative Raman intensities of the v_{12} mode to the v_1 mode decrease on Ag but remain almost unchanged on Au, Cu, and Pt electrodes. The three modes v_{6a}, v_{9a} and v_{8a} become stronger than at OCP.

In order to understand the observed phenomena and extract meaningful information on adsorption properties, Tian *et al.* defined the pyridine adsorption configuration and binding interaction with metals through theoretical simulations, as shown in Figure 8.1b. In this case, it was concluded that when the surface coverage reached a monolayer or the potential was either close to or negative to the potential of zero charge (PZC), the ring-breathing mode would clearly blue-shift from free to adsorbed molecule, which suggested that the pyridine had interacted with the surface via the N-atom in an upright or slightly tilted configuration [34]. Density functional theory (DFT) calculations based on the metal cluster model revealed a binding interaction of pyridine with metals through the nitrogen lone-pair of electrons interacting with the conductance band of the surface metal. This interaction weakens in following the order: transition metals (e.g., Pt) > Cu ~ Au > Ag; that is, among the coinage metals pyridine has a stronger binding interaction with Cu and Au than with Ag [35]. The stronger binding interaction of transition metals (e.g., Pt) was due to a lower unoccupied, energy level than was present in coinage metals, and also the participation of d orbitals in the bonding. This is the main reason why the v_{12} mode exists on the Ag surface but reduces on Cu and Au, and almost vanishes on transition metals. In addition, the binding interaction was mainly attributed to the σ-type bonding between the lone pair orbital of the nitrogen atom and metals, and the p-type bonding of pyridine with metals. The latter results in a blue-shift of the frequencies of v_1 and v_{8a} modes. Instead, the v_{6a} mode has a strong coupling with the pyridine–metal adsorption bond [34], also leading to a blue-shift. As a consequence, the frequency blue-shift of the v_{6a} mode sensitively reflects the strength of the pyridine–metal binding interaction. Overall, the chemisorption of pyridine is through the donation of the nitrogen lone pair electrons and the short-range p-type bonding interaction with metal surfaces. By moving the potential to more negative values, the adsorption bond becomes weaker and this results in a red-shift in the vibrational frequencies.

As for the enhancement mechanism, the electromagnetic enhancement is dominant but the different potential dependences of Raman bands at different metal electrodes show that the chemical enhancement must also operate through: (i) a chemical binding-induced enhancement; and (ii) a charge-transfer mechanism [36–38]. On Ag, charge transfer is the main mechanism causing a decrease in the relative Raman intensity of the v_{12} mode with respect to the v_1 mode, via an electron transition from the Fermi level of the Ag electrode to the unoccupied orbitals. On Cu and Au, a larger relative intensity of the v_1 to v_{12} modes can be interpreted as the synergetic effect of the chemisorption and the charge transfer enhancement mechanism. For v_{8a} and v_{9a} modes, both the strong chemisorption and the charge transfer mechanism, which satisfies the resonance condition, enhance their Raman signals.

By conducting systematic studies of pyridine adsorbed at different electrodes, it has been shown that it is important to combine EC-SERS experiments and theoretical calculations on the frequency and intensity of SERS bands to reveal the detailed interaction of the various factors. By employing SERS-active coinage metals as the nanoparticle core and coated inert SiO_2 or MnO_2 as ultrathin shells, Tian's group recently developed a Shell-Isolated Nanoparticle-Enhanced Raman Spectroscopy (SHINERS) technique [39, 40]. In this case, the "isolated mode" prevents the probed molecules from being adsorbed onto the nanoparticle surfaces, and therefore SHINERS was seen to exhibit outstanding potential for a higher detection sensitivity and a broad application to various materials with diverse morphologies, such as atomically flat single-crystal electrode surfaces. Figure 8.2a shows the SHINERS spectra of pyridine on Au(110) and SERS spectra of pyridine on bare gold nanoparticles obtained under the same experimental conditions [41]. Their comparison led to two surprising results:

- The intensities of SHINERS on Au(110) was comparable to or greater than that of SERS on bare gold nanoparticles; the enhancement factors (EFs) were ca. 10^6 for SHINERS and 10^5 for SERS.
- The potential-dependent intensity profiles were distinctly different for SHINERS and SERS; the maximum intensities of the bands at 1014 and 1039 cm^{-1} were observed at – 0.20 V and – 0.60 V in the SHINERS and SERS experiments, respectively.

The data in Figure 8.2b show that, in SHINERS, pyridine is adsorbed only onto the single-crystal surface as the pinhole-free, silica-coated nanoparticles are chemically and electrically inert. In the SERS experiment, a smooth polycrystalline gold surface is used instead of a single crystal because the citrate-stabilized nanoparticles are polycrystalline. Pyridine is adsorbed onto both the "bare" gold nanoparticles and onto the smooth gold surface. As the electromagnetic field-strength is maximized in the gap between adjacent nanoparticles, most of the signal arises from the pyridine molecules located there, and thus the Raman intensity from adsorbed species will not be very strongly enhanced. Consequently, bare gold nanoparticles cannot provide enhanced Raman information from

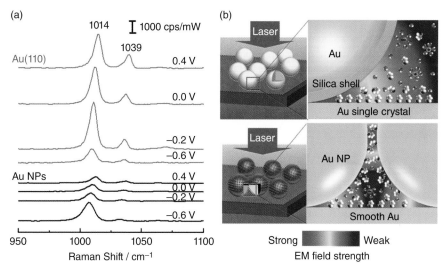

Figure 8.2 *(a) SHINERS spectra of pyridine on Au(110) (red) and SERS spectra of pyridine on "bare" 55 nm Au nanoparticles (black) at different potentials. Electrolyte: 10 mM pyridine + 0.1 M NaClO$_4$; (b) Schematic diagrams of the SHINERS and SERS experiments.*

adsorbed species and/or surface reactions on well-defined single-crystal surfaces. Yet, the SHINERS method overcomes this problem and has the potential to emerge as a powerful tool in (electrochemical) surface science.

8.2 Interfacial Water on Different Metals

Water is the most important solvent in electrochemistry. The adsorption configuration and orientation of water molecules within the electrochemical double layer can significantly affect interfacial electrochemical processes, and therefore the use of SERS to investigate surface water is both interesting and promising. However, two obstacles emerged when obtaining the SERS spectra of water: (i) water is a poor Raman scatterer with a very small cross-section; and (ii) bulk water in solution (~55 M concentration) makes a tremendous contribution to the overall spectra, as compared to surface water.

The SERS spectra of surface water were first acquired in solutions containing concentrated halide ions in 1981, by Fleischmann *et al.* [5] and Pettinger *et al.* [15]. In early studies, silver electrodes roughened by oxidation/reduction cycles in (pseudo-)halide solutions had been employed, and these provided information on the influence of halide ions and electrolyte cations, metal surfaces and electrode potential on the structure of interfacial water [5, 9, 15, 42–47]. The (pseudo-)halide ions were found specifically to be adsorbed onto the surface to stabilize the SERS-active sites and disrupt hydrogen bonds between the surface water and the network of water molecules in the bulk solution. Further studies using NaClO$_4$ solutions confirmed that the cleavage of hydrogen bonds between the water molecules favoured the observation of SERS of water [46]. However, the intensity

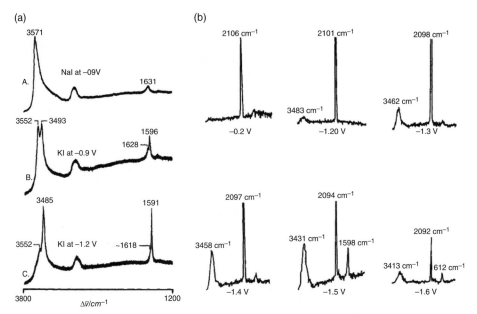

Figure 8.3   *(a) Comparison of the SERS of water, in the OH stretching and bending regions, for molar NaI and KI solutions; (b) The abnormally enhanced δ(HOH) spectra at −1.5 V with CN⁻ adsorbed on roughened Ag.*

of such SERS spectra was irreversibly diminished at negative potentials where the halide ions were completely desorbed, which implied that the spectra corresponded to surface complexes involving "metal ad-atoms (clusters), halide ions, cations, and water," and that those spectra could not therefore be attributed to "normal" water present at the interface. Furthermore, information on surface water was not accessible at the negative potential where the hydrogen evolution reaction occurred.

Some important progress was made in 1987 by Funtikov *et al.* who reported unusual SERS spectra of water, whereby the δ(HOH) bending mode around 1600 cm⁻¹ but not the ν(OH) stretching band around 3500 cm⁻¹ was detected [44]. These authors roughened Ag using a unique *in-situ* oxidation/reduction cycle, without (pseudo-)halide ions, which could be performed only in a narrow −1.10 to −1.45 V potential range. In fact, Fleischmann *et al.* had earlier observed such a dramatic enhancement of the δ(HOH) band, in a potential region where I⁻ [9] and CN⁻ [13] started to desorb (Figure 8.3a and b, respectively), and interpreted this phenomenon as due to quasi-crystal-like water molecules. These results indicated that "normal" SERS spectra of interfacial water could be obtained in the potential region for H₂ evolution.

Since 1990, Tian's group has adopted three experimental approaches to the systematic investigation of surface water structure:

- A special thin-layer spectroelectrochemical cell with a working electrode perpendicular to the solution surface was designed, which allows the hydrogen bubbles generated on the electrode surface to escape smoothly from the thin layer, without severely altering the incident and scattered light.

- The special Funtikov's oxidation/reduction cycle method was used to maintain stable SERS-active sites with no adsorbed (pseudo-)halide ions.
- Difference spectra were recorded to remove the contribution of Raman signals from the bulk water. Thus, a high intensity ratio between bending and stretching modes (I_b/I_s) at very negative potentials was successfully obtained, in clear contrast to the fact that I_b/I_s is normally 0.05 and 0.01 in the Raman spectra of bulk and water vapour, respectively [48].

SERS studies on interfacial water, initially limited to roughened coinage metal electrodes, were recently extended with the help of nanoparticles and transition metal-coated nanoparticles; this approach allowed a clearer interpretation of the potential-dependent spectra to be provided [49–51]. Figure 8.4a shows the SERS spectra of water adsorbed onto Ag, Au, Au@Pt and Au@Pd nanoparticle film electrodes, at different potentials. Core–shell nanoparticles provide an enhancement of two orders of magnitude larger than pure transition metal nanoparticles, thus allowing the detection of extremely weak SERS signals of water. Globally, the SERS intensity increases and the $v(OH)$ stretching band is red-shifted as the potential is made more negative. The red-shift varies at different electrodes; the Stark tuning rate is 80, 14, 30, and 76 cm^{-1} V^{-1} for Au, Pt, Ag, and Pd, respectively (Figure 8.4b). A model for interfacial water adsorbed onto different metals was also proposed (Figure 8.4c) [49].

A broad band at around 2000 cm^{-1} was observed only on the Pt surface, and assigned to the Pt–H stretching vibration mode; this might have been the first Raman spectrum of surface hydrogen ever recorded. With regards to the appearance of the Pt–H band, a full monolayer of on-top adsorbed hydrogen was proposed (see also Chapter 10), above which a second layer of water molecules is located. The stretching band on the Pt surface has the smallest stark tuning rate because water molecules are not directly in contact with Pt. Instead, water molecules can adhere to the bare Pd surface, as implied by the larger stark tuning rate. Recent DFT calculations have suggested that the H-down configuration of water adsorbed onto negatively charged metal clusters can enhance the bending mode more strongly than the stretching mode, and the coupling of the water bending mode with the hydrogen bond is a further enhancement mechanism [50]. Thus, the enhancement in the intensities of the bending vibrations indicate that the water is binding to metal atoms through its H-atom. Accordingly, the abnormally high enhancement at negative potentials could be caused by an electronic enhancement. As the metal conduction electrons have a high polarizability, the surface electronic tail penetrates into the solution to a distance of several Angstroms [52], and thus the combined effects of "the electronic tail" and local optical electric field further enhance the SERS signal of interfacial water [53].

8.3 Coadsorption of Thiourea with Inorganic Anions

Martin Fleischmann and coworkers were the first to recognize that the high sensitivity and high energy resolution of EC-SERS, combined with an analysis of the potential dependence of frequency and intensity of the bands, could be used to characterize the details of layers involving the coadsorption of two species. Two types of electrochemical coadsorption

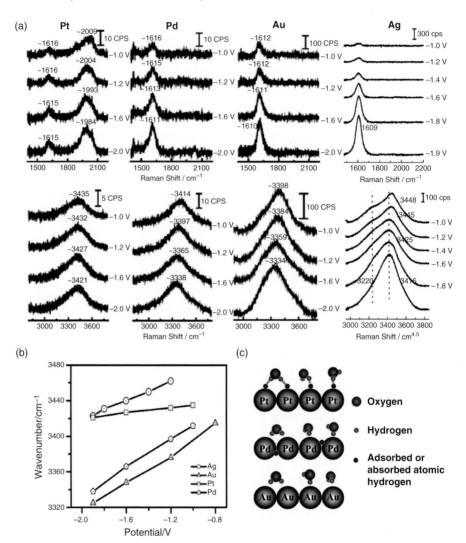

Figure 8.4 *(a) High-resolution SERS spectra of water adsorbed on film electrodes of Au@Pt, Au@Pd, Au and Ag nanoparticles at potentials negative to the potential of zero charge (PZC) with respect to the SCE; (b) Vibrational Stark effect for the O–H stretching vibrations in SERS of interfacial water on different materials of electrodes of Ag (circle), Au (triangle), Au@Pt (square), and Au@Pd (pentagon) nanoparticles; (c) Suggested models of interfacial water adsorbed on Pt, Pd, and Au electrode surfaces.*

termed "parallel" (i.e., competitive) coadsorption and "induced" coadsorption were recognized, and the transition between them was observed [54]. Subsequently, Fleischmann and colleagues studied the coadsorption of thiourea (a common additive in electroplating and refining processes) with inorganic ions such as ClO_4^- and SCN^-. The results of early investigations suggested that thiourea adsorbed onto the metal surface, forming a sulfur–metal bond, and that ClO_4^- might coadsorb with thiourea, through the NH_2 group in neutral

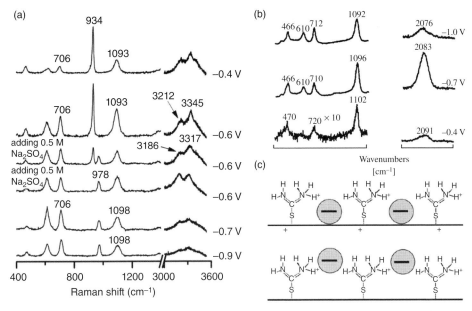

Figure 8.5 *(a) SERS from an Ag surface of the coadsorption process of thiourea with ClO$_4^-$ and SO$_4^{2-}$ as a function of the electrode potential; (b) SERS spectra of C–N stretching vibration band of SCN$^-$ coadsorbed with thiourea on an Ag electrode; (c) Adsorption model showing the coadsorption of anions (SCN$^-$, ClO$_4^-$, or SO$_4^{2-}$) with thiourea in acidic solution at different potentials.*

solution and through the NH$_3^+$ group in acidic solution [10, 13, 14]. As ClO$_4^-$ does not adsorb onto Ag electrodes, its SERS signal cannot be obtained; however, after the addition of thiourea, SERS of ClO$_4^-$ can be detected – together with that of thiourea – indicating an induced coadsorption. The fact that the frequency of the coadsorbed ClO$_4^-$ is identical to that of the free ClO$_4^-$ in solution, and independent of the electrode potential, implies that ClO$_4^-$ interacts only indirectly with the surface.

Tian *et al.* observed that, upon addition of SO$_4^{2-}$, the SERS bands from ClO$_4^-$ were completely replaced by those of SO$_4^{2-}$ (Figure 8.5a). The frequencies of the thiourea bands attributed to NH$_2$ (3345 cm^{-1}) and N–C–N groups (1093 cm^{-1}) changed, further suggesting the coadsorption of thiourea with anions through the NH$_3^+$ group [55, 56]. Unlike ClO$_4^-$, SCN$^-$ strongly interacts with the Ag electrode surface, and thus at relatively positive potentials, SCN$^-$ and thiourea give rise to parallel coadsorption (competitive coadsorption) (Figure 8.5b). In the presence of thiourea, SCN$^-$ can adsorb at more negative potential than without thiourea, and its CN band red-shifts to approach that of free SCN$^-$, which indicates that the coadsorption of SCN$^-$ and thiourea changes from parallel to induced [56] (Figure 8.5c).

Tian's group also studied the thiourea + ClO$_4^-$ system by using time-resolved- SERS (TR-SERS) measurements [55]. The intensities of the different bands were shown to change on different time scales following a potential step (e.g., from –0.9 V to –0.3 V).

The C–S stretching vibrational mode $(710 \ cm^{-1})$ and the SCNN symmetric stretching mode $(1496 \ cm^{-1})$ underwent more than 80% intensity changes within the first 0.5 s, and then slowly approached a constant value. Instead, the intensity of other bands, including the vibrational modes of the amino group $(468, 1094 \ and \ 3350 \ cm^{-1})$ and that of ClO_4^{-}, underwent only 50% of the overall change at the initial stage of the potential step, but then decreased slowly over more than 10 min. Possibly, in order to form a well-structured coadsorbed layer at –0.3 V, the ClO_4^{-} ions interacting with protonated amino groups needed time to rearrange and move to appropriate sites, at some distance from the electrode surface. When the potential was stepped back to –0.9 V, the coadsorbed layer became loose and disordered, and so the intensity change of all bands rapidly followed the change in potential. The TR-SERS result showed that it is possible to focus on individual bonds of the adsorbate in surface dynamic studies, particularly in some complicated coadsorption processes.

8.4   Electroplating Additives

The electroplating of SERS-active coinage metals is technologically important. Consequently, SERS has been used extensively to examine the adsorption behaviors of electroactive complexes, ligands, leveling agents and brighteners present in the common Au, Ag and Cu electrodeposition baths. Although there appears to be a conflict between the roughened substrates needed for SERS and the smooth surfaces sought from electroplating, interesting information was obtained. In most of the studies conducted, systems containing thiourea [10, 13, 14, 57–60] or cyanides [61–65] were addressed.

Several groups agreed that thiourea was adsorbed via the sulfur atom [10, 13, 14, 57, 58], with weakening of the C=S bond order, and coadsorbed with oxyanions [14, 59, 60]. Fleischmann *et al.* [13] reported that, at the silver electrode, thiourea – but not Ag^+–thiourea complexes – was detected by SERS, and concluded that the bonding of thiourea to the Ag surface was stronger than that of the complexes. Thus, Ag plating occurs through a strongly adsorbed thiourea layer, and this phenomenon may be the reason why smooth, reflecting Ag deposits are formed.

When these studies were extended to cyanide-containing baths, they revealed a complex surface chemistry. It was found [13] that at least one Ag^+–cyanide complex [probably $Ag(CN)_2^-$] was adsorbed at a moderately negative potential, while CN^- remained adsorbed at even more negative values. Bozzini *et al.* [62] identified different species adsorbed onto Au; at progressively less-negative potential values, the detected species were: (i) Au–H and Au–NC$^-$ in the HER range; (ii) Au–NC$^-$; (iii) Au–NC$^-$, Au–CN$^-$ and $Au(CN)_2^-$; and (iv) OCN$^-$ in the anodic range. An even higher complexity was reported in a study of Ag–Au alloy deposition [63]. Bozzini *et al.* [65] found that CN^- ions liberated in the cathodic reduction of cyanide complexes of Au, Ag and Cu, were reduced to $CH_2=NH$ and CH_3NH_2 on Au, were polymerized (possibly to polycyanogens) on Ag, but underwent no reaction on Cu.

SERS investigations were carried out using either preroughened electrodes [10, 57–59] or smooth electrodes (e.g., vitreous carbon or Pt) which became SERS-active upon the deposition of Ag or Au nuclei (islands) [13, 61, 63, 64]. SERS was used to study the nucleation of metal deposits and the evolution of deposit morphology during electrodeposition;

however, only weak SERS signals could be obtained from smooth deposits of technological interest [13]. In a study of Ag electrodeposition onto rough Ag, from solutions containing 2-hydroxypyridine, Lin *et al.* [66] observed the intensity of SERS signals to be decreased at potentials slightly negative to that for Ag deposition, but to be markedly increased at more negative potentials where rough deposits were formed. The addition of polyethyleneimine stabilized the SERS intensity at a low level, due to the polymer inducing the formation of smooth deposits.

Electrodeposition on smooth, SERS-inactive substrates has been described as a method for obtaining good-quality SERS spectra of pyridine adsorbed onto Ru, Rh and Pd [67]. However, to date no SERS investigations on the electrodeposition of metals other than Au, Ag and Cu have been reported.

8.5 Inhibition of Copper Corrosion

When, during the early 1980s, SERS was first used to achieve fundamental information about the metal/electrolyte interface in corrosion science, only Ag, Au and Cu were considered to be SERS-active. Consequently, most early studies were focused on the inhibition of Cu corrosion by benzotriazole and other inhibitors. In the same way that pyridine on Ag was used as the model system for many fundamental investigations of SERS, benzotriazole on Cu became the object of most early SERS studies on corrosion inhibition. Benzotriazole is used to inhibit Cu corrosion either through a pretreatment of the Cu sample with the inhibitor ahead of its exposure to the corrosive environment, or by its addition at low concentration to the corrosive environment (when this is of finite volume). SERS studies have focused on both situations in order to elucidate inhibitor adsorption and film formation. The first potential-dependent SERS spectra of benzotriazole adsorbed onto Cu were reported by Kester *et al.* [19] who found that photoalteration of the benzotriazole surface complex induced strongly enhanced Raman signals and increased protection. However, these authors did not discuss the benzotriazole adsorption mode, nor did they assign the observed bands. Later, Rubim *et al.* [20] investigated Cu plates pretreated with benzotriazole, and benzotriazole adsorption from both neutral and acid solutions, in the presence of halides. For this, the SERS spectra were compared with ordinary Raman spectra of aqueous benzotriazole solutions and with those of solid Cu(I) and Cu(II) complexes. Based on results obtained, it was concluded that the Cu pretreatment produced a polymeric Cu(I) complex, $[Cu(I)BTA]_n$, (where BTAH = benzotriazole), confirming previous studies performed with *ex-situ* spectroscopic techniques. $[Cu(I)BTA]_n$ was also identified as the species formed when benzotriazole was absorbed onto Cu from neutral solutions and the electrode was roughened *in-situ*. In acid media, either $[Cu(I)BTA]_n$ or halide-containing complexes $([Cu(I)XBTAH]_4$, X = Cl or Br) were identified, the former being favored at less positive potential values. Adsorption of the benzotriazole molecule was observed upon the reduction of Cu(I) complexes at potential values more negative than the PZC.

Between 1985 and 1987, Fleischmann *et al.* [21–23] continued to perform SERS experiments that led to corrosion rates which could be compared with conventional electrochemical corrosion rate (I_{cor}) measurements. The results of these studies provided a dynamic description of the Cu/electrolyte interface, as a function of potential, concentrations of inhibitor and anions, pH, and time of exposure. Cu was roughened *ex-situ*, in inhibitor-free KCl solutions, to avoid the anodic formation of Cu–inhibitor complexes. When BTAH was compared with 2-mercaptobenzoxazole (MBO), benzimidazole (BIM) and 1-hydroxybenzotriazole (BTAOH) in neutral chloride media [21], I_{cor} measurements showed the following order of inhibition efficiency: BTAH ~ MBO > BIM > BTAOH. In SERS experiments, two strong inhibitors – benzotriazole and MBO – prevented chloride adsorption (i.e., the Cu–Cl band at 286 cm^{-1} was not detected), and their spectra were stable at $E \leq E_{COR}$ (E_{COR} = corrosion potential). It was concluded that these inhibitors had stabilized the ad-atoms necessary to induce SERS. Weaker inhibitors were coadsorbed with chloride, at $E < E_{COR}$. On moving to the open-circuit situation, a steady decrease in the inhibitor bands was observed while the intensity of the Cu–Cl band was initially increased but then decreased; this suggested that chloride had displaced the inhibitor and then induced the corrosion of ad-atoms, with an irreversible loss in SERS signal intensity.

Fleischmann *et al.* [22] compared benzotriazole and 2-mercaptobenzoxazole as inhibitors of copper corrosion in KCl solutions containing low concentrations of cyanide. Benzotriazole proved to be an ineffective inhibitor in cyanide media, while 2-mercaptobenzoxazole remained effective. SERS showed that cyanide, revealed by a broad band centred at 2090 cm^{-1}, displaced benzotriazole from the Cu surface, whereas 2-mercaptobenzoxazole displaced adsorbed cyanide. A synergetic inhibition of Cu corrosion by benzotriazole and benzylamine, both in chloride and chloride/cyanide media, was also shown [22]. As SERS showed that benzylamine had not been adsorbed, its beneficial effect was ascribed to an improved film formation. Subsequent I_{cor} measurements showed that benzotriazole, MBO, 2-mercaptobenzothiazole and 2-mercaptobenzimidazole were all effective inhibitors of copper corrosion in neutral chloride solutions, but the inhibition efficiency of benzotriazole was decreased at pH 1–2 [23]. SERS spectra showed that, at pH 7, benzotriazole and its anionic form were coadsorbed and Cl$^-$ was excluded from the interface. However, at pH $\leq$ 2 undissociated benzotriazole and Cl$^-$ were coadsorbed, such that Cu underwent corrosion. In contrast, the anion from 2-mercaptobenzothiazole was the only adsorbed species at pH between 7 and 2; only at pH 1 was the neutral 2-mercaptobenzothiazole molecule detected. Competitive adsorption experiments showed that the inhibitive action of benzotriazole and 2-mercaptobenzothiazole in neutral/acid media could be explained in terms of adsorption strength.

When Aramaki *et al.* [68–70] also ranked benzotriazole derivatives, they concluded that their inhibition efficiency agreed with their complex-forming ability. The spectra of the neutral inhibitor molecules and those of their Cu(I) polymeric complexes were distinguished, and the stability range of each species was summarized using potential–pH diagrams. The adsorption of neutral molecules was favored by a low pH and a negative potential, while copper complex formation was favored by a high pH and a positive potential. Diagrams based on SERS results agreed, at least qualitatively, with those based on capacity data.

In 1990, Fleischmann *et al.* used NIR Fourier transform SERS (NIR FT-SERS) [71] to study the Cu corrosion inhibitors, benzotriazole and tolyltriazole. Diagnostic bands were identified for each compound, whereby benzotriazole and tolyltriazole were coadsorbed and found to displace each other, depending on their relative concentrations. The NIR FT-SERS investigations were extended to an antifreeze (ethylene glycol) mixture [24], in which both benzotriazole and tolyltriazole were identified by SERS as being adsorbed onto the copper surface, although ordinary Raman spectroscopy showed only the bands of adsorbed ethylene glycol.

The Cu/ benzotriazole system was revisited by Chan and Weaver [72], who explored a wide pH range (2 to 13) using both H_2O and D_2O as solvent. It was shown that, at pH 2 and $E < -0.3$ V (vs SCE), benzotriazole in both H and D forms were adsorbed through their azole ring, with two N atoms interacting with Cu. At about $E = -0.2$ V, the spectra of both forms became identical due to formation of the copper(I) complexes with the deprotonated benzotriazole, $[Cu(I)BTA]_n$. It was also shown that adsorbed benzotriazole was converted to $[Cu(I)BTA]_n$ upon exposure to air.

With no need of any roughening pretreatment, a SHINERS investigation on polycrystalline Cu and single-crystal Cu(100) or Cu(111), immersed in a benzotriazole solution, was carried out to clarify whether $[Cu(I)BTA]_n$ film formation was reversible with potential [73]. For both polycrystalline and single-crystal Cu, identical SERS spectra of benzotriazole were obtained at –0.7 V (Ag/AgCl), which converted to those of $[Cu(I)BTA]_n$ when the potential was swept to more positive values. However, when the potential was swept back, polycrystalline Cu showed a reversible interconversion between $[Cu(I)BTA]_n$ and adsorbed benzotriazole, while single-crystal Cu did not. Through this investigation, SERS results (normally obtained with polycrystalline Cu) were reconciled with infrared reflection-absorption spectroscopy (IRRAS) and sum difference generation spectroscopy (SFG) results (obtained with single crystals).

During recent years, SERS spectra have been used to acquire basic information on new inhibitors [74–77], and also on the adsorption mode of benzotriazole in new solvent environments such as ionic liquids [78]. Thus, three decades after the pioneering studies [19–21], SERS spectroscopy is becoming a standard technique in studies of Cu corrosion inhibition.

8.6   Extension of SERS to the Corrosion of Fe and Its Alloys: Passivity

Extending SERS studies to the corrosion of Fe and its alloys, was a major goal pursued by several groups. Fleischmann *et al.* [17,18] initially proposed an "indirect strategy" whereby, after showing that corrosion protection processes were identical (or very similar) on Fe (or its alloys) and Ag, the latter was used in SERS investigations. The report [25] that opened the way to SERS studies of Fe and its alloys emerged in 1987, when Fe monolayers were deposited onto SERS-active Ag from solutions containing $FeSO_4 \cdot (NH_4)_2SO_4$ and pyridine (pH 3.5). In other experiments, deposits estimated to be approximately 100 monolayers thick were obtained from pyridine-free solutions, and pyridine was added afterwards. In both cases, when SERS spectra of pyridine were obtained, the frequencies and relative intensities of the pyridine bands were different from those obtained with bare Ag and, as a consequence, it was concluded that pyridine had been adsorbed onto Fe. Pyridine spectra were also obtained from Fe electrodes after deposition by galvanic displacement of a low density of Ag nuclei; the spectra obtained from different areas corresponded to pyridine adsorbed onto either Ag or Fe. During the following years the "Fe-on-Ag approach" was used extensively to investigate Fe corrosion inhibitors, while the "Ag-on-Fe approach" was employed in studies on Fe passivity and extended to other metals (e.g., Ni, Cr, Zn) and alloys (steel).

8.6.1   Fe-on-Ag

Aramaki and Uehara [26, 79] confirmed that SERS spectra of pyridine and pyridinium bromide were obtained through the Fe-on-Ag approach, and reported that the pyridine spectra had changed from that of the molecule adsorbed onto Ag to that of the molecule adsorbed onto Fe as the thickness of the Fe deposit exceeded five monolayers. The intensity of the main pyridine peaks and inhibition efficiency of Fe corrosion were linearly correlated. Aramaki's group also carried out systematic investigations on various classes of Fe corrosion inhibitors, including aromatic N- and S-containing compounds and propargylic alcohol. Several positive correlations, but some discrepancies, were identified when comparing SERS results and corrosion inhibition efficiencies.

8.6.2   Ag-on-Fe

The Ag-on-Fe approach was preferred in investigations on passive films because it allowed the study of alloys such as carbon steel and stainless steel, which cannot be effectively electrodeposited. The first investigations on passive Fe using the Ag-on-Fe approach were made by Rubim *et al.* [27] in 1989, while other reports were made during the following years [80–92], with the aim of assessing the chemical nature of surface films on Fe, Ni, carbon steel and stainless steel as a function of electrolyte composition and electrode potential. The technique used to obtain SERS spectra of passive Fe in borate buffer, as described by Devine *et al.* [80, 81], involved the deposition of Ag onto passive Fe, the transfer of Ag-plated Fe to borate buffer, the exhaustive reduction of oxide films, and the formation of passive films under potential control. The optimum Ag deposition charge that maximized the enhancement was evaluated, and compounds proposed as passive film components are briefly summarized in Table 8.1. The identification of passive films with well-defined solid compounds was not always possible as their nature depended heavily on the solution chemistry; typically, the (mainly amorphous) films were both potential- and history-dependent, and underwent changes if removed from solution. The multiplicity of films justified the conflicting findings of various authors who used different techniques. Very recently, a SHINERS investigation of Ni-based alloys in H_2SO_4 showed that heat treatment caused an improved corrosion resistance through the formation of a mixed Cr(III)/(VI) surface oxide [93].

8.7   SERS of Corrosion Inhibitors on Bare Transition Metal Electrodes

The Fe-on-Ag and Ag-on-Fe approaches were subject to some drawbacks. In the former case, pinhole-free metal layers of a few monolayers thickness were difficult to deposit, they were unlikely to have the same crystal structure as bulk Fe, and they had insufficient stability over a wide potential range. With the Ag-on-Fe approach, it could not be excluded that species adsorbed at the Ag/Fe boundary contributed to the overall SERS response in a major fashion. As noted in Chapter 7, however, these concerns were minimized by advances in Raman instrumentation (i.e., the advent of confocal microscopy and the holographic notch filter) and the development of special oxidation/reduction treatments for roughening the electrode surfaces.

Table 8.1 SERS studies of passive metals with the Ag-on-metal approach.

Year	Metal	Medium	Passive film composition	Reference(s)
1989	Fe	H_2SO_4 or Borate or Carbonate	$Fe(OH)_2 + \delta$-FeOOH in H_2SO_4 or borate; $FeCO_3/Fe(OH)_3$ in carbonate	[27]
1991	Fe	Borate	Active region: $Fe(OH)_2 + FeO$ + unknown species From passivation to Flade potential: $Fe_3O_4 + Fe(II,III)(OH)_x{}^a$ Positive of Flade potential: $Fe_3O_4 + Fe(II,III)(OH)_x{}^a + \gamma$-FeOOH.	[80, 81]
1991	Fe	Borate	At -0.6 V(SCE): $Fe_3O_4 + Fe(OH)_2$. At $E = 0.0$ V: α-$Fe_3O_4 + FeOOH + Fe_2O_3$.	[82]
1992	304 SS[b] 316 SS[b]	NaCl	Similar to that of passive films on Fe	[83]
1992	Ni	NaCl or NaOH	$Ni(OH)_2 + NiO$	[84]
1994	Fe	Sulfate or Sulfate + borate	In mildly acid media, sulfate becomes part of the passive film	[85]
1995	Fe, Ni, Cr and 308 SS[b]	Borate	Fe: $Fe(OH)_2$-like species + γ-Fe_2O_3 or Fe_3O_4 Ni: β-$Ni(OH)_2 + NiO$ Cr: $Cr(OH)_3$ + species similar to $Cr(OH)_2$ or CrOOH SS[b]: same species as on alloy components	[86]
1995	Fe	Nitrate or Carbonate	$Fe(OH)_2 + Fe_2O_3/\gamma$-$Fe_3O_4$	[87]
1996	Fe	NaOH or Borate Up to 95 °C	At 25–90 °C: $Fe(OH)_2 + Fe_3O_4$ in prepassive region; Fe_3O_4 + FeOOH in passive region At 95 °C $Fe(OH)_2$	[88]
2001	Carbon steel	NaCl + O_2	Passive film enriched in γ-Fe_2O_3	[90]
2003	Zn	KOH	ZnO	[91]
2010	Fe	Borate	Prepassive region: Fe(II) oxide/hydroxide Passive region: duplex film; inner layer Fe_3O_4 or γ-Fe_2O_3 or both; outer layer Fe(III)-oxide/hydroxide.	[92]

[a]The species $Fe(II,III)(OH)_x{}^*$ is described by Gui and Devine [80,81] as a mixed-oxidation changing substance that initially was $Fe(OH)_2$.
[b]SS, stainless steel.

The first report on SERS from bulk Fe was made in 2000 by Cao *et al.* [94], who obtained spectra of both pyridine and pyrazine. Enhancement factors of two to three orders of magnitude were produced through a sequence of two oxidation/reduction cycles performed in H_2SO_4 and in KCl + pyridine solutions, respectively. An intense band at 280 cm^{-1} in the pyrazine spectrum was assigned to the Fe–N stretching vibration, while its higher intensity compared to Ag–N stretching confirmed that pyrazine had interacted more strongly with Fe than with noble metals.

The corrosion inhibition of Fe by benzotriazole (BTAH) was studied in NaCl and H_2SO_4 solutions [95]. Electrochemical tests showed BTAH to be an effective anodic inhibitor in neutral NaCl solution, and a weaker mixed inhibitor in H_2SO_4 solution. In acid solution, bands for N–H in-plane bending and sulfate were detected which suggested that, at potentials positive to the PZC, sulfate ions were adsorbed. At the PZC, neutral BTAH was adsorbed and sulfate was coadsorbed with a protonated form ($BTAH_2^+$), while at potentials negative to the PZC, $BTAH_2^+$ was physisorbed (see Figure 8.6a–c). In neutral solution, the presence of a Fe(II)-BTA film was proposed. In a further study [96], the inhibition efficiency of BTAH was found to decrease in the order Cu > Fe > Ni. At sufficiently positive potentials, Cu and Fe were coated by $[Cu(I)BTA]_n$ and $[Fe(II)BTA_2]_n$, respectively; at a more negative potential, BTAH was adsorbed. Similar conclusions were reached for Ni [97].

Roughening procedures based on oxidation/reduction cycles followed by the exhaustive reduction of ZnO, in neutral $NaClO_4$ solutions, were used to obtain spectra of pyridine adsorbed onto solid Zn in a narrow potential window [98]. At more positive potentials, the pyridine response disappeared and ZnO was detected. Yang *et al.* [99] found that the thiolate form of 2-mercaptobenzothiazole was adsorbed on Zn through both of its S atoms, and moved from a quasi-flat to a quasi-perpendicular orientation as the potential was taken negative.

These few examples show that SERS has become a versatile technique for the study of corrosion inhibition of those technologically important metals which were once considered SERS-inactive. However, the recording of high-quality SERS spectra requires the use of roughening procedures that can cause profound modifications of surface morphology.

8.8   Lithium Batteries

With their high energy and power density, lithium batteries have attracted wide research interest and have been used to power various portable consumer electronic devices, electric vehicles, and implantable medical applications. It is well known that the electrochemical properties of lithium batteries (e.g., specific capacity, reversibility, cycling behavior) are heavily dependent on the changes of structures and composition of the electrodes, electrolyte, and solid–electrolyte interphase (SEI) during the charge–discharge process. Therefore, Raman spectroscopy is appropriate to understanding such changes as it allows the sensitive detection of structural variations at the atomic level. Indeed, unique molecular and crystalline information is then accessible [100].

The SEI is an inhomogeneous film which is composed of various reduction products and results from the chemical reaction between lithium and an electrolyte solution. The SEI plays the key role in lithium ion insertion–deinsertion and the electrochemical processes at the interface. However, Raman spectroscopy is rarely employed to investigate SEI layers

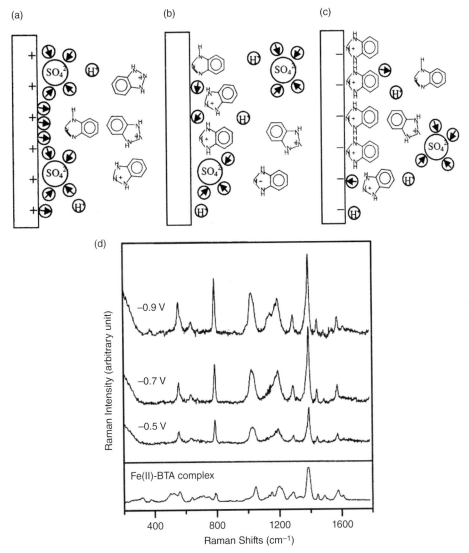

Figure 8.6 *Schematic of adsorption mode of benzotriazole on an iron electrode in sulfuric acid at potentials. (a) Positive to potential of zero charge (PZC); (b) PZC; (c) Negative to PZC; (d) Normal Raman spectrum of the Fe(II)–BTA complex and potential-dependent surface Raman spectra of BTAH on an iron electrode in saline water.*

as they are usually very thin and their normal Raman scattering is too weak to be detected. Because of its high sensitivity to surface structures, SERS can serve as a powerful tool to study the interfacial phenomena in lithium ion batteries and to understand the mechanism of formation of the SEI film at its early stages [101–107].

Silver, which is one of the best SERS-active substrates, can be used as the positive electrode anode for lithium batteries since, when the battery is discharged, it is converted to a LiAg alloy. When in contact with the electrolyte, it forms an SEI that would be expected to have a composition and structure similar to that formed at a lithium negative electrode during battery operation. When Li *et al.* [101–103] discharged a Li/Ag cell with electrolytes of both ethylcarbonate/diethylcarbonate and propylene carbonate/diethylcarbonate containing lithium salts, they obtained SERS spectra from the surface of the Ag positive electrode after washing off the electrolyte and mounting in a vacuum cell. The intercalation of lithium into the silver lattice alters the surface morphology of the Ag electrode, making it SERS-active. An analysis of the spectra suggested that, in dry conditions, the SEI layer was mainly composed of alkyl carbonates, $ROCO_2Li$ (R = alkyl groups) and amorphous Li_2CO_3. However, the $ROCO_2Li$ disappeared such that $LiOH \cdot H_2O$ and Li_2CO_3 became the main components of the SEI layer when a trace of water was present in the electrolyte or the atmosphere.

Recently, Schmitz *et al.* [107] investigated the SEI film on a copper electrode after lithium plating in ethylcarbonate/diethylcarbonate. By using a specially designed *in-situ* Raman cell, these authors obtained the enhanced Raman signal of lithium carbonate, Li_2CO_3, and lithium acetylide, Li_2C_2, which were identified as SEI components, on the electrically roughened lithium-covered copper electrode. The Raman mapping images, based respectively on the bands of Li_2CO_3 and Li_2C_2, revealed the Li_2CO_3 to be distributed homogeneously over the copper substrate, while the Li_2C_2 was mainly located on the plated lithium surface.

Nowadays, the Li-air (O_2) battery is attracting a great deal of interest because of its (theoretical) high energy density [108]. Bruce's group was the first to define the O_2 reduction process in the Li-O_2 battery by *in-situ* SERS studies on a roughened Au electrode [109]. In this case, lithium superoxide (LiO_2) was found to be an intermediate in O_2 reduction, which then disproportionated to the final product, lithium peroxide (Li_2O_2). In contrast, the LiO_2 is not an intermediate in oxidation; that is, oxidation does not follow the reverse pathway to reduction. Operation of the rechargeable Li-O_2 battery depends critically on the repeated and highly reversible formation/decomposition of Li_2O_2 at the positive electrode upon cycling. SERS provides evidence of Li_2O_2 formation/decomposition also in improved Li-O_2 batteries, which suffer less capacity loss upon cycling [110, 111].

Although difficulties in studying electrodes for lithium batteries have hindered a wider application of SERS, time-resolved Raman (or SERS) measurements of the changes of bulk materials and interfacial composition, the spatial-resolved confocal Raman observation of inhomogeneous surfaces could be of great interest. See [112] for a recent review.

8.9   Intermediates of Electrocatalysis

The characterization of reaction mechanisms and intermediates of complex chemical reactions forms a central topic in electrochemistry [113–118]. The electrochemical reductive cleavage of carbon–halogen bonds is an important process in electro-organic synthesis, waste stream treatment and electron-transfer mechanisms. Benzyl chloride $(PhCH_2Cl)$ reduction in organic solvents has been widely investigated following debate as to whether the carbon–chlorine bond would be reductively cleaved via a concerted or a nonconcerted reaction pathway. Recent studies have shown that Pd, Cu – and especially Ag – cathodes

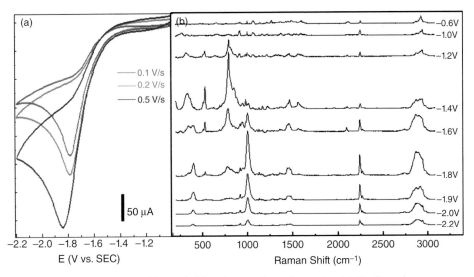

Figure 8.7 *The reduction of benzyl chloride at a silver electrode. (a) Cyclic voltammogram;*
(b) SERS spectra as a function of potential.

possess unexpectedly high electrocatalytic activity towards these reactions. Since voltammetry is insufficient to identify reaction intermediates and unravel the exact origin of the
strong catalysis by Ag, Tian and Amatore performed an *in-situ* SERS-DFT study on the electrochemical reduction of $PhCH_2Cl$ at the Ag electrode in acetonitrile, exploiting for the first
time the potential of EC-SERS in the identification of transient reaction intermediates at the
electrode surface [119]. Figure 8.7a shows the cyclic voltammograms of $PhCH_2Cl$, which
exhibit a single irreversible reduction peak at the Ag electrode. These voltammograms dis-
close little regarding the reduction mechanism, although the peak potential is substantially
positive to that at vitreous carbon, confirming that Ag is an electrocatalyst for the reaction.
SERS spectra were recorded at potentials from –0.6 to –2.2 V, and suggested three regions
within this range (Figure 8.7b). At potentials from –0.6 to –1.2 V, where the voltammogram
shows little current, the spectral features resembled those of soluble $PhCH_2Cl$, indicating
a weak interaction between benzyl chloride and silver. When the potential was scanned to
–1.2 V, the spectra were changed dramatically and the peaks reached a maximum intensity
at –1.4 V, suggesting the formation of a new surface species. At even more negative potentials, spectra with other new features grew gradually and reached a maximum at –1.8 V, the
peak potential observed in the voltammogram. These features are most likely due to the final
reaction product(s), which then decay because of desorption at more negative potentials.
Further insight was achieved by coupling DFT evaluations with SERS [120], allowing a precise description of the main mechanistic features leading to the exceptional electrocatalytic
properties of silver cathodes. First, the DFT simulation suggested that the SERS spectra
at –1.2 V to –1.4 V had most likely originated from the combined adsorption of benzyl
radical and benzyl anion, as the reaction intermediates. The final products – assumed to be
mostly toluene, 1, 4-diphenylethane and 3-phenylpropanenitrile – then contributed to the
overall spectra at –1.8 V. The results of this study highlighted the need to combine cyclic

voltammography and EC-SERS methods with DFT and thermodynamic calculations, in order to clarify interfacial structures and complex mechanisms. Clearly, this is a future direction for EC-SERS.

Acknowledgments

Whenever work from the Chinese groups is mentioned in this chapter, it is the great contribution of the self-motivated and hard-working students of Xiamen University.

References

(1) Fleischmann, M., Hendra, P.J. and McQuillan, A.J. (1974) Raman spectra of pyridine adsorbed at a silver electrode. *Chemical Physics Letters*, **26**, 163–166.

(2) Albrecht, M.G. and Creighton, J.A. (1977) Anomalously intense Raman spectra of pyridine at a silver electrode. *Journal of the American Chemical Society*, **99**, 5215–5217.

(3) Jeanmaire, D.L. and Van Duyne, R.P. (1977) Surface Raman spectroelectrochemistry: part I. Heterocyclic, aromatic, and aliphatic amines adsorbed on the anodized silver electrode. *Journal of Electroanalytical Chemistry*, **84**, 1–20.

(4) Paul, R.L., McQuillan, A.J., Hendra, P.J. and Fleischmann, M. (1975) Laser Raman spectroscopy at the surface of a copper electrode. *Journal of Electroanalytical Chemistry*, **66**, 248–249.

(5) Fleischmann, M., Hendra, P.J., Hill, I.R. and Pemble, M.E. (1981) Enhanced Raman spectra from species formed by the coadsorption of halide ions and water molecules on silver electrodes. *Journal of Electroanalytical Chemistry*, **117**, 243–255.

(6) Fleischmann, M., Hill, I.R. and Pemble, M.E. (1982) Surface-enhanced Raman spectroscopy of $^{12}CN^-$ and $^{13}CN^-$ adsorbed at silver electrodes. *Journal of Electroanalytical Chemistry*, **136**, 361–370.

(7) Fleischmann, M., Graves, P.R., Hill, I.R. and Robinson, J. (1983) Simultaneous Raman spectroscopic and differential double-layer capacitance measurements of pyridine adsorbed on roughened silver electrodes. *Chemical Physics Letters*, **98**, 503–506.

(8) Fleischmann, M., Graves, P.R., Hill, I.R. and Robinson, J. (1983) Raman spectroscopy of pyridine adsorbed on roughened β-palladium hydride electrodes. *Chemical Physics Letters*, **95**, 322–324.

(9) Fleischmann, M. and Hill, I.R. (1983) The observation of solvated metal ions in the double layer region at silver electrodes using surface enhanced Raman scattering. *Journal of Electroanalytical Chemistry*, **146**, 367–376.

(10) Fleischmann, M., Hill, I.R. and Sundholm, G. (1983) A Raman spectroscopic study of thiourea adsorbed on silver and copper electrodes. *Journal of Electroanalytical Chemistry*, **157**, 359–368.

(11) Fleischmann, M. and Hill, I.R. (1983) Surface-enhanced Raman scattering from silver electrodes: formation and photolysis of chemisorbed pyridine species. *Journal of Electroanalytical Chemistry*, **146**, 353–365.

(12) Fleischmann, M., Graves, P.R. and Robinson, J. (1985) Enhanced and normal Raman scattering from pyridine adsorbed on rough and smooth silver electrodes. *Journal of Electroanalytical Chemistry*, **182**, 73–85.

(13) Fleischmann, M., Sundholm, G. and Tian, Z.Q. (1986) An SERS study of silver electrodeposition from thiourea and cyanide containing solutions. *Electrochimica Acta*, **31**, 907–916.

(14) Tian, Z.Q., Lian, Y.Z. and Fleischmann, M. (1990) In-situ Raman spectroscopic studies on coadsorption of thiourea with anions at silver electrodes. *Electrochimica Acta*, **35**, 879–883.

(15) Pettinger, B., Philpott, M.R. and Gordon, J.G. (1981) Contribution of specifically adsorbed ions, water, and impurities to the surface-enhanced Raman spectroscopy (SERS) of Ag electrodes. *The Journal of Physical Chemistry*, **74**, 934–940.

(16) Fleischmann, M., Tian, Z.Q. and Li, L.J. (1987) Raman spectroscopy of adsorbates on thin film electrodes deposited on silver substrates. *Journal of Electroanalytical Chemistry*, **217**, 397–410.

(17) Fleischmann, M., Hill, I.R., Mengoli, G. and Musiani, M.M. (1983) A Raman spectroscopic investigation of the electropolymerization of phenol on silver electrodes. *Electrochimica Acta*, **28**, 1545–1553.

(18) Mengoli, G., Musiani, M.M., Pelli, B. *et al.* (1983) The effect of triton on the electropolymerization of phenol; an investigation of the adhesion of coatings using surface-enhanced Raman scattering (SERS). *Electrochimica Acta*, **28**, 1733–1740.

(19) Kester, J.J., Furtak, T.E. and Bevolo, A.J. (1982) Surface-enhanced Raman scattering in corrosion science: benzotriazole on copper. *Journal of The Electrochemical Society*, **129**, 1716–1719.

(20) Rubim, J., Gutz, I.G.R., Sala, O. and Orville-Thomas, W.J. (1983) Surface-enhanced Raman spectra of benzotriazole adsorbed on a copper electrode. *Journal of Molecular Structure*, **100**, 571–583.

(21) Fleischmann, M., Hill, I.R., Mengoli, G. *et al.* (1985) A comparative study of the efficiency of some organic inhibitors for the corrosion of copper in aqueous chloride media using electrochemical and surface-enhanced Raman scattering techniques. *Electrochimica Acta*, **30**, 879–888.

(22) Fleischmann, M., Mengoli, G., Musiani, M.M. and Pagura, C. (1985) An electrochemical and Raman spectroscopic investigation of synergetic effects in the corrosion inhibition of copper. *Electrochimica Acta*, **30**, 1591–1602.

(23) Musiani, M.M., Mengoli, G., Fleischmann, M. and Lowry, R.B. (1987) An electrochemical and SERS investigation of the influence of pH on the effectiveness of some corrosion inhibitors of copper. *Journal of Electroanalytical Chemistry*, **217**, 187–202.

(24) Sockalingum, D., Fleischmann, M. and Musiani, M.M. (1991) Near-infrared Fourier transform surface-enhanced Raman scattering of azole copper corrosion inhibitors in aqueous chloride media. *Spectrochimica Acta Part A: Molecular and Biomolecular*, **47**, 1475–1485.

(25) Mengoli, G., Musiani, M.M., Fleischman, M. *et al.* (1987) Enhanced Raman scattering from iron electrodes. *Electrochimica Acta*, **32**, 1239–1245.

(26) Aramaki, K. and Uehara, J. (1989) A SERS study on adsorption of some organic compounds on iron. *Journal of The Electrochemical Society*, **136**, 1299–1303.

(27) Rubim, J.C. and Dünnwald, J. (1989) Enhanced Raman scattering from passive films on silver-coated iron electrodes. *Journal of Electroanalytical Chemistry*, **258**, 327–344.

(28) Fleischmann, M., Hill, I.R. and Sundholm, G. (1983) A Raman spectroscopic study of quinoline and isoquinoline adsorbed on copper and silver electrodes. *Journal of Electroanalytical Chemistry*, **158**, 153–164.

(29) Fleischmann, M., Graves, P.R. and Robinson, J. (1985) The Raman spectroscopy of the ferricyanide/ferrocyanide system at gold, β-palladium hydride and platinum electrodes. *Journal of Electroanalytical Chemistry*, **182**, 87–98.

(30) Korzeniewski, C., Severson, M.W., Schmidt, P.P. *et al.* (1987) Theoretical analysis of the vibrational spectra of ferricyanide and ferrocyanide adsorbed on metal electrodes. *The Journal of Physical Chemistry*, **91**, 5568–5573.

(31) Tian, Z.Q., Ren, B. and Wu, D.Y. (2002) Surface-enhanced Raman scattering: from noble to transition metals and from rough surfaces to ordered nanostructures. *The Journal of Physical Chemistry B*, **106**, 9463–9483.

(32) Wu, D.Y., Li, J.F., Ren, B. and Tian, Z.Q. (2008) Electrochemical surface-enhanced Raman spectroscopy of nanostructures. *Chemical Society Reviews*, **37**, 1025–1041.

(33) Tian, Z.Q., Ren, B., Li, J.F. and Yang, Z.L. (2007) Expanding generality of surface-enhanced Raman spectroscopy with borrowing SERS activity strategy. *Chemical Communications*, **0**, 3514–3534.

(34) Wu, D.Y., Ren, B., Jiang, Y.X. *et al.* (2002) Density functional study and normal-mode analysis of the bindings and vibrational frequency shifts of the Pyridine-M (M = Cu, Ag, Au and Pt) Complexes. *The Journal of Physical Chemistry A*, **106**, 9042–9052.

(35) Wu, D.Y., Ren, B., Xu, X. *et al.* (2003) Periodic trends in the bonding and vibrational coupling: Pyridine interacting with transition metals and noble metals studied by surface-enhanced Raman spectroscopy and density-functional theory. *The Journal of Physical Chemistry*, **119**, 9.

(36) Otto, A., Mrozek, I., Grabhorn, H. and Akemann, W. (1992) Surface-enhanced Raman scattering. *Journal of Physics: Condensed Matter*, **4**, 1143.

(37) Gao, P., Gosztola, D., Leung, L.W.H. and Weaver, M.J. (1987) Surface-enhanced Raman scattering at gold electrodes: dependence on electrochemical pretreatment conditions and comparisons with silver. *Journal of Electroanalytical Chemistry*, **233**, 211–222.

(38) Wu, D.Y., Ren, B. and Tian, Z.Q. (2006) A theoretical study on SERS intensity of pyridine adsorbed on transition metal electrodes. *Israel Journal of Chemistry*, **46**, 317–327.

(39) Li, J.F., Huang, Y.F., Ding, Y. *et al.* (2010) Shell-isolated nanoparticle-enhanced Raman spectroscopy. *Nature*, **464**, 392–395.

(40) Li, J.F., Tian, X.D., Li, S.B. *et al.* (2013) Surface analysis using shell-isolated nanoparticle-enhanced Raman spectroscopy. *Nature Protocols*, **8**, 52–65.

(41) Li, J.F., Ding, S.Y., Yang, Z.L. *et al.* (2011) Extraordinary enhancement of Raman scattering from pyridine on single crystal Au and Pt electrodes by shell-isolated Au nanoparticles. *Journal of the American Chemical Society*, **133**, 15922–15925.

(42) Owen, J.F., Chen, T.T., Chang, R.K. and Laube, B.L. (1983) Irreversible loss of adatoms on Ag electrodes during potential cycling determined from surface-enhanced Raman intensities. *Surface Science*, **131**, 195–220.

(43) Chen, T.T. and Chang, R.K. (1985) Surface-enhanced Raman scattering of interfacial DOD, HOD, and OD$^-$. *Surface Science*, **158**, 325–332.

(44) Funtikov, A.M., Sigalaev, S.K. and Kazarinov, V.E. (1987) Surface-enhanced Raman scattering and local photoemission currents on the freshly prepared surface of a silver electrode. *Journal of Electroanalytical Chemistry*, **228**, 197–218.

(45) Gui, J. and Devine, T.M. (1989) Influence of hydroxide on the SERS of water and chloride. *Surface Science*, **224**, 525–542.

(46) Tian, Z.Q., Lian, Y.Z. and Lin, T.Q. (1989) SERS studies on interfacial water in concentrated NaClO$_4$ solutions. *Journal of Electroanalytical Chemistry*, **265**, 277–282.

(47) Kwon, M.Y. and Kim, J.J. (1990) Surface-enhanced Raman scattering of water in ethanol: Electrolytic concentration dependence. *Chemical Physics Letters*, **169**, 337–341.

(48) Tian, Z.Q., Sigalaev, S.K., Zou, S.Z. *et al.* (1994) The observation of SERS of water in a wide potential range from the Ag/NaClO$_4$ system. *Electrochimica Acta*, **39**, 2195–2196.

(49) Jiang, Y.X., Li, J.F., Wu, D.Y. *et al.* (2007) Characterization of surface water on Au core Pt-group metal shell nanoparticles coated electrodes by surface-enhanced Raman spectroscopy. *Chemical Communications*, **0**, 4608–4610.

(50) Wu, D.Y., Duan, S., Liu, X.M. *et al.* (2008) Theoretical study of binding interactions and vibrational Raman spectra of water in hydrogen-bonded anionic complexes: $(H_2O)^{n-}$ ($n = 2$ and 3), $H_2O\ldots X\text{-}(X = F, Cl, Br, and I)$, and $H_2O\ldots M\text{-}(M = Cu, Ag, and Au)$. *The Journal of Physical Chemistry A*, **112**, 1313–1321.

(51) Li, J.F., Huang, Y.F., Duan, S. *et al.* (2010) SERS and DFT study of water on metal cathodes of silver, gold and platinum nanoparticles. *Physical Chemistry Chemical Physics*, **12**, 2493–2502.

(52) Kornyshev, A.A., Schmickler, W. and Vorotyntsev, M.A. (1982) Nonlocal electrostatic approach to the problem of a double layer at a metal–electrolyte interface. *Physical Review B*, **25**, 5244–5256.

(53) Persson, B.N.J., Zhao, K. and Zhang, Z. (2006) Chemical contribution to surface-enhanced Raman scattering. *Physical Review Letters*, **96**, 207401.

(54) Tian, Z.Q. and Ren, B. (2007) Raman Spectroscopy of electrode surfaces, in *Encyclopedia of Electrochemistry*, vol. 3 (ed. P. Unwin), Wiley-VCH Verlag GmbH & Co. KGaA.

(55) Tian, Z.Q., Li, W.H., Mao, B.W. and Gao, J.S. (1994) Surface-enhanced Raman spectroscopic studies on structural dynamics of coadsorption of thiourea and ClO_4^- at Ag electrodes. *Journal of Electroanalytical Chemistry*, **379**, 271–279.

(56) Tian, Z.Q., Li, W.H., Qiao, Z.H. *et al.* (1995) Molecular-level investigations of different types of coadsorption at Ag electrodes by Raman spectroscopy. *Russian Journal of Electrochemistry*, **31**, 935–940.

(57) Loo, B.H. (1982) Molecular orientation of thiourea chemisorbed on copper and silver surfaces. *Chemical Physics Letters*, **89**, 346–350.

(58) Macomber, S.H. and Furtak, T.E. (1982) The short-range component of surface-enhanced Raman scattering: thiourea adsorbed on a silver electrode. *Chemical Physics Letters*, **90**, 59–63.

(59) Brown, G.M., Hope, G.A., Schweinsberg, D.P. and Fredericks, P.M. (1995) SERS study of the interaction of thiourea with a copper electrode in sulphuric acid solution. *Journal of Electroanalytical Chemistry*, **380**, 161–166.

(60) Reents, B., Plieth, W., Macagno, V.A. and Lacconi, G.I. (1998) Influence of thiourea on silver deposition: spectroscopic investigation. *Journal of Electroanalytical Chemistry*, **453**, 121–127.

(61) Lacconi, G., Reents, B. and Plieth, W. (1992) Raman spectroscopy of silver plating from a cyanide electrolyte. *Journal of Electroanalytical Chemistry*, **325**, 207–217.

(62) Bozzini, B. and Fanigliulo, A. (2002) An in situ spectroelectrochemical Raman investigation of Au electrodeposition and electrodissolution in $KAu(CN)_2$ solution. *Journal of Applied Electrochemistry*, **32**, 1043–1048.

(63) Bozzini, B., Pietro De Gaudenzi, G. and Mele, C. (2004) A SERS investigation of the electrodeposition of Ag–Au alloys from free-cyanide solutions. *Journal of Electroanalytical Chemistry*, **563**, 133–143.

(64) Bozzini, B., Romanello, V. and Mele, C. (2007) A SERS investigation of the electrodeposition of Au in a phosphate solution. *Surface and Coatings Technology*, **201**, 6267–6272.

(65) Bozzini, B., D'Urzo, L., Mele, C. and Romanello, V. (2008) A SERS investigation of cyanide adsorption and reactivity during the electrodeposition of gold, silver, and copper from aqueous cyanocomplexes solutions. *The Journal of Physical Chemistry C*, **112**, 6352–6358.

(66) Lin, Z.B., Tian, J.H., Xie, B.G. *et al.* (2009) Electrochemical and *in situ* SERS studies on the adsorption of 2-hydroxypyridine and polyethyleneimine during silver electroplating. *The Journal of Physical Chemistry C*, **113**, 9224–9229.

(67) Gao, J.S. and Tian, Z.Q. (1997) Surface Raman spectroscopic studies of ruthenium, rhodium and palladium electrodes deposited on glassy carbon substrates. *Spectrochimica Acta Part A: Molecular and Biomolecular*, **53**, 1595–1600.

(68) Youda, R., Nishihara, H. and Aramaki, K. (1988) A SERS study on inhibition mechanisms of benzotriazole and its derivatives for copper corrosion in sulphate solutions. *Corrosion Science*, **28**, 87–96.

(69) Youda, R., Nishihara, H. and Aramaki, K. (1990) SERS and impedance study of the equilibrium between complex formation and adsorption of benzotriazole and 4-hydroxybenzotriazole on a copper electrode in sulphate solutions. *Electrochimica Acta*, **35**, 1011–1017.

(70) Aramaki, K., Kiuchi, T., Sumiyoshi, T. and Nishihara, H. (1991) Surface enhanced Raman scattering and impedance studies on the inhibition of copper corrosion in sulphate solutions by 5-substituted benzotriazoles. *Corrosion Science*, **32**, 593–607.

(71) Fleischmann, M., Sockalingum, D. and Musiani, M.M. (1990) The use of near infrared Fourier transform techniques in the study of surface-enhanced Raman spectra. *Spectrochimica Acta Part A: Molecular and Biomolecular*, **46**, 285–294.

(72) Chan, H.Y.H. and Weaver, M.J. (1999) A vibrational structural analysis of benzotriazole adsorption and phase film formation on copper using surface-enhanced Raman spectroscopy. *Langmuir*, **15**, 3348–3355.

(73) Honesty, N.R. and Gewirth, A.A. (2012) Shell-isolated nanoparticle enhanced Raman spectroscopy (SHINERS) investigation of benzotriazole film formation on Cu(100), Cu(111), and Cu(poly). *Journal of Raman Spectroscopy*, **43**, 46–50.

(74) Sun, Y., Song, W., Zhu, X. *et al.* (2009) Electrochemical and in situ SERS spectroelectro-chemical investigations of 4-methyl-4*H*-1,2,4-triazole-3-thiol monolayers at a silver electrode. *Journal of Raman Spectroscopy*, **40**, 1306–1311.

(75) Liao, Q.Q., Yue, Z.W., Yang, D. *et al.* (2011) Inhibition of copper corrosion in sodium chloride solution by the self-assembled monolayer of sodium diethyldithiocarbamate. *Corrosion Science*, **53**, 1999–2005.

(76) Pan, Y.C., Wen, Y., Xue, L.Y. *et al.* (2012) Adsorption behavior of methimazole monolayers on a copper surface and its corrosion inhibition. *The Journal of Physical Chemistry C*, **116**, 3532–3538.

(77) Pan, Y.C., Wen, Y., Zhang, R. *et al.* (2012) Electrochemical and SERS spectroscopic investigations of 4-methyl-4*H*-1,2,4-triazole-3-thiol monolayers self-assembled on copper surface. *Applied Surface Science*, **258**, 3956–3961.

(78) Costa, L.A.F., Breyer, H.S. and Rubim, J.C. (2010) Surface-enhanced Raman scattering (SERS) on copper electrodes in 1-*n*-butyl-3-methylimidazolium tetrafluorborate (BMI.BF$_4$): The adsorption of benzotriazole (BTAH). *Vibrational Spectroscopy*, **54**, 103–106.

(79) Aramaki, K. and Uehara, J. (1990) The surface-enhanced Raman scattering spectra of pyridine adsorbed on an iron surface. *Journal of The Electrochemical Society*, **137**, 185–187.

(80) Gui, J. and Devine, T.M. (1991) Obtaining surface-enhanced Raman spectra from the passive film on iron. *Journal of The Electrochemical Society*, **138**, 1376–1384.

(81) Gui, J. and Devine, T.M. (1991) In situ vibrational spectra of the passive film on iron in buffered borate solution. *Corrosion Science*, **32**, 1105–1124.

(82) Melendres, C.A., Acho, J. and Knight, R.L. (1991) On the breakdown of passivity of iron by thiocyanate. *Journal of The Electrochemical Society*, **138**, 877–878.

(83) Ferreira, M.G.S., Mourae Silva, T., Catarino, A. *et al.* (1992) Electrochemical and laser Raman spectroscopy studies of stainless steel in 0.15 M NaCl solution. *Journal of The Electrochemical Society*, **139**, 3146–3151.

(84) Melendres, C.A. and Pankuch, M. (1992) On the composition of the passive film on nickel: a surface-enhanced Raman spectroelectrochemical study. *Journal of Electroanalytical Chemistry*, **333**, 103–113.

(85) Gui, J. and Devine, T.M. (1994) The influence of sulfate ions on the surface enhanced Raman spectra of passive films formed on iron. *Corrosion Science*, **36**, 441–462.

(86) Oblonsky, L.J. and Devine, T.M. (1995) A surface enhanced Raman spectroscopic study of the passive films formed in borate buffer on iron, nickel, chromium and stainless steel. *Corrosion Science*, **37**, 17–41.

(87) Gui, J. and Devine, T.M. (1995) A SERS investigation of the passive films formed on iron in mildly alkaline solutions of carbonate/bicarbonate and nitrate. *Corrosion Science*, **37**, 1177–1189.

(88) Simpson, L.J. and Melendres, C.A. (1996) Temperature dependence of the surface-enhanced Raman spectroelectrochemistry of iron in aqueous solutions. *Electrochimica Acta*, **41**, 1727–1730.

(89) Oblonsky, L.J., Virtanen, S., Schroeder, V. and Devine, T.M. (1997) Surface-enhanced Raman spectroscopy of iron oxide thin films: comparison with the passive film on iron. *Journal of The Electrochemical Society*, **144**, 1604–1609.

(90) Baek, W.C., Kang, T., Sohn, H.J. and Kho, Y.T. (2001) In situ surface-enhanced Raman spectroscopic study on the effect of dissolved oxygen on the corrosion film on low carbon steel in 0.01 M NaCl solution. *Electrochimica Acta*, **46**, 2321–2325.

(91) Cai, W.B. and Scherson, D.A. (2003) In-situ Raman spectroscopy of zinc electrodes in alkaline solutions. *Journal of The Electrochemical Society*, **150**, B217–B223.

(92) Harrington, S.P., Wang, F. and Devine, T.M. (2010) The structure and electronic properties of passive and prepassive films of iron in borate buffer. *Electrochimica Acta*, **55**, 4092–4102.

(93) Honesty, N.R. and Gewirth, A.A. (2013) Investigating the effect of aging on transpassive behavior of Ni-based alloys in sulfuric acid with shell-isolated nanoparticle enhanced Raman spectroscopy (SHINERS). *Corrosion Science*, **67**, 67–74.

(94) Cao, P.G., Yao, J.L., Ren, B. *et al.* (2000) Surface-enhanced Raman scattering from bare Fe electrode surfaces. *Chemical Physics Letters*, **316**, 1–5.

(95) Cao, P., Gu, R. and Tian, Z. (2002) Electrochemical and surface-enhanced Raman spectroscopy studies on inhibition of iron corrosion by benzotriazole. *Langmuir*, **18**, 7609–7615.
(96) Cao, P.G., Yao, J.L., Zheng, J.W. *et al.* (2002) Comparative study of inhibition effects of benzotriazole for metals in neutral solutions as observed with surface-enhanced Raman spectroscopy. *Langmuir*, **18**, 100–104.
(97) Gu, W., Fan, X.M., Yao, J.L. *et al.* (2009) Investigation on surface-enhanced Raman scattering activity on an ex-situ ORC roughened nickel electrode. *Journal of Raman Spectroscopy*, **40**, 405–410.
(98) Gu, R.A., Shen, X.Y., Liu, G.K. *et al.* (2004) Surface-enhanced Raman scattering from bare Zn Electrode. *The Journal of Physical Chemistry B*, **108**, 17519–17522.
(99) Yang, H., Sun, Y., Ji, J. *et al.* (2008) 2-Mercaptobenzothiazole monolayers on zinc and silver surfaces for anticorrosion. *Corrosion Science*, **50**, 3160–3167.
(100) Baddour-Hadjean, R. and Pereira-Ramos, J.P. (2009) Raman microspectrometry applied to the study of electrode materials for lithium batteries. *Chemical Reviews*, **110**, 1278–1319.
(101) Li, G., Li, H., Mo, Y. *et al.* (2000) Surface-enhanced resonance Raman spectroscopy of rhodamine 6G adsorbed on silver electrode in lithium batteries. *Chemical Physics Letters*, **330**, 249–254.
(102) Li, H., Mo, Y., Pei, N. *et al.* (2000) Surface-enhanced Raman scattering study on passivating films of Ag electrodes in lithium batteries. *The Journal of Physical Chemistry B*, **104**, 8477–8480.
(103) Li, G., Li, H., Mo, Y. *et al.* (2002) Further identification to the SEI film on Ag electrode in lithium batteries by surface-enhanced Raman scattering (SERS). *Journal of Power Sources*, **104**, 190–194.
(104) Itoh, T., Abe, K., Mohamedi, M. *et al.* (2001) In-situ SERS spectroscopy of Ag-modified pyrolytic graphite in organic electrolytes. *Journal of Solid State Electrochemistry*, **5**, 328–333.
(105) Matsuo, Y., Kostecki, R. and McLarnon, F. (2001) Surface layer formation on thin-film LiMn$_2$O$_4$ electrodes at elevated temperatures. *Journal of The Electrochemical Society*, **148**, A687–A692.
(106) Naoi, K., Suematsu, S., Komiyama, M. and Ogihara, N. (2002) An electrochemical study of poly(2,2'-dithiodianiline) in acidic aqueous media. *Electrochimica Acta*, **47**, 1091–1096.
(107) Schmitz, R., Muller, R.A., Schmitz, R.W. *et al.* (2013) SEI investigations on copper electrodes after lithium plating with Raman spectroscopy and mass spectrometry. *Journal of Power Sources*, **233**, 110–114.
(108) Girishkumar, G., McCloskey, B., Luntz, A.C. *et al.* (2010) Lithium–air battery: promise and challenges. *The Journal of Physical Chemistry Letters*, **1**, 2193–2203.
(109) Peng, Z., Freunberger, S.A., Hardwick, L.J. *et al.* (2011) Oxygen reactions in a non-aqueous Li+ electrolyte. *Angewandte Chemie-International Edition*, **50**, 6351–6355.
(110) Peng, Z., Freunberger, S.A., Chen, Y. and Bruce, P.G. (2012) A reversible and higher-rate Li-O$_2$ battery. *Science*, **337**, 563–566.
(111) Chen, Y., Freunberger, S.A., Peng, Z. *et al.* (2013) Charging a Li-O$_2$ battery using a redox mediator. *Nature Chemistry*, **5**, 489–494.
(112) Stancovski, V. and Badilescu, S. (2014) In-situ Raman spectroscopic-electrochemical studies of lithium battery materials: a historical overview. *Journal of Applied Electrochemistry*, **44**, 23–43.
(113) Weaver, M.J., Gao, P., Gosztola, D. *et al.* (1986) Surface-enhanced Raman spectroscopy, in *Excited States and Reactive Intermediates, ACS Symposium Series* (ed. A.B.P. Lever), American Chemical Society.
(114) Gao, P., Gosztola, D. and Weaver, M.J. (1988) Surface-enhanced Raman spectroscopy as a probe of electroorganic reaction pathways. 1. Processes involving adsorbed nitrobenzene, azobenzene, and related species. *The Journal of Physical Chemistry*, **92**, 7122–7130.
(115) Birke, R. and Lombardi, J. (1994) Investigation of radical ions with time-resolved surface enhanced Raman spectroscopy. *Molecular Engineering*, **4**, 277–310.
(116) Zhang, W., Vivoni, A., Lombardi, J.R. and Birke, R.L. (1995) Time-resolved SERS study of direct photochemical charge transfer between FMN and a Ag electrode. *The Journal of Physical Chemistry*, **99**, 12846–12857.

(117) Mrozek, M.F., Luo, H. and Weaver, M.J. (2000) Formic acid electrooxidation on platinum-group metals: is adsorbed carbon monoxide solely a catalytic poison? *Langmuir*, **16**, 8463–8469.

(118) Ren, B., Li, X.Q., She, C.X. *et al.* (2000) Surface Raman spectroscopy as a versatile technique to study methanol oxidation on rough Pt electrodes. *Electrochimica Acta*, **46**, 193–205.

(119) Wang, A., Huang, Y.F., Sur, U.K. *et al.* (2010). In-situ identification of intermediates of benzyl chloride reduction at a silver electrode by SERS coupled with DFT calculations. *Journal of the American Chemical Society*, **132**, 9534–9536.

(120) Huang, Y.F., Wu, D.Y., Wang, A. *et al.* (2010) Bridging the gap between electrochemical and organometallic activation: benzyl chloride reduction at silver cathodes. *Journal of the American Chemical Society*, **132**, 17199–17210.

9

In-Situ Scanning Probe Microscopies: Imaging and Beyond

Bing-Wei Mao
Xiamen University, State Key Laboratory of Physical Chemistry of Solid Surfaces, China

One of the most important tasks of modern electrochemistry is to develop microscopic pictures of solid–liquid interfaces and thus to provide a basis for the detailed understanding of electrochemical processes. To fulfill this task, the development of surface-specific and structure-sensitive *in-situ* methods to characterize electrochemical interfacial processes is indispensable. As early as 1970, Professor Martin Fleischmann was one of the pioneers in exploring *in-situ* methods that included surface-enhanced Raman spectroscopy [1], surface X-ray diffraction [2] and nuclear magnetic resonance [3] to characterize electrochemical interfaces. Nowadays, nontraditional electrochemical methods that include spectroscopic and microscopic as well as diffraction techniques have been extensively applied, and this has promoted an understanding of electrochemical interfaces at both atomic and molecular levels.

Scanning probe microscopy (SPM) incorporates a family of microscopic techniques that utilize small probes which are located closely above a surface and are scanned across the surface to image the physical and chemical properties of the surface by detecting local tunneling currents, weak forces, light or electrochemical current as a function of *x-y* position. The most important members of the SPM family are scanning tunneling microscopy (STM) and atomic force microscopy (AFM), both of which allow surface imaging and manipulation down to atomic scale resolution. STM was introduced in 1981 by Binnig and Rorher [4], who were awarded the Nobel Prize in Physics in 1986. Based on localized tunneling current detection, STM has an extremely high spatial resolution of 0.1

Developments in Electrochemistry: Science Inspired by Martin Fleischmann, First Edition.
Edited by Derek Pletcher, Zhong-Qun Tian and David E. Williams.
© 2014 John Wiley & Sons, Ltd. Published 2014 by John Wiley & Sons, Ltd.

and 0.01 nm in lateral and vertical directions, respectively, providing real space structural arrangement and electronic properties of the surface atoms of conducting substrates. AFM was first described in 1986 and can operate with insulating as well as conducting substrate surfaces [5]. More importantly, because of their flexibility to working environments, STM and AFM have become indispensable tools in fields ranging from surface science, materials science, life sciences to nanoscience. In particular, the features that STM and AFM can be used to characterize surfaces immersed in electrolytes have led to the emergence of *in-situ* means for direct observations of electrochemical interfacial processes [6–10], which are of profound significance in surface electrochemistry. By making use of the precise probe positioning ability and various probe–surface interactions, STM and AFM are used in electrochemistry beyond the scope of imaging [11–13]. The facile potential control of not only the surface but also the probe in electrochemical environments facilitates tip-assisted local surface processes, providing opportunities for new and far-reaching interdisciplinary researches within nanoelectrochemistry.

In this chapter, attention is focused on *in-situ* STM and AFM, and recent advances of *in-situ* SPM in surface electrochemistry and nanoelectrochemistry are introduced, with applications that include surface characterization, nanostructuring, and molecular electronics. First, a brief discussion of the principles and features of STM and AFM is provided, and this is followed by some selected examples of the capabilities of both techniques in the study of surface and nanoelectrochemistry, mostly acquired in recent studies conducted by the present author's group. Emphasis is placed on the roles of *in-situ* STM and AFM from a methodological point of view. Finally, the prospects for the further development of *in-situ* SPM are reviewed.

9.1   Principle of *In-Situ* STM and *In-Situ* AFM

9.1.1   Principle of *In-Situ* STM

According to quantum tunneling theory, when two conductors are brought sufficiently close to each other, the electron wavefunctions near the Fermi energy (E_F) of the two conductors overlap to a certain degree in the energy barrier region. If a bias voltage V_b is applied between the two conductors, as shown in Figure 9.1a, electrons will tunnel continuously through the barrier from the lower electrode to the upper electrode, forming a detectable flow of current that is proportional to the inverse exponential of the distance between the two conductors. The invention of STM was based on making one of the conductors – that is, the top one in Figure 9.1a – a probe with an atomically sharp tip apex, and located above the surface of the other conductor (sample) within a close enough separation [14]. The approximated expression of tunneling current between the tip and surface is given by Bardeen's perturbation theory [15]:

$$I_t \approx V_b \rho_s(0, E_F) \exp(-2\kappa s) \tag{9.1}$$

where $\rho_s(0, E_F)$ refers to the local density of state (LDOS) near E_F of the sample surface, $\kappa = (2m\varphi_0)^{1/2}/\hbar$ is known as decay constant (or $1/\kappa$ = decay length), where m and $\hbar$ are the electron mass and normalized Plank constant, respectively, φ_0 is the effective tunnel barrier between the sample and tip (often approximated by the averaged value of the work functions of the two conductors), and s is the tip–sample distance. It can be seen from Equation (9.1)

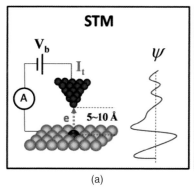
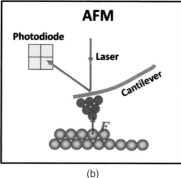

(a)   (b)

Figure 9.1 *Schematic illustrations of the working principle of (a) scanning tunneling microscopy (STM) and (b) atomic force microscopy (AFM).*

that the tunneling current I_t is proportional to the inverse exponential of the tip–surface distance, s. Assuming an effective tunnel barrier $\varphi_0 = 4$ eV, which gives $1/\kappa \sim 0.1$ nm, I_t would increase by almost one order of magnitude upon a decrease of s by 0.1 nm at a specified position of the surface where $\rho_s(0,E_F)$ is fixed. When the tip is scanned laterally relative to the surface, I_t varies depending primarily on the tip–sample distance, generating an image of tunneling current that can, with care, be converted to surface topography. In other words, very small variation of the tip–sample distance would lead to a large tunneling current variation, and it is this exponential relationship that forms the basis of the atomic resolution imaging ability of STM. In practice, however, an atomically flat single-crystalline surface is also necessary to achieve atomic resolution.

The tunneling current is also proportional to the exponential of $\varphi_0^{1/2}$, and linearly proportional to $\rho_s(0,E_F)$. As the tip is scanned laterally over the surface, the variation of tunneling current is actually a combined contribution of surface morphology and electronic properties. It should be recognized that the bias voltage and the tunneling current should be kept low so as not to induce processes other than tunneling. Care must be taken to distinguish the origins of features of an STM images. Several cases are worthy of mention:

 (i) For metals, the surface distribution of the local density of states is usually consistent with the location of surface atoms, and the interpretation of atomic resolution STM images of pure metals is relatively straightforward;
 (ii) For semiconductors, the directional covalent bonds between surface atoms contribute to the distribution of surface local density of states, resulting in more complicated STM images. In this case, the distribution of surface atoms is not exactly equivalent to the distribution of the local density of states. The interpretation of STM images of semiconductors should be based on the analysis and understanding of the surface electronic properties in order to separate spatial from electronic features;
(iii) For surfaces with adsorbed molecules, it is generally accepted that the tunneling through molecules may be classified into "through-space," "through-bond" as well as resonant mechanisms [16]. For weakly adsorbed molecules, the electrons tunnel "through-space", and the frontier orbital of the molecules serve as the antennas which can feel the path of tunneling electrons and are thus imaged by STM. For strongly

adsorbed molecules, factors such as surface state, bias and tunneling current can promote "through-bond" tunneling with an enhanced tunneling probability of certain specific groups of the molecules so that STM images do not necessarily reflect the original molecular structural characteristics. Coadsorption of molecules places even more complexity on the interpretation as one of the components or all of the components may contribute to the STM image. For redox molecules whose equilibrium potential is reachable within the Fermi level of the tip or surface electrode, resonant tunneling can occur when the Fermi level of the electrode is in alignment with the HOMO or LOMO state of the redox species.

In order for STM to work with electrochemical interfaces, the instrument is equipped with a bipotentiostat for independent potential control of both the tip and surface with respect to a chosen reference electrode in a four-electrode cell, so that both electrodes are under well-defined electrochemical conditions. Furthermore, since Faradaic current could also flow through the tip electrode, this would be superimposed on the tunneling current and interfere severely with the detection of tunneling current, and even destabilize the geometry of the tip apex. It is therefore essential to insulate the side wall of the metallic tip electrode to suppress the Faradaic current while leaving a small tip apex for tunneling [7, 8].

9.1.2   Principle of *In-Situ* AFM

AFM functions with a fine cantilever below which there is an apex with a sharp tip above the surface of interest so that weak forces (10^{-11} to 10^{-6} N) are exerted to the tip. As shown in Figure 9.1b, the cantilever is bent away from the surface under study, and this leads to the deflection of a reflected laser beam at the back of the cantilever. The deflection can be read out and converted to force via the spring constant of the cantilever. The force can range from long-range magnetic and electrostatic forces to short-range van de Waal's attractive forces and contact repulsive forces. A cantilever with an appropriate spring constant is necessary to match the different force ranges of interest. The short-range forces are measured and mapped laterally to yield AFM images of topography or friction of the surface, while the long-range forces can be probed as a function of tip–sample distance to yield so-called "force curves." There are basically two types of operating mode for AFM imaging: constant force and constant height mode. When scanning the probe laterally across a rigid surface in the contact mode under constant force, the tip will always follows the contour of the surface under the presence of repulsive force, so that the surface topography is imaged [14].

AFM relies on the detection of force, not current, and therefore the substrate does not need to be conductive. AFM can also work with an electrochemical interface [6, 17–21]. As force measurements are insensitive to the current flow at the substrate surface, AFM can be used to study electrochemical processes accompanied by faradaic currents. However, gas evolution at the surface should be avoided as bubbles interfere with the light path of the laser beam.

It should be noted that the force measured in AFM is proportional to the inverse of the tip–sample distance to the power of *n*. This is less sensitive than the exponential variation of tunneling current as a function of tip–sample distance in STM. In addition, due to the restricted sharpness of the tip apex of an AFM probe (presently, the finest commercial probes have a radius $\sim$10 nm), the convolution effect of the tip shape may be obvious – that is, the surface characteristic features may be influenced by the tip shape. This limits the imaging resolution for surfaces that do not have periodicity at the atomic scale and has restricted AFM applications for atomic resolution investigations of surface structures. However, AFM force curves are well-suited to investigations of the electric double layer [22–25].

9.2   *In-Situ* STM Characterization of Surface Electrochemical Processes

9.2.1   *In-Situ* STM Study of Electrode–Aqueous Solution Interfaces

A great number of *in-situ* STM studies have been carried out in aqueous solutions relating to topics that include potential-induced surface reconstruction [9, 23–26], ionic adsorption [27–30], molecular adsorption [31–33] and metal electrodeposition [34–38] on a variety of metal single-crystal substrates, including coinage metals (Au, Ag, Cu) and precious metals (Pt, Ru, Rh and Ir). Among the substrates, Au(111) and Au(100) single-crystal surfaces are most frequently employed because they are relatively inert and easy to prepare and pretreat yet are rich in structural-sensitive surface processes. The potential-induced reconstruction of Au(111)-($\sqrt{3} \times 22$) and Au(100)-hex and the lifting of the reconstructions have been extensively studied [9]. It was found that reconstructions are promoted at negatively charged surfaces at a more negative potential, while lifting of the reconstruction is promoted by anion adsorption towards more positive potentials. The shift of potential of zero charge of reconstructed and unreconstructed Au surfaces has been found to be responsible for the charging current observed on the cyclic voltammograms [9]. These reconstructed and unreconstructed Au single-crystal surfaces provide a playground for ionic and molecular adsorption and electrodeposition of metals and semiconductors. In sulfuric acid solution, adsorption of SO_4^{2-} anions on Au(111)-(1×1) leads to a disordered–ordered structural transition upon positive potential excursion, resulting in the formation of an Au(111)-($\sqrt{3} \times \sqrt{7}$) lattice structure [9, 27]. The Cu underpotential deposition (UPD) on Au(111) in sulfuric acid solution [39] is a classic example showing coadsorption of SO_4^{2-} anions with the Cu UPD submonolayer. The adsorption of SO_4^{2-} on top of the interstices of the Cu honeycomb structure gives an apparent ($\sqrt{3} \times \sqrt{3}$) lattice structure, but this structure had been misleadingly attributed to Cu UPD structure. This shows that care must be taken when interpreting STM images [16, 39], and independent experiments should be designed and performed in order to clarify ambiguities.

   The *in-situ* STM characterization of electrochemical interfaces in aqueous solutions has provided invaluable information for elucidating potential-dependent surface structural details at atomic or submolecular resolution. The above-mentioned studies represent only a few examples of the technique and several reviews of *in-situ* STM studies of the electrochemical interface in aqueous solutions provide further information [6, 8, 9, 24, 27, 30, 34].

9.2.2   *In-Situ* STM Study of Electrode–Ionic Liquid Interface

Room temperature ionic liquids (RTILs) are compounds that are composed of organic cations and multiatom anions in the liquid state near or below room temperature. They are well suited as solvents for electrochemistry [40] as they have wide electrochemical windows, good thermal stability, and low vapor pressure. Notably, RTILs provide opportunities for investigating electrode processes that take place outside the potential range limited by water electrolysis and with improved safety and performance. However, electrochemical interfaces in ionic liquids are rather complicated because of the involvement of strong and multinatured interactions between the highly concentrated, large solvent ions and the electrode surface. Currently, the understanding of such interfaces is still in its infancy, and issues such as double-layer structures and various ion–ion and ion–electrode interactions require adequate experimental and theoretical investigations.

Although the low vapor pressure of ionic liquids is advantageous for *in-situ* STM characterization in these media, to obtain STM images with a high resolution in ionic liquids is more challenging than in aqueous solutions, because of the intrinsically high viscosity and consequent large tunneling barrier of ionic liquids. The tunneling currents decrease so that shorter tip–sample distance is required to maintain a given current, while mechanical vibration introduced by the glovebox or similar equipment pose additional difficulties. There are also favorable aspects for *in-situ* STM measurements in ionic liquids, for example, tip-insulation is a less crucial step in ionic liquids than in aqueous solutions because of the lower leakage currents.

Among a spectrum of cations and anions that have been documented to form ionic liquids, imidazolium-based ionic liquids are of particular interest because the imidazolium cations have localized positive charges at the ring and neutral hydrophobic tails of variable length as the alkyl side chain is changed, and they are therefore suitable for fundamental investigations by *in-situ* STM [10, 41–44]. Figure 9.2 shows potential-dependent STM images of Au(100) surfaces in neat 1-butyl-3-methylimidazolium tetrafluoroborate (BMIBF$_4$). Upon a positive potential excursion, the BF$_4^-$ anions are adsorbed at around -0.05 V in an ordered $\begin{pmatrix} 3 & -1 \\ -1 & 3 \end{pmatrix}$ structure. This is followed by the observation of a clear atomic-resolution STM image of bare Au(100)-(1 × 1) at -0.5 V. The fourfold square lattice of the Au(100) surface is clearly identified from the images, with periodicities close to the inter-atomic distance of the corresponding surface. The adsorption of BMI$^+$ cations successively forms disordered and then ordered micelle-like structures at potentials negative of -0.7 V [42]. The potential region where the adsorption of anions and cations are observed are just at the two sides of the potential of capacitance maxima in a differential capacitance curve for the system, providing an experimental basis to validate theoretical prediction that potential of zero charge (PZC) is located close to the maxima [45].

The disordered adsorption of BMI$^+$ cations causes etching of the surface due to effective interaction of the cations with the surface. As the adsorption coverage increases upon a negative potential excursion, the cations can form an ordered adlayer structure in the form of double rows; these are referred to as a micelle-like structure (Figure 9.2d). Once a stable micelle-like structure has been achieved, the surface etching is relieved because of the constraint of the BMI$^+$ molecules by the intermolecular interaction in the ordered adsorption structure [42, 43]. At potentials more negative than -1.25 V, the surface reconstruction of Au(100)-hex is promoted. The reconstruction involves an accumulation of 24% of the surface Au atoms in a form of sixfold, hexagonal arrangement. The crystallographic mismatches of the topmost surface structure with the corresponding second topmost surface structure create the unique corrugations seen in the STM images, with orthogonal features along the two characteristic directions of Au(100) [42, 43]. Similar surface processes of Au(111) surface in BMI$^+$-based ionic liquids have also been observed, which show a distinct feature of worm-like adsorption structure of BMI$^+$ and Au(111)-($\sqrt{2} \times 22$) reconstruction

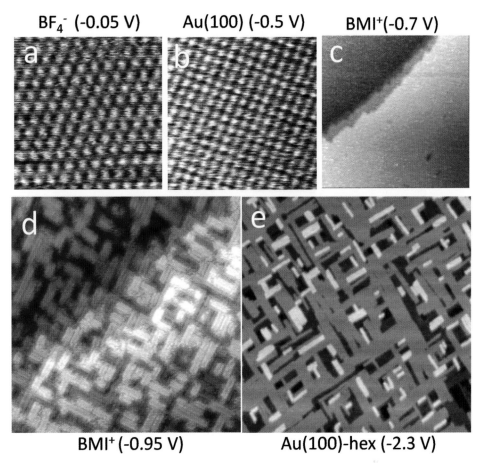

BF$_4^-$ (-0.05 V)   **Au(100) (-0.5 V)**   **BMI$^+$(-0.7 V)**

BMI$^+$ (-0.95 V)   **Au(100)-hex (-2.3 V)**

Figure 9.2 In-situ *STM images of Au(100) in BMIBF$_4$, showing potential-dependent surface processes. Scan areas: (a) 5 × 5 nm^2; (b) 4 × 4 nm^2; (c) 100 × 100 nm^2; (d) 60 × 60 nm^2; (e) 150 × 150 nm^2. A Pt wire was used as the quasi-reference electrode.*

[41, 43]. The observed features of Au(100)-hex and Au(111)-($\sqrt{2} \times 22$) reconstruction are in full agreement with those observed in aqueous solutions [9, 23, 24].

Ionic liquids have unusual bulk and interfacial properties that are brought about by the intrinsic strong interactions between the highly concentrated solvent ions and their interaction with the surface. These properties make ionic liquids a new type of medium for investigations of electrode processes, including those that can occur within the electrochemical window of aqueous solutions. The adsorption of neutral inorganic molecules SbCl$_3$ and BiCl$_3$ [46–48], and the electrodeposition of ferromagnetic metals (Fe, Co) [49, 50] and Ge [51], have been investigated systematically using *in-situ* STM. Sb(III)/Bi(III) from SbCl$_3$/BiCl$_3$ are prone to hydrolysis in aqueous solutions to form the oxygen-containing species, SbO$^+$/BiO$^+$, which then are irreversibly adsorbed at the metal surfaces. However, such hydrolysis of metal precursors is largely suppressed in ionic liquids and SbCl$_3$/BiCl$_3$ remains in covalent form. The molecule–surface interaction between SbCl$_3$/BiCl$_3$ and

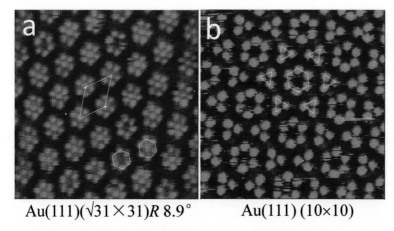

$$Au(111)(\sqrt{31} \times 31)R\,8.9° \qquad\qquad Au(111)\,(10 \times 10)$$

Figure 9.3 *Ordered arrays of supramolecular structures of SbCl₃ at 0 V (a) and BiCl₃ at −0.30 V (b) on Au(111) in the BMIBF₄ ionic liquid. Scan size: 10 nm × 10 nm. Potentials are versus a Pt quasi-reference electrode.*

Au(111) is created by a potential-dependent partial charge transfer from the Au to the *d* empty orbital of $SbCl_3/BiCl_3$; this results in the molecular adsorption of $SbCl_3/BiCl_3$ as unique supramolecular 3- and 6-/7-member clusters in a lattice structure of Au(100)-$(\sqrt{31} \times \sqrt{31})R8.9°$ in the case of $SbCl_3$, and Au(100)-(10×10) in the case of $BiCl_3$, as shown in Figure 9.3 [47]. However, such supramolecular structures are destroyed as the ionic strength is increased by the addition of $NaClO_4$ up to 6 M. Such behavior results from the delicate balance of various interactions at the electrified electrode–ionic liquid interfaces.

The electrodeposition of ferromagnetic metals in ionic liquids also exhibits a dramatically different behavior compared to that in aqueous solutions. By using $FeCl_3$ as the precursor, *in-situ* STM results have shown that Fe is deposited with the formation of shape-ordered structures, namely pseudo-rods on Au(111) and pseudo-square rings on Au(100) [49]. These interesting structures are explained as results of magnetostatic interactions under crystallographic constraints. The roles of ionic adsorption of the ionic liquid at the electrode surface and the Fe grains, as well as the local enrichment of magnetic reactant, cannot be underestimated. On the other hand, Co electrodeposition from an ionic liquid is influenced by the cobalt source in the ionic liquid electrolyte [50]. On the reconstructed Au(111) surface, when using $CoCl_2$ in $BMIBF_4$, Co nucleates preferentially at structure imperfections of the Au(111)-$(\sqrt{2} \times 22)$ reconstruction at a surprisingly negative potential (−2.05 V versus Pt). However, when using $Co(BF_4)_2$ as the precursor in $BMIBF_4$, Co deposition takes place at a much less negative potential and proceeds in a three-dimensional progressive nucleation and growth mode, without preference in nucleation sites.

9.3   *In-Situ* AFM Probing of Electric Double Layer

Electrochemical interfaces are charged interfaces. In isotropic media such as aqueous solutions, a metal–electrolyte interface has charge and potential distribution mainly across the diffuse electric double layer on the solution side. The associated electrostatic forces can be detected using a conductive AFM probe. Early *in-situ* AFM measurements of electrostatic forces in aqueous solutions relied on the use of a negatively charged silica sphere (10–20 μm in diameter) attached to the cantilever of an AFM probe [52, 53]. In studies performed by Bard and coworkers [53], a set of potential-dependent long-range electrostatic forces curves were measured. The large tip area and well-defined tip geometry of the silica sphere allows direct comparisons to be made with Derjaguin–Landau–Verwey–Overbeek (DLVO) theory, which is necessary to calibrate the surface potential of the silica and to treat the force curves. Charge and potential distribution within the diffuse layer of the electric double layer is obtained, and this is compared to those calculated from the Poisson–Boltzmann equation suitable for the Gouy–Chapman–Stern (GCS) model of the double layer. Surprisingly, it was found that the effective charge determined from the AFM force curves was smaller than that injected electrochemically. The authors suggested that classical GCS theory would be inadequate to describe the diffuse electrical double layer, and that ion correlation and ion condensation effects might need to be considered to account for the reduced surface charge.

AFM force measurement is also an important means for investigating the electric double layers of electrode–ionic liquid interfaces. Ionic liquid molecules are asymmetric in structure with alkyl side chains on the cations, making them an anisotropic media. In the past, there has been a general consensus that ionic liquids form a layered structure near solid surfaces [22, 54, 55], with the layering being caused by van de Waals molecular interactions among the hydrophobic alkyl side chains, balanced by the molecule–surface interaction. A layered structure can resist an external increasing pressing force until a threshold force value is reached. Previously, AFM [22, 54] or surface force apparatus (SFA) [55] have been used to directly probe the layered structures of the interfaces. In the AFM force measurements, by steadily driving the piezo tube towards a layered structure, the AFM tip attached to the piezo tube can be brought to compress the layered structure with an increasing force until the layered is ruptured. This process is accompanied by a continuous bending of the AFM cantilever when the tip has made contact with the layer, and this is followed by a (partial) relief of the cantilever bending after rupture of the layered structure. This leaves a transition of force appearing in the repulsive region of the approach curve. If more than one layered structure exists at the interface, then an approach force curve with more than one force transition will be observed, where each transition corresponds to one layered structure. The number, thickness and rupture forces can be determined from the force curves. Subsequently, from the potential-dependency of these values much information is obtained that is valuable when elucidating the electric double layer at the electrode–ionic liquid interface.

Ionic liquids have a high ionic strength, and therefore the electric double layer at the metal–ionic liquid interface is rather compact. In addition, the rupture forces of the layered structure are small (in the range of a few nanonewtons). To probe the weak forces associated with the layered structures within the compact region, the micrometer silica sphere employed in the aqueous solution is no longer applicable. Instead, the use of cantilevers with nanoscale and usually an atomically rough tip apex, as well as a low spring constant (on the scale of 0.1 N m^{-1}) are necessary.

In the present author's group, the electric double layer of Au(111)-BMIPF$_6$ interface and its dependence on potential has been investigated systematically using *in-situ* AFM force curve measurements [22]. Figure 9.4 shows a high-quality AFM force curve measured at

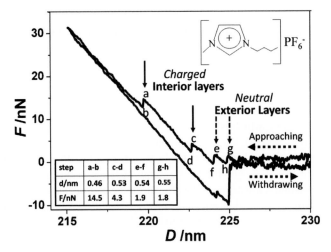

Figure 9.4   *AFM force curve from Au(111)-BMIPF$_6$ interface, showing clear stepwise force changes at −0.8 V versus Pt quasi-reference electrode. The horizontal axis is the relative reading of Z-piezo tube displacement.*

−0.8 V, a potential which is far more negative than the PZC at −0.08 V. The presence of four saw-toothed force transitions marked in Figure 9.4 by g–h, e–f, c–d and a–b in the approach direction are the results of a *partial* relief of the cantilever bending due to the sequential rupture of four layered structures under increasing levels of compression force by the AFM tip. The rupture force of each layered structure may be read directly from the height of the corresponding force at the transition points, that is, points g, e, c, and a. The layer thickness can also be calculated, based on the piezo tube travel distance and the cantilever deflection upon rupture of the layered structure. The similarity in thickness of the different layers indicates an ordered arrangement of BMIPF$_6$ molecules containing both cations and anions. The first layer from the surface shows that the largest rupture force peaked at 11.4 nN, while the second layer had a remarkably reduced force of 3.2 nN. More importantly, the third and fourth layers had very close force values of 1.8 nN; in other words, the rupture forces of the two layers are essentially distance-insensitive. Based on these feature, the four-layered structure can be classified into two groups, namely *distance-sensitive* interior layers (i.e., the first and second layers) and *distance-insensitive* exterior layers (i.e., the other two layers). When a detailed analysis of the potential-dependency of the interior and exterior layers was carried out across the PZC, it was found that the interior layers existed only at potentials away from the PZC and the number of layers increased as the surface charge increased. On the other hand, the exterior layers existed at all potentials, including the PZC. In other words, the presence of the exterior layers is *potential-independent*, and should be related to the general and intrinsic property of the layering behavior of the ionic liquid at solid–ionic liquid interfaces. The above results suggest that the electrode–ionic liquid interface is composed of an electric double-layer region formed by a charged layered structure, and an electric neutral region formed by neutral layered structures. No diffuse layer of conventional meaning appeared to be present. For future investigations of electrode–ionic liquid interfaces by AFM force measurements, tip modification, temperature control and a systematic investigation of varying the alkane side chain length of the imidazolium cations or types of ionic liquids would be highly desirable. A deeper insight into the structure and property of the electric double layer would also be possible.

9.4 Electrochemical STM Break-Junction for Surface Nanostructuring and Nanoelectronics and Molecular Electronics

For STM, if the tip and sample are brought to a distance where tunneling is not the only mechanism, then various localized processes can occur. By utilizing the precise three-dimensional tip positioning ability and tip-assisted localized processes, STM has been extended beyond imaging and has today become a platform for the manipulation, nanos-tructuring, and construction of nanoscale junctions.

One successful application of STM beyond imaging has been the STM break junction (STM-BJ) technique [56] and modification thereof [13, 57, 58], which are becoming very popular experimental platforms for nanoelectronics and molecular electronics [59–61]. In this technique, Figure 9.5a, a tip (usually Au) is brought into mechanical contact to a defined depth (crash-to-contact) with a single-crystalline surface of the same material. The tip is then withdrawn at a suitable rate such that a metal nanoconstruction is formed, elongated, and eventually broken (break-of-contact). During tip withdrawal, the conductance is recorded as a function of piezo displacement. The procedure can be repeated many thousands of

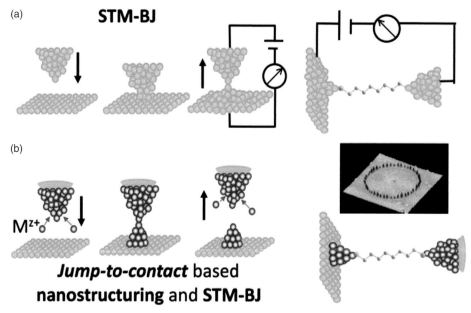

Figure 9.5 *Schematic illustration of working principle of STM-BJ (a) and jump-to-contact-based nanostructuring and STM-BJ (b). The STM image in (b) is a ring of 48 Fe clusters created by jump-to-contact-based nanostructuring.*

times to collect statistically large numbers of conductance traces for the construction of a conductance histogram. In fact, each break of contact process creates a pair of electrodes. If appropriate molecules are present in the system, there is a probability for a nearby molecule to "clip" to the pair of electrodes via the anchoring groups at each end of the molecule, so as to construct a metal–molecule–metal molecular junction.

The STM-BJ technique is reliable, flexible and efficient, and has been widely used for constructing atomic wires and single-molecule junctions for electron quantum transport studies. Gold is particularly suitable as the tip material and surface in STM-BJ techniques because of its good elasticity. A variety of Au–molecule–Au molecular junctions have been tested, among which saturated *n*-alkane with varying chain length, polyphenylene-based π-conjugate molecules [61], as well as redox molecules [13, 57, 62], have been the main focal points of these investigations. Overall, these studies have led to significant advances in understanding electron transport through molecules and their dependence on various factors, including distance and configuration of the molecule, metal–molecule interaction, and the contact geometry of the anchoring group.

By employing the STM-BJ technique in aqueous solutions, studies have been conducted by Mao *et al.* of the single-molecule conductance of 4,4′-bipyridine (BPY), 1,2-di(pyridin-4-yl)ethene (BPY-EE) and 1,2-di(pyridin-4-yl)ethane (BPY-EA), the latter two molecules being equivalent to two pyridine rings that sandwich the conjugated ethene group and the nonconjugated ethane group, respectively [63]. Furthermore, employing STM-BJ in electrochemical environments provides a new dimension to nanomolecular and single-molecule science by introducing notions such as interfacial charge transfer and electrochemical "gating." Additional studies have been conducted, in collaboration with the Amatore research team in Paris, of the single-molecule conductance of three different redox systems that have been self-assembled onto Au by using the STM-BJ method, and their electrochemical heterogeneous rate constants compared with ultrafast voltammetry [64]. The results showed that fast systems did, indeed, provide a higher conductance, and consequently electronic coupling factors for both experimental approaches were evaluated based on super-exchange mechanism theory. The results suggested, rather surprisingly, that coupling was on the same order of magnitude or even larger in conductance measurements, whereas electron transfer occurred over larger distances than in transient electrochemistry.

Modifying the tip approach procedure towards the surface can bring a paradigm change to the tip–sample interaction mechanism. Using the so-called "jump-to-contact process", a STM tip-induced nanostructuring technique has been established by Kolb and coworkers [12] whereby clusters of the smallest achievable size in solution can be created. In this technique (see Figure 9.5b), a thin film metal is electrodeposited on a tip, while the substrate surface is maintained at a potential which prevents bulk deposition of the metal. Under constant current operation mode of STM, and with low settings of proportional and integrated gains, an external voltage pulse is added to the z-piezo tube to drive the metal-loaded tip towards the surface until a jump-to-contact process occurs. When a transfer of the metal atoms from the tip to the surface occurs, a metal nanoconstruction is formed; the STM feedback circuit then responds to the rising current upon jump-to-contact by reducing the internal voltage applied to the z-piezo tube. This causes the tip to be withdrawn from the surface when the nanoconstruction is elongated and finally broken, so as to create a cluster at the surface. Repeating the above procedure allows the generation of large-scale arrays or patterns of the metal clusters. One primary requirement for the successful creation of metal clusters on the substrate surface is that the cohesive energy of the metal that has been electrodeposited on the STM tip should be smaller than that of the substrate, so that atoms can be transferred from the tip to the surface rather than in the reverse direction. Furthermore, the negative potential limit for the system must be such that solvent electrolysis does not interfere with the electrodeposition of the metal or semiconductor of interest. Because of these two constraints, only Cu [12], Ag [65], Cd [66] and Pd [67] clusters have been created on Au(111) in aqueous solutions.

In order to extend the nanostructuring systems to metals and semiconductors whose Nernst potentials are more negative than the hydrogen evolution potential, the use of ionic liquids is necessary. In fact, the cohesive properties of the surface atoms on both the tip-loaded metal and the substrate surface, as well as various interfacial interactions, may be modified in ionic liquids due to the adsorption of solvent ions. Hence, the nanostructuring of metals with larger bulk cohesive energies might be possible. The generality of tip-induced nanostructuring has been demonstrated by the nanostructuring of the reactive metals/semiconductor Zn [68], Fe [69] and Ge [51], in alkylimidazolium-based ionic liquids. Figure 9.5b (upper right) shows a ring of 48 Fe clusters with a ring diameter of 120 nm [69], and where the clusters are about 8–10 nm in diameter and 1–1.5 nm in height.

An electrochemically assisted jump-to-contact process has also been applied in the development of a modified STM-BJ approach by the present author's group [70]. This technique has the advantage that it allows the construction of chemically well-defined, atomic-size contacts, and is possible with both chemically active and/or soft metals. The metal contacts created in this way have been confirmed as having a well-defined structure, suitable for studying the mechanical properties of the nanocontacts [62, 71]. As in the conventional STM-BJ approach, the pair of metal electrodes created after breaking the contact provides the electrodes to construct metal–molecule–metal molecular junctions. These improvements have extended the capability of conventional STM-BJ to create a variety of metal nanocontacts and single-molecule junctions beyond the Au–molecule–Au junctions.

The electrochemically assisted jump-to-contact based STM-BJ approach has been used successfully for investigating the electron transport of metals and elemental semiconductors with widely differing physical and chemical properties, including Ag [62] and Cu [72] (coinage metals), Pd [62] (transition metal), Fe [71] (ferromagnetic) and Ge [51]; the latter two systems were measured in an ionic liquid, and such clusters have been shown to have well-defined structures [62, 71]. As shown in Figure 9.6, novel statistical distributions of the last-step length from the conductance traces can be observed with up to five for Fe and three for Cu peaks at integral multiples close to 0.075 nm, a subatomic distance [71]. The ultrafine structural rearrangement of the contact region is responsible for such a feature, and most likely one of the {111}-equivalent planes in the contact region, upon tip stretching with displacement, is glided in the direction from fcc-hollow-site to hcp-hollow-site at one-third of the regular layer spacing of the bulk metal. The observation can provide a new understanding of, and promote further efforts for, the experimental investigation of the mechanical properties of metal nanocontacts.

The conductance of molecular junctions formed with succinic acid using Cu, Ag and Au as metal electrodes has also been studied systemically using the jump-to-contact STM-BJ approach, with values of 18.6, 13.2 and 5.6 nS being found for Cu, Ag and Au electrodes, respectively [73, 74]. The observed decrease in conductance indicates a weakening of

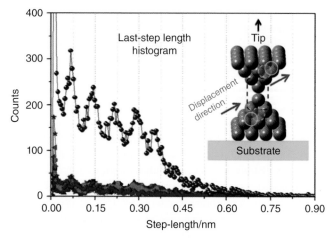

Figure 9.6   *Last-step length histogram of 16 000 conductance curves (from ten sets of 1600 curves) of Fe atomic contacts using Au(111) (blue). Histograms of the ten sets of conductance curves are also provided in the bottom part. The inset shows a schematic representation of gliding of atomic planes in the contact region with displacement of atoms from for example, fcc-hollow-site to the nearest hcp-hollow-site.*

electronic coupling efficiency at the electrode–molecule contacts in the order Cu > Ag > Au, and this should be taken into account when evaluating molecular conductance in the junctions.

9.5   Outlook

The application of *in-situ* SPM to electrode–electrolyte interfaces has not only led to enormous progress in the fundamental investigations of surface electrochemistry and nano-electrochemistry but has also catalyzed the merging of new interdisciplinary topics with electrochemistry. Specially designed video STM instrumentation [75, 76] that provides images at about 20 frames per second, and has been available in some laboratories for several years, allows the kinetics of electrochemical processes to be followed with increased time resolution. However, there remains much room for further developments in the instrumentation as well as applications in *in-situ* SPM.

From a methodological point of view, of particularly interest have been improvements in the chemical sensitivity of STM and AFM characterization. This is especially desirable for electrochemists, as electrochemical environments prevent the combined characterization by other surface techniques, as are frequently used for composition determinations in vacuum. Tunneling spectroscopy measurements to obtain $I{\sim}V$ and $dI/dV{\sim}V$ relationships may provide a certain degree of information regarding the electronic structure of the substrate surface and adsorbed molecules [77], and the use of ionic liquids of large electrochemical windows is favorable in this respect. One major enhancement would be to complement SPM with other spatial, time- and energy-resolved surface *in-situ* techniques. For example, a combination of scanning electrochemical microscopy and atomic force microscopy

(SECM-AFM) has been developed that enables mapping of the electrochemical properties of interfaces at high spatial resolution [70]. A more exciting approach would be to incorporate spectroscopy into an SPM system. Tip-enhanced Raman spectroscopy (TERS) [78,79], which utilizes the localized surface plasmon resonance from an STM or AFM tip to enhance the Raman signals of the surface species under the tip, is already available in either atmospheric or vacuum conditions. Vibrational spectroscopy of spatial resolution down to a few nanometers has also been demonstrated, and STM or AFM images could perhaps be interpreted with the support of vibrational spectroscopic information in assigning the composition and interaction. The successful application of TERS in electrochemistry in the near future would be highly desirable.

From an applications point of view, the incorporation of light and/or magnetic fields into the STM-BJ platform has already attracted attention, and both nanoelectronics and molecular electronics are currently being extended towards molecular optoelectronics [80] and molecular spintronics [81]. When studying these topics, apart from an appropriate electrochemical control, the correlation of electron conduction through the metal atomic wire and molecular junction with the electrochemical electron transfer may be profoundly significant. Furthermore, both *in-situ* STM and AFM from and beyond imaging are expected to be directed towards important practical systems such as electrochemical energy devices to elucidate key issues in the field. There is no doubt that with the improvement of time resolution, spatial resolution and also energy resolution, the investigation and understanding of electrochemical interfacial structures and processes will be brought to a new phase.

References

(1) Fleischmann, M., Hendra, P.J. and McQuillan, A.J. (1973) Raman spectra from electrode surfaces. *Chemical Communications*, **80**.
(2) Fleischmann, M., Hendra, P.J. and Robinson, J. (1980) X-ray diffraction from adsorbed iodine on graphite. *Nature*, **288**, 152.
(3) Newmark, R.D., Fleischmann, M. and Pons, S. (1988) The observation of surface species on platinum colloids using ^{195}Pt NMR. *Journal of Electroanalytical Chemistry*, **255**, 235.
(4) Binnig, G., Rohrer, H., Gerber, C. and Weibel, E. (1981) Tunneling through a controllable vacuum gap. *Applied Physics Letters*, **40**, 178–180.
(5) Binnig, G., Quate, C.F. and Gerber, C. (1986) Atomic force microscopy. *Physical Review Letters*, **56**, 930–933.
(6) Gewirth, A.A. and Niece, B.K. (1997) Electrochemical applications of in situ scanning probe microscopy. *Chemical Reviews*, **97**, 1129–1162.
(7) Moffat, T.P. (1999) Scanning tunneling microscopy studies of metal electrodes. *Electroanalytical Chemistry*, **21**, 211–316.
(8) Itaya, K. (1998) In-situ scanning tunneling microscopy in electrolyte solutions. *Progress in Surface Science*, **58**, 121–248.
(9) Kolb, D.M. (2001) Electrochemical surface science. *Angewandte Chemie International Edition*, **40**, 1162–1181.
(10) Su, Y.Z., Fu, Y.C., Wei, Y.M. *et al.* (2010) Electrode-ionic liquid interface: electric double layer and metal electrodeposition. *ChemPhysChem*, **11**, 2764–2778.
(11) Tao, N.J., Li, C.Z. and He, H.X. (2000) Scanning tunneling microscopy applications in electrochemistry - beyond imaging. *Journal of Electroanalytical Chemistry*, **492**, 81–93.
(12) Kolb, D.M., Ullmann, R. and Will, T. (1997) Nanofabrication of small copper clusters on gold(111) electrodes by a scanning tunneling microscope. *Science*, **275**, 1097–1099.

(13) Li, C., Mishchenko, A. and Wandlowski, T. (2012) Charge transport in single molecular junctions at the solid/liquid interface. *Topics in Current Chemistry*, **313**, 121–188.

(14) Bhushan, B. and Marti, O. (2005) Scanning probe microscopy - principle of operation, instrumentation, and probes, in *Nanotribology and Nanomechanics* (ed. B. Bhushan), Springer, Berlin, Heidelberg, New York, pp. 41–115.

(15) Chen, C.J. (1993) *Introduction to Scanning Tunneling Microscopy*, Oxford University Press, New York.

(16) Giancarlo, L.C. and Flynn, G.W. (1998) Scanning tunneling and atomic force microscopy probes of self-assembled, physisorbed monolayers: Peeking at the Peaks. *Annual Review of Physical Chemistry*, **49**, 297–336.

(17) Manne, S., Hansma, P.K., Massie, J. *et al.* (1991) Atomic-resolution electrochemistry with the atomic force microscope: copper deposition on gold. *Science*, **251**, 183.

(18) Nishizawa, T., Nakada, T., Kinoshita, Y. *et al.* (1996) AFM observation of a sulfate adlayer on Au(111) in sulfuric acid solution. *Surface Science*, **367**, L73–L78.

(19) Kubo, K., Hirai, N., Tanaka, T. and Hara, S. (2003) In situ observation on Au(100) surface in molten EMImBF(4) by electrochemical atomic force microscopy. *Surface Science*, **546**, L785–L788.

(20) Uosaki, K. and Koinuma, M. (1993) In-situ and real-time monitoring of the inset surface by atomic-force microscopy with atomic-resolution. *Journal of Electroanalytical Chemistry*, **357**, 301–306.

(21) Cai, X.W., Gao, J.S., Xie, Z.X. *et al.* (1998) Nanomodification of polypyrrole and polyaniline on highly oriented pyrolytic graphite electrodes by atomic force microscopy. *Langmuir*, **14**, 2508–2514.

(22) Zhang, X., Zhong, Y.X., Yan, J.W. *et al.* (2012) Probing double layer structures of Au (111)– BMIPF$_6$ ionic liquid interfaces from potential-dependent AFM force curves. *Chemical Communications*, **48**, 582–584.

(23) Haiss, W. (2001) Surface stress of clean and adsorbate-covered solids. *Reports on Progress in Physics*, **64**, 591–648.

(24) Kolb, D.M. (1996) Reconstruction phenomena at metal-electrolyte interfaces. *Progress in Surface Science*, **51**, 109–173.

(25) Zei, M.S. and Ertl, G. (1999) On the structural transformation of the reconstructed Pt(100) in electrolyte solution. *Surface Science*, **442**, 19–26.

(26) Wu, Q., Shang, W.H., Yan, J.W. and Mao, B.W. (2003) Metal adlayer-induced relaxation of Au(111) reconstruction under electrochemical control. *Journal of Physical Chemistry B*, **107**, 4065–4069.

(27) Lipkowski, J., Shi, Z., Chen, A. *et al.* (1998) Ionic adsorption at the Au(111) electrode. *Electrochimica Acta*, **43**, 2857–2888.

(28) Itaya, K., Batina, N., Kunitake, M. *et al.* (1997) In situ scanning tunneling microscopy of organic molecules adsorbed on iodine-modified Au(111), Ag(111), and Pt(111) electrodes. *Solid-Liquid Electrochemical Interfaces. ACS Symposium Series*, **656**, 171–188.

(29) Kim, Y.G., Yau, S.L. and Itaya, K. (1996) Direct observation of complexation of alkali cations on cyanide-modified Pt(111) by scanning tunneling microscopy. *Journal of the American Chemical Society*, **118**, 393–400.

(30) Magnussen, O.M. (2002) Ordered anion adlayers on metal electrode surfaces. *Chemical Reviews*, **102**, 679–725.

(31) Wan, L.J. and Itaya, K. (1997) In situ scanning tunnelling microscopy of benzene, naphthalene, and anthracene adsorbed on Cu(111) in solution. *Langmuir*, **13**, 7173–7179.

(32) Cunha, F. and Tao, N.J. (1995) Surface-charge induced order-disorder transition in an organic monolayer. *Physical Review Letters*, **75**, 2376–2379.

(33) Dretschkow, T., Dakkouri, A.S. and Wandlowski, T. (1997) In-situ scanning tunneling microscopy study of uracil on Au(111) and Au(100). *Langmuir*, **13**, 2843–2856.

(34) Herrero, E., Buller, L.J. and Abruña, H.D. (2001) Underpotential deposition at single crystal surfaces of Au, Pt, Ag and other materials. *Chemical Reviews*, **101**, 1897–1930.

(35) Schmidt, U., Vinzelberg, S. and Staikov, G. (1996) Pb UPD on Ag(100) and Au(100) - 2D phase formation studied by in situ STM. *Surface Science*, **348**, 261–279.

(36) Esplandiu, M.J., Schneeweiss, M.A. and Kolb, D.M. (1999) An in situ scanning tunneling microscopy study of Ag electrodeposition on Au(111). *Physical Chemistry Chemical Physics*, **1**, 4847–4854.

(37) Randler, R., Dietterle, M. and Kolb, D.M. (1999) The initial stages of Cu deposition on Au(100) as studied by in situ STM: The epitaxial growth of bcc Cu. *Zeitschrift Fur Physikalische Chemie*, **208**, 43–56.

(38) Mao, B.W., Tang, J. and Randler, R. (2002) Clustering and anisotropy in monolayer formation under potential control: Sn on Au(111). *Langmuir*, **18**, 5329–5332.

(39) Wu, S., Lipkowski, J., Tyliszczak, T. and Hitchcock, A.P. (1995) Effect of anion adsorption on early stages of copper electrocrystallization at Au(111) surface. *Progress in Surface Science*, **50**, 227–236.

(40) Buzzeo, M.C., Evans, R.G. and Compton, R.G. (2004) Non-haloaluminate room temperature ionic liquids in electrochemistry - A review. *ChemPhysChem*, **5**, 1106–1120.

(41) Lin, L.G., Wang, Y., Yan, J.W. *et al.* (2003) An in situ STM study on the long-range surface restructuring of Au(111) in a non-chloroaluminumated ionic liquid. *Electrochemistry Communications*, **5**, 995–999.

(42) Su, Y.Z., Fu, Y.C., Yan, J.W. *et al.* (2009) Double layer of Au(100) – ionic liquid interface and its stability in imidazolium-based ionic liquids. *Angewandte Chemie*, **121**, 5250–5253.

(43) Su, Y.Z., Yan, J.W., Li, M.G. *et al.* (2012) Adsorption of solvent cations on Au(111) and Au(100) in alkylimidazolium-based ionic liquids – worm-like versus micelle-like structures. *Zeitschrift fur Physikalische Chemie*, **226**, 979–994.

(44) Su, Y.Z., Yan, J.W., Li, M.G. *et al.* (2013) Electric double layer of Au(100)/imidazolium-based ionic liquids interface: effect of cation size. *Journal of Physical Chemistry C*, **117**, 205–212.

(45) Kornyshev, A.A. (2007) Double-layer in ionic liquids: paradigm change. *Journal of Physical Chemistry B*, **111**, 5545–5557.

(46) Fu, Y.C., Yan, J.W., Wang, Y. *et al.* (2007) In situ STM studies on the underpotential deposition of antimony on Au(111) and Au(100) in a BMIBF$_4$Ionic liquid. *Journal of Physical Chemistry C*, **111**, 10467–10477.

(47) Fu, Y.C., Su, Y.Z., Wu, D.Y. *et al.* (2009) Supramolecular aggregation of inorganic molecules at Au(111) electrodes under a strong ionic atmosphere. *Journal of the American Chemical Society*, **131**, 14728–14737.

(48) Fu, Y.C., Su, Y.Z., Zhang, H.M. *et al.* (2010) An in situ scanning tunneling microscopic study of electrodeposition of bismuth on Au(111) in a 1-butyl-3-methylimidazolium tetrafluoroborate ionic liquid: Precursor adsorption and underpotential deposition. *Electrochimica Acta*, **55**, 8105–8110.

(49) Wei, Y.M., Fu, Y.C., Yan, J.W. *et al.* (2010) Growth and shape-ordering of iron nanostructures on Au single crystalline electrodes in an ionic liquid: a paradigm of magnetostatic coupling. *Journal of the American Chemical Society*, **132**, 8152–8157.

(50) Lin, L.G., Yan, J.W., Wang, Y. *et al.* (2006) An in situ STM study of cobalt electrodeposition on Au(111) in BMIBF4 ionic liquid. *Journal of Experimental Nanoscience*, **1**, 269–278.

(51) Xie, X.F., Yan, J.W., Liang, J.H. *et al.* (2013) Measurement of the quantum conductance of germanium by an electrochemical scanning tunneling microscope break junction based on a jump-to-contact mechanism. *Chemistry - An Asian Journal*, **8**, 2401–2406.

(52) Ishino, T., Hieda, H., Tanaka, K. and Gemma, N. (1994) Electrical double-layer forces measured with an atomic-force microscope while electrochemically controlling surface-potential of the cantilever. *Japanese Journal of Applied Physics Part 2*, **33**, L1552–L1554.

(53) Wang, J. and Bard, A.J. (2001) Direct atomic force microscopic determination of surface charge at the gold/electrolyte interface – the inadequacy of classical GCS theory in describing the double-layer charge distribution. *Journal of Physical Chemistry B*, **105**, 5217.

(54) Atkin, R. (2007) Structure in confined room-temperature ionic liquids. *Journal of Physical Chemistry C*, **111**, 5162.

(55) Perkin, S., Crowhurst, L., Niedermeyer, H. *et al.* (2011) Self-assembly in the electrical double layer of ionic liquids. *Chemical Communications*, **47**, 6572–6574.

(56) Xu, B.Q. and Tao, N.J. (2003) Measurement of single molecule conductance by repeated formation of molecular junctions. *Science*, **301**, 1221–1223.

(57) Haiss, W., van Zalinge, H., Higgins, S.J. *et al.* (2003) Redox state dependence of single molecule conductivity. *Journal of the American Chemical Society*, **125**, 15294–15295.

(58) Sabater, C., Untiedt, C., Palacios, J. and Caturla, M. (2012) Mechanical annealing of metallic electrodes at the atomic scale. *Physical Review Letters*, **108**, 205502.

(59) Agraït, N., Yeyati, A.L. and Van Ruitenbeek, J.M. (2003) Quantum properties of atomic-sized conductors. *Physics Reports*, **377**, 81–279.

(60) Chen, F., Hihath, J., Huang, Z.F. *et al.* (2007) Measurement of single-molecule conductance. *Annual Review of Physical Chemistry*, **58**, 535–564.

(61) Salomon, A., Cahen, D., Lindsay, S. *et al.* (2003) Comparison of electronic transport measurements on organic molecules. *Advanced Materials*, **15**, 1881–1890.

(62) Liang, J.H., Liu, L., Gao, Y.J. *et al.* (2013) Correlating conductance and structure of silver nano-contacts created by jump-to-contact STM break junction. *Journal of Electroanalytical Chemistry*, **688**, 257–261.

(63) Zhou, X.S., Chen, Z.B., Liu, S.H. *et al.* (2008) Single molecule conductance of dipyridines with conjugated ethene and nonconjugated ethane bridging group. *Journal of Physical Chemistry C*, **112**, 3935–3940.

(64) Zhou, X.S., Liu, L., Fortgang, P. *et al.* (2011) Do molecular conductances correlate with electrochemical rate constants? Experimental insights. *Journal of the American Chemical Society*, **133**, 7509–7516.

(65) Kolb, D.M., Engelmann, G.E. and Ziegler, J.C. (2000) Electrochemical fabrication of large arrays of metal nanoclusters. *Solid State Ionics*, **131**, 69–78.

(66) Zhang, Y., Maupai, S. and Schmuki, P. (2004) EC-STM tip induced Cd nanostructures on Au (111). *Surface Science*, **551**, L33–L39.

(67) Engelmann, G.E., Ziegler, J.C. and Kolb, D.M. (1998) Nanofabrication of small palladium clusters on Au (111) electrodes with a scanning tunnelling microscope. *Journal of the Electrochemical Society*, **145**, L33–L35.

(68) Wang, J.G., Tang, J., Fu, Y.C. *et al.* (2007) STM tip-induced nanostructuring of Zn in an ionic liquid on Au(111) electrode surfaces. *Electrochemistry Communications*, **9**, 633–638.

(69) Wei, Y.M., Zhou, X.S., Wang, J.G. *et al.* (2008) The creation of nanostructures on an Au (111) electrode by tip-induced iron deposition from an ionic liquid. *Small*, **4**, 1355–1358.

(70) Holder, M.N., Gardner, C.E., Macpherson, J.V. and Unwin, P.R. (2005) Combined scanning electrochemical-atomic force microscopy (SECM-AFM): Simulation and experiment for flux-generation at un-insulated metal-coated probes. *Journal of Electroanalytical Chemistry*, **585**, 8–18.

(71) Wei, Y.M., Liang, J.H., Chen, Z.B. *et al.* (2013) Stretching single atom contacts at multiple subatomic step-length. *Physical Chemistry Chemical Physics*, **15**, 12459–12465.

(72) Zhou, X.S., Wei, Y.M., Liu, L. *et al.* (2008) Extending the capability of STM break junction for conductance measurement of atomic-size nanowires: An electrochemical strategy. *Journal of the American Chemical Society*, **130**, 13228–13230.

(73) Zhou, X.S., Liang, J.H., Chen, Z.B. and Mao, B.W. (2011) An electrochemical jump-to-contact STM-break junction approach to construct single molecular junctions with different metallic electrodes. *Electrochemistry Communications*, **13**, 407–410.

(74) Peng, Z.L., Chen, Z.B., Zhou, X.Y. *et al.* (2012) Single molecule conductance of carboxylic acids contacting Ag and Cu electrodes. *Journal of Physical Chemistry C*, **116**, 21699–21705.

(75) Schitter, G. and Rost, M.J. (2008) Scanning probe microscopy at video-rate. *Materials Today*, **11**, 40–48.

(76) Tansel, T., Taranovskyy, A. and Magnussen, O.M. (2010) In situ video-STM studies of adsorbate dynamics at electrochemical interfaces. *ChemPhysChem*, **11**, 1438–1445.

(77) Hipps, K.W. (2012) Tunneling spectroscopy of organic monolayers and single molecules. *Topics in Current Chemistry*, **313**, 189–216.

(78) Schmid, T., Opilik, L., Blum, C. and Zenobi, R. (2013) Nanoscale chemical imaging using tip-enhanced Raman spectroscopy: a critical review. *Angewandte Chemie International Edition*, **52**, 5940–5954.

(79) Liu, Z., Ding, S.Y., Chen, Z.B. *et al.* (2011) Revealing the molecular structure of single-molecule junctions in different conductance states by fishing-mode tip-enhanced Raman spectroscopy. *Nature Communications*, **2**, 305.
(80) Galperin, M. and Nitzan, A. (2012) Molecular optoelectronics: the interaction of molecular conduction junctions with light. *Physical Chemistry Chemical Physics*, **14**, 9421–9438.
(81) Sanvito, S. (2011) Molecular spintronics. *Chemical Society Reviews*, **40**, 3336–3355.

10

In-Situ Infrared Spectroelectrochemical Studies of the Hydrogen Evolution Reaction

Richard J. Nichols

University of Liverpool, Department of Chemistry, UK

Early on, Martin Fleischmann had recognized the need for techniques such as *in-situ* spectroscopies to provide information to supplement that from voltammetry and hence to gain a more detailed knowledge about the mechanisms of electrode reactions. By the 1970s, techniques such as external reflection *in-situ* infrared (IR) spectroscopy and surface-enhanced Raman spectroscopy [1] were providing truly molecular-level views of electrochemical interfaces and were also inspiring the development of other *in-situ* spectroscopies. The mechanism of the hydrogen evolution reaction has long been a challenge and the application of IR techniques to the study of this reaction is the subject of this chapter.

10.1 The H⁺/H₂ Couple

The hydrogen evolution reaction (HER) and hydrogen oxidation reaction (HOR) have been the focus of numerous scientific studies; indeed, they probably represent the most studied reactions in electrochemistry. The demonstration of the electrolysis of water dates back to 1800, when the chemist and inventor William Nicholson teamed up with the surgeon Anthony Carlisle to decompose water into its constituent gases with a voltaic pile [2]. Today, the HER and HOR have become exemplary reactions for understanding and developing

Developments in Electrochemistry: Science Inspired by Martin Fleischmann, First Edition.
Edited by Derek Pletcher, Zhong-Qun Tian and David E. Williams.

methodology in electrochemical kinetics and electrocatalysis. In fact, the Tafel equation was first published in 1905 in the context of the cathodic HER [3].

Interest in the HER and HOR extends to a range of important technologies that include H_2 oxidation at fuel cell anodes, the electrolytic production of hydrogen, and the role of adsorbed and absorbed hydrogen in a number of other important technologies, including electrocatalytic hydrogenations, hydrogen storage metals and alloys, and hydrogen permeation membranes for hydrogen separation. Without doubt, there are many reasons for studying the electrochemical HER and HOR reactions.

At first sight, an uninitiated view of the HER (and HOR) might be that it is "uncomplicated" given the apparent "simplicity" of the reactants and products, namely H^+ and H_2. But this view could not be further from the truth, with the extensive literature being beset with speculation, controversy and incongruent opinions. This has been to a large part due to the historical difficulty in identifying and thoroughly characterizing the reaction intermediates and mechanisms, even on well-defined platinum single crystals, and also the very fast reaction kinetics for both the HER and HOR, particularly on Pt in acidic conditions at room temperature. However, considerable advances have been made during the past 30 years in electrochemical surface science and these have contributed towards a new comprehension of the molecular-level details, the mechanism and reaction intermediates, and an understanding of the surface structural sensitivity of these reactions. In this respect, relevant advances in electrochemical surface science include well-defined platinum single-crystal electrodes, careful electrokinetic measurements at such surfaces, and *in-situ* spectroscopic detection of adsorbed reaction intermediates, the latter of which will be the focus of this chapter.

The HER on platinum proceeds through two, successive, elementary reaction steps; H_2 oxidation takes place by the reverse sequence. The first common step in the HER is the Volmer step, in which a proton is discharged to form adsorbed hydrogen atoms:

$$H^+ + e \rightarrow H_{ads} \tag{10.1}$$

This is followed by either the Tafel step (10.2) or the Heyrovsky step (10.3). The former step involves the combination of two adsorbed hydrogen atoms, while the latter envisages the combination of a proton in solution, an electron, and adsorbed hydrogen to give dihydrogen:

$$2H_{ads} \rightarrow H_2 \tag{10.2}$$

$$H^+ + e + H_{ads} \rightarrow H_2 \tag{10.3}$$

The nature of the adsorbed intermediate (H_{ads}) and its impact on the mechanism of hydrogen evolution has been a major concern in HER electrocatalysis. Platinum has long been known to form adsorbed hydrogen [4], and the adsorption of hydrogen holds a particular significance in the overall development of modern electrochemical surface science.

10.2 Single-Crystal Surfaces

With the development of well-defined single-crystal electrode surfaces, it was seen that hydrogen adsorption has unique and distinctive features between the three low-index planes of Pt. Studies conducted during the late 1970s by the groups of Yeager, Hubbard and Ross involved the preparation of platinum single crystals in ultrahigh-vacuum (UHV), followed

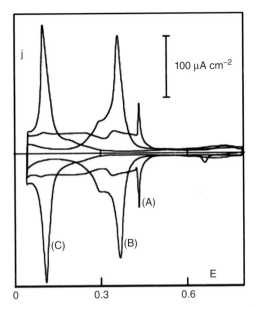

Figure 10.1 *Cyclic voltammograms of flame-cleaned (H_2 + Ar atmosphere) (A) Pt(111), (B) Pt(100) and (C) Pt(110), all in 0.5 M H_2SO_4 at 50 mV s^{-1}. Reproduced with permission from Ref. [9], © Société Chemique de France.*

by transfer into an electrochemical environment for the cyclic voltammetric analysis of hydrogen adsorption [5,6]. Later, Clavilier developed a method by which properly oriented single-crystal beads could be prepared in a flame, followed by rapid quenching in ultrapure water and transfer to the electrochemical cell for voltammetric analysis [7,8]. This method for the preparation of single-crystal platinum electrodes alleviates the need for expensive UHV apparatus, and has been widely adopted by the electrochemical community and has undoubtedly contributed greatly to the understanding of adsorption processes on platinum electrode surfaces, and the platinum–electrolyte interface in general. It is clearly apparent that the voltammetric response differs greatly between the three low-index planes of platinum, as seen in the overlaid voltammograms from Clavilier *et al.* in Figure 10.1 [8]. These voltammograms cover the "hydrogen adsorption" region, although there is also charge flow for anion adsorption. With Pt(111), the hydrogen adsorption peak (from about 0–0.35 V versus reversible hydrogen electrode; RHE) is mostly separated from the anion adsorption (ca. 0.3–0.45 V). On the other hand, for Pt(100) in 0.5 M aqueous H_2SO_4 there is a significant overlap between hydrogen and anion adsorption/desorption. Nevertheless, it is clear that there is great structural sensitivity to the formation of adsorbed hydrogen on these platinum single crystals, while the substantial charges involved indicate a high coverage of adsorbed hydrogen at potentials appreciably positive of the reversible potential for hydrogen gas evolution. This hydrogen has been referred to as "underpotentially deposited hydrogen" (H_{UPD}), due to its formation at potentials positive of the Nernst potential for the HER.

This high coverage of adsorbed hydrogen on both single-crystal Pt surfaces and polycrystalline Pt prior to the onset of hydrogen evolution is at clear odds with the electrokinetic data which requires a low coverage at zero current followed by an increasing coverage as the hydrogen evolution current increases. This apparent disparity has been discussed by Schuldiner [9], and indicates that H_{UPD} is not the intermediate in the HER. On the contrary, the intermediate is another adsorbed species which is formed at very low or vanishing coverage at significant underpotentials, but then shows increasing coverage at progressively higher hydrogen evolution overpotentials. This intermediate has been termed overpotentially deposited hydrogen (H_{OPD}).

10.3 Subtractively Normalized Interfacial Fourier Transform Infrared Spectroscopy

The direct spectroscopic detection of any form of adsorbed hydrogen at platinum metal electrode surfaces eluded detection for many years until 1988, when Nichols and Bewick observed a clear spectroscopic band at ~2090 cm^{-1} which was only apparent at potentials of hydrogen gas evolution [10]. This has been followed by a collection of further studies by several groups worldwide, continuing up to 2013 [11–17], which have aimed at further examining this spectroscopic feature for H_{OPD} and its relation to HER and HOR electrokinetics. The original spectra published by Nichols and Bewick in 1988 are shown in Figure 10.2 [10, 18, 19]. These are potential difference spectra recorded with the so-called SNIFTIRS (Subtractively Normalised Interfacial Fourier Transform Infrared Spectroscopy) technique. This FTIR-based *in-situ* IR spectroscopy method involves recording sets of IR spectra at two electrode potentials and then subtracting (and normalizing) them to produce potential difference spectra with upwards-pointing spectroscopic bands corresponding to species in abundance at one of the potential limits, and downwards-pointing bands corresponding to abundance at the other limit. By using this potential difference method, the tiny spectral absorbance of adsorbed species can be distinguished from the large spectroscopic background which includes electrolyte and other absorbance in the path of the optical beam. The spectra in Figure 10.2 were obtained with the polycrystalline Pt electrode in aqueous 1 M H_2SO_4 using a fixed base potential, E_1, of +442 mV (RHE) and progressively varying the second potential, E_2, to more negative potentials, first corresponding to H_{UPD} and then to potentials of hydrogen evolution. No spectral bands were observed in either the double-layer region or at potentials corresponding to the so-called weakly or strongly bonded H_{UPD} regions of polycrystalline Pt electrodes. Only when the electrode potential, E_2, was extended into region of hydrogen evolution did the spectral band at ~2090 cm^{-1} become apparent, as can be seen in Figure 10.2a. The upwards-pointing band corresponds to spectral absorbance at E_2. Figure 10.2 also shows the cathodic linear sweep voltammogram for polycrystalline platinum with the rapidly rising current for hydrogen evolution at $E < 100$ mV (versus RHE). The HER commences at positive potentials under these conditions, as the working electrode compartment is not saturated with H_2 gas at 1 atm, which simply displaces the E^{θ}_{HER} to positive potentials according to the Nernst equation [20, 21]. It is most notable that the rapid rise in band intensity for the spectroscopic band is concomitant with the rising hydrogen evolution current in the linear sweep voltammogram (Figure 10.2b). It is also noted that the ~2090 cm^{-1} band is seen on rhodium, iridium [10] and Ir(111) cathodes [22] at potentials of hydrogen evolution.

The correlation between the band intensity for the 2090 cm^{-1} band and the rate of hydrogen formation in Figure 10.2b and c is a strong indication that it results from the

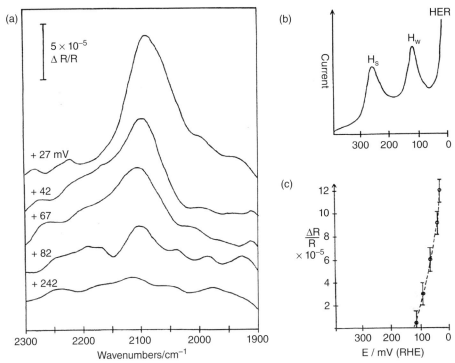

Figure 10.2 *(a) SNIFTIR difference spectra from a polycrystalline Pt electrode in 1 M H_2SO_4 at potentials E_2 as marked in the figure, and with a reference potential E_1 = +442 mV (versus RHE); (b) A linear sweep voltammogram over the hydrogen adsorption and hydrogen evolution region for the same system, showing the historically so-called "strongly bound" hydrogen (H_s), "weakly bound" hydrogen (H_w) and hydrogen evolution reaction (HER); (c) The potential dependence of the intensity of the ~2090 cm^{-1} absorption band. Reprinted from Ref. [10] with permission from Elsevier.*

H_{OPD} adsorbed hydrogen species, which is a reaction intermediate in the hydrogen evolution reaction described by Equations (10.1–10.3). However, the nature of the species which gives rise to the 2090 cm^{-1} absorption remains unclear. There should be concern that this arises from adsorbed CO which gives spectroscopic absorption in the ~2100–1700 cm^{-1} range, depending on the surface coordination. The IR band of adsorbed CO can be so intense that bands can be readily apparent even at levels significantly below 1% coverage. Ultrapure electrolytes and appropriate degassing were applied, but in addition an electrode potential program was used to guard against CO accumulation during the spectral accumulation [10]. This involved a positive potential pulse to potentials of CO electro-oxidation which was applied to the polycrystalline electrode periodically during an interlude between blocks of spectral accumulation [10]. This is conveniently done using the SNIFTIRS method, which involves the accumulation of N spectra at E_1 followed by a short waiting time, and then an accumulation of N spectra at E_2, this cycle being repeated M times (this results in the recording of $N \times M$ spectra at each potential). The electro-oxidation pulse is simply applied

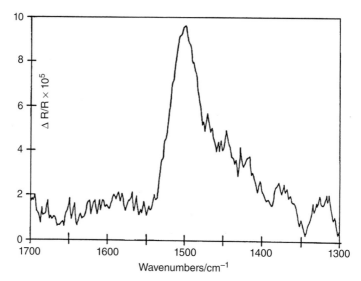

Figure 10.3   *A SNIFTIR difference spectrum showing the on-top Pt–D vibration for polycrystalline Pt. The electrolyte was 1 M H_2SO_4, $E_1 = 442$ mV and $E_2 = +12$ mV (versus RHE). Reprinted from Ref. [10] with permission from Elsevier.*

in the intermission between blocks. Nevertheless, this alone is not sufficient proof that the band arises from a Pt–H vibrational mode since, as discussed later, CO_2 can be reduced to CO on Pt at potentials of hydrogen adsorption and evolution on Pt electrodes. Firm confirmation that the band is indeed a Pt–H vibrational mode was derived from isotopic substitution experiments [10]. The spectrum in Figure 10.3, recorded in D_2O with 1 M H_2SO_4, shows a clear band at 1500 cm^{-1}. This represents a factor of 1.39 by which the 2090 cm^{-1} band is shifted to a lower frequency, which is precisely what is expected for the isotopic effect and the influence of anharmonicity [10].

Given that this IR band arises from a Pt–H vibration from the isotopic substitution experiments, can more be deduced about the likely surface coordination? The surface coordination of this intermediate has been discussed by Nichols and Bewick [10], and also subsequently by other authors. Infrared bands at around 2120/2060 cm^{-1} have been observed in early gas-phase experiments on alumina-supported platinum in a hydrogen atmosphere [23,24], and have been attributed to hydrogen bound to a single surface platinum atom (on-top). By contrast, a much lower frequency of ~1230 cm^{-1} is exhibited by the symmetric dipole active vibrational mode of hydrogen atoms in three-coordinate sites on Pt(111) in UHV [25]. On the other hand, the 2090 cm^{-1} is also too high in frequency for hydrogen in a two-coordinate bridging site [16]. Clearly, the 2090 cm^{-1} absorbance is out of the range of hydrogen in multifold sites but is congruent with atomic hydrogen bound to single Pt atoms; hence, Nichols and Bewick attributed the observed peak, and consequently also H_{OPD}, to on-top hydrogen [10]. As this spectroscopic band correlates qualitatively with the rate of H_2 formation and not with H_{UPD} coverage, the implication is that H_{UPD} cannot be hydrogen atoms singly coordinated on-top of a surface Pt atom. Although, as discussed later, this is contrary to the conclusions of other subsequent studies, this finding is strongly supported by others.

10.4   Surface-Enhanced Raman Spectroscopy

The vibrational band near 2090 cm^{-1} was later confirmed with Raman spectroscopy at the polycrystalline Pt electrode, as seen in Figure 10.4 [26–29]. High-quality Raman spectra could be obtained at platinum by using a sensitive confocal Raman spectrometer and appropriately roughened electrode surfaces. As for the SNIFTIRS measurements, the ca. 2090 cm^{-1} band was detected only in the electrode potential regions near and within hydrogen evolution. Indeed, the rise in the band does not correlate with the rise of the voltammetric peaks for either strongly bound or weakly bound hydrogen adsorption on the polycrystalline electrode (a weak band is only apparent at potentials negative to that for the weakly bound hydrogen peak [27, 29]). As with earlier SNIFTIRS measurements, this band could be attributed to an adsorbed hydrogen species through the use of isotopic substitution, and a band was observed at ~1500 cm^{-1} in D$_2$O electrolyte [26]. This corresponded to a factor of 1.39, in exact agreement with the SNIFTIRS measurements. A strong potential dependence of about 60 cm^{-1} V^{-1} was also observed [27, 29], with a frequency shift towards lower values with a decrease in electrode potential. This has been discussed in terms of increasing lateral interactions with increasing coverage of the on-top hydrogen [29, 30].

The aforementioned *in-situ* IR spectroscopy measurements were conducted by SNIFTIRS, which requires the platinum working electrode to be placed against the IR transmitting window to form an electrolyte thin layer between the window and the electrode. This thin electrolyte layer configuration leads to an ohmic drop when sustained currents are drawn during electrochemical reactions, but also, in the case of the HER, this

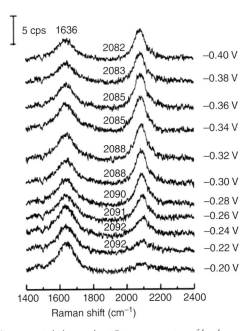

Figure 10.4   *Electrode potential-dependent Raman spectra of hydrogen adsorption at a roughened polycrystalline platinum electrode surface. Reprinted from Ref. [27] with permission from Elsevier.*

can lead to the accumulation of gas bubbles if the electrode potential is extended too far into the region of hydrogen evolution. For this reason the band could only be observed by SNIFTIRS over a region of about 100 mV, but nevertheless it increased in intensity by about 10-fold over this relatively narrow potential window.

10.5   Surface-Enhanced IR Absorption Spectroscopy

The advent of surface-enhanced IR absorption spectroscopy (SEIRAS) removed this requirement for a thin-layer configuration. SEIRAS uses a Kretschmann-type attenuated total reflection (ATR) configuration [13], and for these experiments a thin layer of platinum is deposited onto the top face of the ATR prism and forms the working electrode surface. A typical thickness is about 100 nm, which is sufficiently thin to allow penetration of the evanescent optical field to the electrochemical interface. As the evanescent light penetrates into the electrolyte to only a limited extent, there is much less interference from the bulk electrolyte than in the external reflection methods, such as SNIFTIRS. Important, also, is the significant enhancement of spectral absorbance which can be between 10- to 10000-fold when comparing adsorbed species with their equivalents in electrolyte solution. The surface enhancement is believed to result from the nanoscale roughness of the polycrystalline metal film; such surface enhancement and avoidance of the thin layer are clear benefits when studying the HER, where the band intensities are relatively low and gas bubble formation is problematic. On the other hand, the SEIRAS technique can be applied only to polycrystalline Pt electrodes and not to well-defined Pt single crystals.

Figure 10.5 shows the potential dependence of ATR-SEIRA spectra recorded on poly-crystalline Pt in 0.5 M H_2SO_4 [15]. These are fully consistent with the previous SNIFTIRS

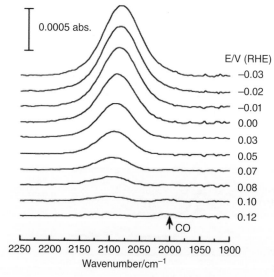

Figure 10.5   *SEIRA spectra recorded as a function of electrode potential, showing the rise of the band corresponding to on-top Pt–H at potentials of hydrogen evolution at ~2100 cm⁻¹. Spectra recorded on polycrystalline Pt (ATR substrates) in 0.5 M H_2SO_4. Reprinted from Ref. [15] with permission from Elsevier.*

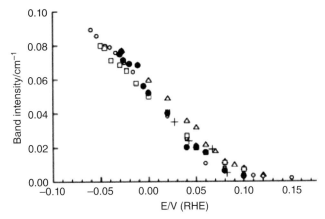

Figure 10.6   *The electrode potential dependence of the integrated band intensity of terminal H observed in 0.01 M H₂SO₄ (□), 0.5 M H₂SO₄ (•), 3 M H₂SO₄ (○) and 1 M HClO₄ (□). The data points shown by (+) were taken from Ref. [10] and multiplied by a factor of 8.5 to compensate for the difference in sensitivity between SEIRAS and IRAS. Reprinted from Ref. [15] with permission from Elsevier.*

studies [10], and show a band rising rapidly in intensity at potentials of the HER [13–15]. This spectroscopic signature was only observed at potentials of hydrogen evolution, and it was not for the underpotentially deposited hydrogen. When D_2O was used instead of H_2O the spectroscopic band shifted to $1500\ cm^{-1}$, which is consistent with the SNIFTIRS [10] and SERS [26] observations. This clearly reinforces the assignment of this feature to H_{OPD}. Figure 10.6 shows the intensity of the terminal Pt–H spectroscopic band versus the electrode potential in various concentrations of H_2SO_4 solutions and in $HClO_4$ [15]. The data points shown as (+) in this figure are from the SNIFTIR spectra of Nichols and Bewick [10]. The intensities of the latter have been adjusted by a factor of 8.5 to account for the greater sensitivity of the SEIRA method. Nevertheless, it can be seen that the external and internal reflection spectroscopy methods agree very well. This figure also shows that the SEIRA method can record IR spectra deeper into the hydrogen evolution region; this means that the band intensity can be more reliably followed over a greater potential window, which also means that there are enhanced possibilities to analyze the mechanism of the HER, as discussed by Kunimatsu and coworkers [13–15].

Direct observation of the band intensity of the terminal Pt–H over a relatively wide electrode potential range (as in Figure 10.6) provides a relatively straightforward access to an analysis of the HER mechanism, provided that reasonable assumptions can be made to relate the band intensity to the coverage of adsorbed hydrogen ($\theta_{H(a)}$) or, more precisely, to the activity of the hydrogen evolution surface intermediate ($a_{H(a)}$) [15]. At very low coverage it is reasonable to assume that these are equivalent. On polycrystalline Pt, a Tafel slope of 30 mV per decade at low overvoltage has been obtained for the HER; this value is consistent with the notion that proton discharge to form adsorbed hydrogen atoms [the Volmer step; Equation (10.1)] is in quasi-equilibrium, and that the following step – the recombination of two adsorbed hydrogen atoms [the Tafel step; Equation (10.3)], is rate-determining. The direct observation of the spectroscopic band has provided the opportunity to evaluate whether this species is an adsorbed intermediate in this mechanism [15]. Within the framework of the Volmer–Tafel mechanism, and assuming that $a_{H(a)}$ approximately equals $\theta_{H(a)}$, the following relationships can be straightforwardly derived [15]:

$$\log A_{H(a)} \propto \frac{1}{2.303} \frac{F}{RT} \eta \tag{10.4}$$

and

$$\log i \propto 2 \log A_{H(a)} \tag{10.5}$$

where $\log A_{H(a)}$ is the integrated band intensity of the hydrogen intermediate which is taken to be directly proportional to its surface coverage, i is the current, while η is the overpotential.

From Equation (10.4), a slope of $(60 \text{ mV})^{-1}$ is expected in a plot of log (band intensity) versus overpotential, and this is what is observed in Figure 10.7a, at least in the low-current regime. An analysis of log i versus log (band intensity), as shown in Figure 10.7b, gives the expected slope of 2 [Equation (10.5)], also again at least in the low-current regime [15]; this confirms a Volmer–Tafel mechanism with the Tafel step being rate-determining. However, the relationships in Equations (10.4) and (10.5) are not held at the higher current densities, as shown in Figures 10.7a and b. This has been attributed to the assumption that the activity of the adsorbed H_{OPD} intermediate, $a_{H(a)}$, equals the coverage $\theta_{H(a)}$ [15]. Kunimatsu *et al.* have generalized this assumption by taking a Frumkin adsorption isotherm to express the activity of the terminal hydrogen [15]. The same group then showed that the deviations apparent in Figures 10.7a and b could be removed by adopting a Frumkin isotherm into the analysis. Best-fit linearizations of the data to the relationships in Equations (10.4) and (10.5) were achieved by incorporating the Frumkin isotherm with interaction parameters between about 2 and 6 (the dimensionless parameter g in the Frumkin isotherm) [15]. These positive g-values imply that lateral repulsive interactions exist between adsorbed H atoms.

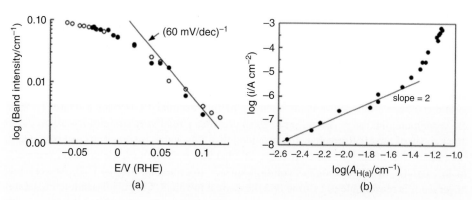

Figure 10.7   *(a) A plot of log (IR band intensity) for on-top hydrogen versus electrode potential and (b) a plot of log (current density) for the HER versus log (IR band intensity). These plots have been used in the analysis of the HER mechanism (see text). Data recorded for polycrystalline Pt (on the ATR prism) in 0.5 M H_2SO_4 (•) or 3 M H_2SO_4 (○). Reprinted from Ref. [15] with permission from Elsevier.*

The clear picture emerging from the SNIFTIRS, Raman and SEIRAS studies of the hydrogen adsorption and evolution regions on polycrystalline Pt electrodes is that on-top hydrogen atoms are only detected at potentials of hydrogen evolution [10, 12–17, 19, 26–29]. The assignment of the 2090 cm^{-1} to on-top hydrogen was supported by isotopic substitution experiments performed for all three methods (SNIFTIRS, Raman and SEIRAS). Moreover, the SEIRAS measurements have been able to link this on-top Pt–H with the intermediate in the Volmer–Tafel mechanism [15]. Plots of log (band intensity) versus over-potential, and also log i versus log (band intensity), were consistent with this mechanism and the recombination of two terminal hydrogen (Tafel step) being rate-determining [15]. These relationships are straightforwardly observed at very low current densities, and then also at higher coverage of the intermediate when lateral repulsive adsorbate interactions were included.

10.6   *In-Situ* Sum Frequency Generation Spectroscopy

On the other hand, vibrational spectra have also been recorded through the hydrogen adsorption region and to potentials of hydrogen evolution using *in-situ* sum frequency generation (SFG) spectroscopy [31–35]. These SFG measurements were at variance with the results of the SNIFTIRS, Raman and SEIRAS studies of the hydrogen adsorption region, and these conflicting results are discussed in the following paragraph.

SFG is a sensitive and selective method for obtaining vibrational spectra at both metal surfaces in UHV and under electrochemical conditions. SFG only detects molecules at the interface due to the nonlinear optical selection rules. In addition, the vibrational modes under study must be both IR- and Raman-active to be detectable in SFG. The SFG studies performed by Tadjeddine *et al.* were conducted with a free electron laser as an IR source, while spectra were recorded throughout the hydrogen UPD region for polycrystalline platinum and also Pt(100), Pt(110) and Pt(111) in a series of investigations [31–35]. Peaks were seen in the SFG spectra in the UPD hydrogen region at the following positions: 1890 and 1970 cm^{-1} for Pt(100), 1900 and 1980 cm^{-1} for Pt(110), and 1945 and 2020 cm^{-1} for Pt(111) [31–33]. Of note was the slight sensitivity of these frequencies to the crystallographic orientation of the surface. The intensity of the peaks increased progressively through the UPD hydrogen region as the electrode potential was lowered, except for Pt(110) where a sudden intensity increase was observed just before the onset of hydrogen evolution [33]. Although the frequency of these peaks in the 1800–2100 cm^{-1} region would seem to indicate on-top hydrogen, no isotopic substitution has been conducted to confirm a Pt–H assignment. The same group also contended that three surface-bound hydrogen atoms were in turn hydrogen-bonded to a water dimer. SFG spectra have also been recorded at potentials of hydrogen evolution, and a new lower wavenumber peak was seen to appear at 1770 cm^{-1} on all Pt surfaces, with the higher wavenumber resonances recorded in the UPD region remaining unperturbed [33]. It has been argued that the 1770 cm^{-1} band arises from adsorbed dihydride, since hydride complexes of the form H_2Pt–R exhibit frequencies around 1735 cm^{-1}. Furthermore, it is contended that this dihydride species is the intermediate for the HER reaction [35]. Clearly, these spectroscopic data from SFG and the appropriate conclusions contradict the SNIFTIRS, Raman and SEIRAS studies. Several groups have offered ideas to explain this contradictory data. For example, Nanbu *et al.*

suggested that vibrational bands around the ~ 2000 cm^{-1} region and ~ 1800 cm^{-1} could arise from adsorbed CO in atop and bridge sites, respectively [12]; this conclusion was reached by recording spectra in the H$_{UPD}$ region in the presence and absence of CO$_2$ [12]. Adsorbed CO can be produced by the electroreduction of CO$_2$ absorbed from surrounding air in the hydrogen adsorption region. It should be noted that even a tiny surface coverage of adsorbed CO can give rise to a detectable IR active vibrational band. In a subsequent review, Jerkiewicz contended that the SFG experiments involve the very time-consuming alignment of laser beams, and that the electrochemical cell used in SFG was prone to contamination [36]. Kunimatsu *et al.* have also suggested that CO derived from impurities in the electrolytes was an issue in the SFG studies [13]. There is, therefore, a body of evidence and opinion that supports the contention that the SFG spectral bands observed in the H$_{UPD}$ region arise from adsorbed CO, possibly generated by CO$_2$ reduction. It has also been noted that such features can be observed in SEIRA spectra [15], and that CO-free spectra can be recorded by prepolarization of the electrode to potentials of CO oxidation before recording sample and reference spectra, as performed in certain SNIFTIRS [10] and SEIRAS [15] studies. In order to fully resolve this issue, it would appear that the SFG studies would benefit from revisiting with a focus on isotopic labelling confirmation and the possible influence of adventitious impurities such as CO$_2$.

10.7   Spectroscopy at Single-Crystal Surfaces

Notwithstanding the results of the SFG studies, there seems to be a good consensus for on-top hydrogen being the active intermediate for the HER on polycrystalline electrodes. Although several studies have been conducted for single-crystal platinum electrodes, the picture there is not clear. As stated earlier, the number of *in-situ* spectroscopic techniques which can be currently applied to reasonably characterize single crystalline platinum electrodes is more limited. SNIFTIRS can be deployed, while Raman and SEIRAS can only be currently applied to polycrystalline Pt for these systems. Nichols and Bewick recorded spectra of the on-top intermediate on low-index Pt surfaces at potentials of hydrogen evolution [10, 19]. The spectra recorded on Pt(110) are shown in Figure 10.8. Ogasawara and Ito have studied the low-index Pt surfaces and also the stepped surface, Pt(11 1 1) at potentials of hydrogen UPD and also the HER [11]. For Pt(100), Pt(110) and the stepped Pt(11 1 1) surfaces, spectroscopic bands were observed at 2080–1990 cm^{-1} (although no isotopic substitution was presented) [11]. These bands were only observed in the potential region where the UPD hydrogen was saturated, and therefore they could not be attributed to a H$_{UPD}$ species. A band was seen at potentials of hydrogen evolution on the stepped Pt(11 1 1) surface (6(100) × (111)), but not on Pt(111) [11]. Thus, it was concluded that on-top hydrogen could not be observed on well-oriented Pt(111). Clearly, the results of Ogasawara and Ito showed a strong structural sensitivity of the on-top hydrogen band, which in turn highlighted the quandary arising from this apparent surface sensitivity of the spectral band, while conventional electrochemical data at that time pointed to structural insensitivity [37, 38]. However, since then it has become apparent from renewed electrokinetics measurements that the HER (and HOR) are indeed surface structure-sensitive. Indeed, in acidic electrolytes Markovic has shown that the exchange current densities for the HER/HOR in acid solutions follow the order Pt(111) << (100) < (110) [39]. Retrospectively, this

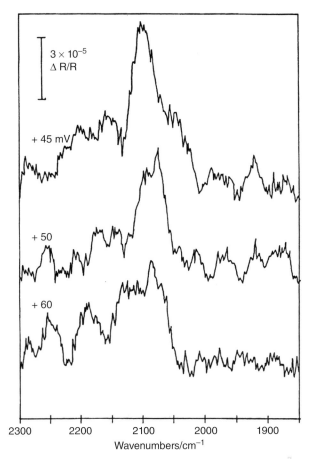

Figure 10.8 *SNIFTIR difference spectra for Pt(110) in 1 M HClO$_4$. E$_1$ = 650 mV and E$_2$ as marked (versus RHE). Reprinted from Ref. [18] R.J. Nichols, PhD thesis, University of Southampton, England (1989).*

point could be used to justify the lack of observation of on-top hydrogen for the HER on Pt(111). More recent studies have been conducted on low-index and stepped Pt surfaces by Nakamura *et al.* [16], and showed a strong surface structural influence for the H$_{OPD}$ intermediate, as detected using *in-situ* IR spectroscopy [16]. A clear band was detected for Pt(110) at ~2080 cm^{-1} at potentials of hydrogen evolution, and not at underpotentials [16]; the band was not detected on Pt(100) and Pt(111) [16], and this might be attributed to the HER reactivity sequence Pt(111) << (100) < (110), as classified by Markovic [39], with on-top hydrogen still being the intermediate on Pt(100) and Pt(111) but being of insufficient coverage (and hence intensity) to readily detect in these SNIFTIRS experiments. However, Nakamura *et al.* sought another explanation for these results [16]. The spectra are shown in Figure 10.9 of high-index planes of Pt in 0.5 M H$_2$SO$_4$ at 0 V versus RHE (the SNIFTIRS reference potential was 0.8 V). A band around 2080 cm^{-1} was observed

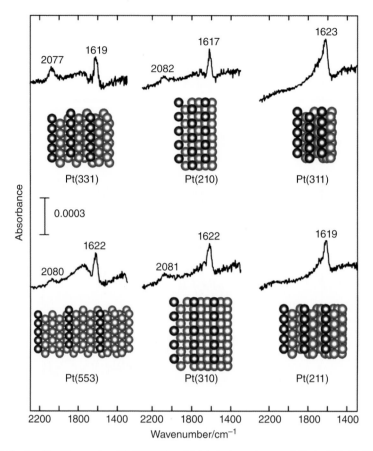

Figure 10.9 In-situ IR spectra (SNIFTIRS) of high-index planes of Pt in 0.5 M H_2SO_4. All spectra recorded at 0 V (RHE) with a reference potential of 0.8 V [16]. Reprinted from Surface Science, Vol. 605, Nakamura, M.; Kobayashi, T.; Hoshi, N.: Structural dependence of intermediate species for the hydrogen evolution reaction on single crystal electrodes of Pt, pages 1462–1465, copyright 2011, with permission from Elsevier.

on Pt(331), Pt(210), Pt(553) and Pt(310), but not on Pt(311) and Pt(211). These high-index planes belong to the following surfaces:

- Pt(331)$n = 3$ and Pt(553)$n = 5$ belong to the $(n-1)(111)–(110)$ series; these have a (111) terrace and (110) step.
- Pt(210)$n = 2$ and Pt(310)$n = 3$ belong to the $n(100)–(110)$ series; these have a (100) terrace and (110) step.
- Pt(311)$n = 2$ and Pt(211)$n = 3$ belong to the $n(111)–(100)$ series; these have a (111) terrace and (100) step.

The 2080 cm^{-1} band is observed for the first two of these series which have a (110) step, but does not appear on the third series, which has a (100) step. Hence, the 2080 cm^{-1} band was associated with atop hydrogen adsorbed onto (110) structures [16]. This may be interpreted as atop hydrogen only being present on (110) structures, or not being sufficiently intense to be spectroscopically observed on other structures but still being present there, albeit at a comparatively low coverage. For the Pt(311) and Pt(211) surfaces (which have a (100) step), where the 2080 cm^{-1} band is not observed, the authors pointed to a small blue-shifted tail on the ~1620 cm^{-1} band (see Figure 10.9). This tail was also observed as a weak feature on the low-index Pt(100) surface. The main band here at 1620 cm^{-1} is simply the δ(HOH) mode of interfacial water. With the aid of density functional theory (DFT) calculations, they assigned the small tail to an asymmetric bridge hydrogen (i.e., a hydrogen bridging two platinum atoms with unequal Pt–H bond lengths) [16]; this was believed to be the H$_{OPD}$ intermediate on (100) structures [16]. Although this was an interesting study, the proximity of the band tail to the main water feature, and the lack of any isotopic substitution call for confirmatory experiments in D$_2$O to indeed confirm that it corresponds to a Pt$_2$H-type adsorbate.

10.8   Overall Conclusions

With the exception of the SFG studies, where many reservations exist that have been discussed earlier in the chapter, the clear implication of the observations of on-top hydrogen only at potentials of hydrogen evolution is that this is not a UPD hydrogen species. In other words, UPD hydrogen must correspond to some surface-bound hydrogen species that is multicoordinated, perhaps bound to two or three Pt surface atoms, or hydrogen embedded below the surface plane of Pt atoms. For example, by using electron energy loss spectroscopy (EELS), an adsorbed monolayer of hydrogen on Pt(111) has been shown in UHV to provide dipole active modes for H in threefold hollow f.c.c. sites at 1234 and 911 cm^{-1} [25]. These have not been spectroscopically detected within the *in-situ* environment, even when HF has been used as the electrolyte to remove anion interference [16]. Indeed, there is no definitive spectroscopic characterization of any UPD hydrogen species, despite such species existing at high coverage at potentials immediately preceding the HER. In the absence of any firm spectroscopic characterization no absolute assignments can be made, although a collection exists of nonetheless very interesting observations and albeit indirect deductions concerning the nature of the H$_{UPD}$ species. This includes an impressive thermodynamic analysis of hydrogen adsorption by the groups of Jerkiewicz and Markovic [20, 36, 39–42]. For example, Jerkiewicz *et al.* have carried out a detailed thermodynamic analysis of the electrochemical adsorption of H$_{UPD}$ on polycrystalline and single-crystal Pt electrodes [20, 36, 40, 41]. From a thermodynamic analysis of cyclic voltammograms and their temperature dependence, it was possible to determine thermodynamic state functions for H$_{UPD}$, including $\Delta G°$, $\Delta H°$ and $\Delta S°$, and also the surface bond energy and Gibbs energies of lateral interactions. For instance, for Pt(111) the surface bond energy for M-H$_{UPD}$ was determined and found to be very similar to hydrogen adsorbed onto Pt(111) from the UHV [36]. This proximity of the UHV and electrochemical values led to the conclusion that both involved similar surface bonds, while the thermodynamic equivalence was taken to indicate that atomic hydrogen occupies the same adsorption sites on Pt(111), despite the very different environments [36]. From UHV EELS data [25] this is presumably the threefold hollow site.

Further evidence of the inequivalence of H$_{UPD}$ and H$_{OPD}$ adsorption sites was obtained from the observations by Protopopoff and Marcus that although adsorbed sulfur atoms on Pt electrodes would block H_{UPD}, the HER could still take place but the rate would be greatly reduced [43, 44]. This led to the conclusion that H_{UPD} occupies different sites from the intermediate responsible for the HER. In a similar vein, cyanide adsorbed onto Pt(111) has been seen to block H_{UPD} but not to change the onset potential for hydrogen evolution in cyclic voltammetry studies [45]. Reflection spectroscopy also showed significant differences for H_{UPD} in different adsorption regions, and/or on different crystal faces. Bewick *et al.* noted that the "strongly bound" H_{UPD} peak on polycrystalline gave rise to strong reflectivity changes, whereas the "weakly bound" species did not [46, 47]. It was suggested that the former corresponded to a proton lying just below the surface plane of the metal, with its electron contributing to the conduction band of the metal and giving rise to large reflectivity changes [46, 47]. In later studies, when Nichols and Bewick recorded the IR reflectivity spectra for Pt(100) and Pt(111) [18, 19], the Pt(100) surface showed large reflectivity changes for H_{UPD} but the Pt(111) surface did not. It was suggested that this was indirect evidence for the formation of subsurface "strongly bound" hydrogen on the (100) surface, probably involving the fourfold hollow sites [19].

References

(1) Fleischmann, M., Hendra, P.J. and McQuillan, A.J. (1974) Raman-spectra of pyridine adsorbed at a silver electrode. *Chemical Physics Letters*, **26**, 163–166.

(2) Golinski, J. (2004) W. Nicholson (1753–1815), in *Oxford Dictionary of National Biography*, Oxford University Press, http://www.oxforddnb.com/view/article/20153. Accessed 11th February 2014.

(3) Tafel, J. (1905) The polarisation of cathodic hydrogen development. *Zeitschrift fur Physikalische Chemie – Stochiometrie und Verwandtschaftslehre*, **50**, 641–712.

(4) Bold, W. and Breiter, M. (1960) Bestimmung der adsorptionswarme von wasserstoff an aktiven platinmetallelektroden in schwefelsaurer losung. *Zeitschrift fur Elektrochemie*, **64**, 897–902.

(5) Ross, P.N. (1981) Hydrogen chemisorption on Pt single-crystal surfaces in acidic solutions. *Surface Science*, **102**, 463–485.

(6) Hubbard, A.T. (1988) Electrochemistry at well-characterized surfaces. *Chemical Reviews*, **88**, 633–656.

(7) Clavilier, J., Faure, R., Guinet, G. and Durand, R. (1980) Preparation of mono-crystalline Pt microelectrodes and electrochemical study of the plane surfaces cut in the direction of the (111) and (110) planes. *Journal of Electroanalytical Chemistry*, **107**, 205–209.

(8) Clavilier, J., Rodes, A., Elachi, K. and Zamakhchari, M.A. (1991) Electrochemistry at platinum single-crystal surfaces in acidic media - hydrogen and oxygen-adsorption. *Journal de Chimie Physique et de Physico-Chimie Biologique*, **88**, 1291–1337.

(9) Schuldiner, S. (1959) Hydrogen overvoltage on bright platinum.3. Effect of hydrogen pressure. *Journal of the Electrochemical Society*, **106**, 891–895.

(10) Nichols, R.J. and Bewick, A. (1988) Spectroscopic identification of the adsorbed intermediate in hydrogen evolution on platinum. *Journal of Electroanalytical Chemistry*, **243**, 445–453.

(11) Ogasawara, H. and Ito, M. (1994) Hydrogen adsorption on Pt(100), Pt(110), Pt(111) and Pt(1111) electrode surfaces studied by in-situ infrared reflection-absorption spectroscopy. *Chemical Physics Letters*, **221**, 213–218.

(12) Nanbu, N., Kitamura, F., Ohsaka, T. and Tokuda, K. (2000) Adsorption of atomic hydrogen on a polycrystalline Pt electrode surface studied by FT-IRAS: the influence of adsorbed carbon monoxide on the spectral feature. *Journal of Electroanalytical Chemistry*, **485**, 128–134.

(13) Kunimatsu, K., Senzaki, T., Tsushima, M. and Osawa, M. (2005) A combined surface-enhanced infrared and electrochemical kinetics study of hydrogen adsorption and evolution on a Pt electrode. *Chemical Physics Letters*, **401**, 451–454.

(14) Kunimatsu, K., Uchida, H., Osawa, M. and Watanabe, M. (2006) In situ infrared spectroscopic and electrochemical study of hydrogen electro-oxidation on Pt electrode in sulfuric acid. *Journal of Electroanalytical Chemistry*, **587**, 299–307.

(15) Kunimatsu, K., Senzaki, T., Samjeske, G. *et al.* (2007) Hydrogen adsorption and hydrogen evolution reaction on a polycrystalline Pt electrode studied by surface-enhanced infrared absorption spectroscopy. *Electrochimica Acta*, **52**, 5715–5724.

(16) Nakamura, M., Kobayashi, T. and Hoshi, N. (2011) Structural dependence of intermediate species for the hydrogen evolution reaction on single crystal electrodes of Pt. *Surface Science*, **605**, 1462–1465.

(17) Dong, Y., Hu, G., Hu, X. *et al.* (2013) Hydrogen adsorption and oxidation on Pt Film: an in situ real-time attenuated total reflection infrared (ATR-IR) spectroscopic study. *Journal of Physical Chemistry C*, **117**, 12537–12543.

(18) Nichols, R.J. (1989) Infrared spectroscopic investigations of electrocatalysis. PhD thesis, University of Southampton, England.

(19) Nichols, R.J. (1992) IR spectroscopy of molecules at the solid-solution interface, in *Adsorption of Molecules at Metal Electrodes,* Frontiers in Electrochemistry, Chapter 7, vol. **1** (eds P.N. Ross and J. Lipkowski), VCH Publ. Inc., New York, pp. 347–389.

(20) Jerkiewicz, G. (1998) Hydrogen sorption at/in electrodes. *Progress in Surface Science*, **57**, 137–186.

(21) Zolfaghari, A., Chayer, M. and Jerkiewicz, G. (1997) Energetics of the underpotential deposition of hydrogen on platinum electrodes. 1. Absence of coadsorbed species. *Journal of the Electrochemical Society*, **144**, 3034–3041.

(22) Senna, T., Ikemiya, N. and Ito, M. (2001) In situ IRAS and STM of adsorbate structures on an Ir(111) electrode in sulfuric acid electrolyte. *Journal of Electroanalytical Chemistry*, **511**, 115–121.

(23) Jayasooriya, U.A., Chesters, M.A., Howard, M.W. *et al.* (1980) Vibrational spectroscopic characterization of hydrogen bridged between metal atoms – a model for the adsorption of hydrogen on low-index faces of tungsten. *Surface Science*, **93**, 526–534.

(24) Dixon, L.T., Barth, R. and Gryder, J.W. (1975) Infrared active species of hydrogen adsorbed by alumina-supported platinum. *Journal of Catalysis*, **37**, 368–375.

(25) Badescu, S.C., Jacobi, K., Wang, Y. *et al.* (2003) Vibrational states of a H monolayer on the Pt(111) surface. *Physical Review B*, **68**, 205401.

(26) Ren, B., Huang, Q.J., Cai, W.B. *et al.* (1996) Surface Raman spectra of pyridine and hydrogen on bare platinum and nickel electrodes. *Journal of Electroanalytical Chemistry*, **415**, 175–178.

(27) Ren, B., Xu, X., Li, X.Q. *et al.* (1999) Extending surface Raman spectroscopic studies to transition metals for practical applications II. Hydrogen adsorption at platinum electrodes. *Surface Science*, **427–28**, 157–161.

(28) Tian, Z.Q., Ren, B., Chen, Y.X. *et al.* (1996) Probing electrode/electrolyte interfacial structure in the potential region of hydrogen evolution by Raman spectroscopy. *Journal of the Chemical Society, Faraday Transactions*, **92**, 3829–3838.

(29) Xu, X., Ren, B., Wu, D.Y. *et al.* (1999) Raman spectroscopic and quantum chemical study of hydrogen adsorption at platinum electrodes. *Surface and Interface Analysis*, **28**, 111–114.

(30) Xu, X., Wu, D.Y., Ren, B. *et al.* (1999) On-top adsorption of hydrogen at platinum electrodes: a quantum-chemical study. *Chemical Physics Letters*, **311**, 193–201.

(31) Peremans, A. and Tadjeddine, A. (1994) Vibrational spectroscopy of electrochemically deposited hydrogen on platinum. *Physical Review Letters*, **73**, 3010–3013.

(32) Peremans, A. and Tadjeddine, A. (1995) Electrochemical deposition of hydrogen on platinum single-crystals studied by infrared-visible sum-frequency generation. *Journal of Chemical Physics*, **103**, 7197–7203.

(33) Tadjeddine, A. and Peremans, A. (1996) Vibrational spectroscopy of the electrochemical interface by visible infrared sum-frequency generation. *Surface Science*, **368**, 377–383.

(34) Tadjeddine, A. and Peremans, A. (1996) Vibrational spectroscopy of the electrochemical interface by visible infrared sum frequency generation. *Journal of Electroanalytical Chemistry*, **409**, 115–121.

(35) Tadjeddine, A., Peremans, A. and Guyotsionnest, P. (1995) Vibrational spectroscopy of the electrochemical interface by visible-infrared sum-frequency generation. *Surface Science*, **335**, 210–220.

(36) Jerkiewicz, G. (2010) Electrochemical hydrogen adsorption and absorption. Part 1: under-potential deposition of hydrogen. *Electrocatalysis*, **1**, 179–199.

(37) Kita, H., Ye, S. and Gao, Y. (1992) Mass-transfer effect in hydrogen evolution reaction on Pt single-crystal electrodes in acid-solution. *Journal of Electroanalytical Chemistry*, **334**, 351–357.

(38) Seto, K., Iannelli, A., Love, B. and Lipkowski, J. (1987) The influence of surface crystallography on the rate of hydrogen evolution at Pt electrodes. *Journal of Electroanalytical Chemistry*, **226**, 351–360.

(39) Markovic, N.M., Grgur, B.N. and Ross, P.N. (1997) Temperature-dependent hydrogen electrochemistry on platinum low-index single-crystal surfaces in acid solutions. *Journal of Physical Chemistry B*, **101**, 5405–5413.

(40) Radovic-Hrapovic, Z. and Jerkiewicz, G. (2001) The temperature dependence of the cyclic-voltammetry response for the Pt(110) electrode in aqueous H_2SO_4 solution. *Journal of Electroanalytical Chemistry*, **499**, 61–66.

(41) Zolfaghari, A. and Jerkiewicz, G. (1999) Temperature-dependent research on Pt(111) and Pt(100) electrodes in aqueous H_2SO_4. *Journal of Electroanalytical Chemistry*, **467**, 177–185.

(42) Markovic, N.M., Schmidt, T.J., Grgur, B.N. *et al.* (1999) Effect of temperature on surface processes at the Pt(111)-liquid interface: hydrogen adsorption, oxide formation, and CO oxidation. *Journal of Physical Chemistry B*, **103**, 8568–8577.

(43) Protopopoff, E. and Marcus, P. (1987) Effect of chemisorbed sulfur on the electrochemical hydrogen adsorption and recombination reactions on Pt(111). *Journal of Vacuum Science & Technology A - Vacuum Surfaces and Films*, **5**, 944–947.

(44) Protopopoff, E. and Marcus, P. (1991) Effects of chemisorbed sulfur on the hydrogen adsorption and evolution on metal single-crystal surfaces. *Journal de Chimie Physique et de Physico-Chimie Biologique*, **88**, 1423–1452.

(45) Cuesta, A. (2011) Atomic ensemble effects in electrocatalysis: the site-knockout strategy. *ChemPhysChem*, **12**, 2375–2385.

(46) Bewick, A. and Tuxford, A.M. (1973) Studies of adsorbed hydrogen on platinum cathodes using modulated specular reflectance spectroscopy. *Journal of Electroanalytical Chemistry*, **47**, 255–264.

(47) Bewick, A., Kunimatsu, K., Robinson, J. and Russell, J.W. (1981) IR vibrational spectroscopy of species in the electrode-electrolyte solution interphase. *Journal of Electroanalytical Chemistry*, **119**, 175–185.

11

Electrochemical Noise: A Powerful General Tool

Claude Gabrielli[1] and David E. Williams[2]
[1]*Université Pierre et Marie Curie, LISE, France*
[2]*University of Auckland, School of Chemical Sciences, New Zealand*

At the end of the 1960s, a few electrochemists began to investigate electrode–solution systems by considering the random fluctuations of either the current passing through the interface or the voltage across this interface. Previously, electrochemical studies had been based only on current–voltage relationships at steady state or under linear voltammetric sweeping, or impedance measurements at relatively high frequencies – that is, averaged or deterministic quantities. The new and original approach was based on the study of the random behavior of the electrode–solution system generated by stochastic processes such as chemical reactions or diffusion occurring at the interface. This approach was founded on the theory of the noise arising in semiconductors [1]. Whether fluctuations from a particular process are observable depends of course on its magnitude in relation to the additional noise generated by the system, including the noise due to the measurement instruments and the experimental set-up.

The minimum value of the noise generated by a system is the thermal or Johnson noise resulting from the random motion of electrons or ions within the system when it is in thermal equilibrium with its surroundings. An expression for the Johnson noise in terms of a fluctuating voltage was first derived by Nyquist [2] in 1928 and takes the form

$$< \Delta V^2 >= 4kT \, \mathrm{Re} \, [Z(f)] \, \Delta f \qquad (11.1)$$

Developments in Electrochemistry: Science Inspired by Martin Fleischmann, First Edition.
Edited by Derek Pletcher, Zhong-Qun Tian and David E. Williams.
© 2014 John Wiley & Sons, Ltd. Published 2014 by John Wiley & Sons, Ltd.

where $< \Delta V^2 >$ is the spectral density of voltage fluctuations, k is Boltzmann's constant, T is the absolute temperature, Re[Z] is the real part of the impedance of the system, and Δf the frequency bandwidth within which the noise is measured. Clearly, any noise that is to be studied must be detectable above the Johnson noise of the system.

When the system under investigation is far from equilibrium – that is, when the current in one direction can be neglected in comparison with the current for the process being studied – the electrical noise may be presented as a random sequence of noncorrelated current pulses. Each of these pulses transfers an electric charge of value γe in the external circuit (shot noise), where e is the elementary electronic charge. The current noise of such a system is described by Schottky's theorem [3]:

$$< \Delta I^2 >= 2\gamma eJ \tag{11.2}$$

where $< \Delta I^2 >$ is the spectral density of current fluctuations and J is the average current.

The first pioneering studies to be reported in electrochemistry in which electrochemical noise was used as the analysis tool mainly dealt with theoretical approaches to generation–recombination noise in electrochemical systems, and its possible use to study homogeneous reaction kinetics. Fleischmann [4] investigated the fluctuating current arising from inherent fluctuations in the number of ionic particles present at the interface caused by the generation and recombination of ions as a result of electrochemical reactions in weak electrolytes. Barker [5] studied the noise theory and related irreversible reactions to shot noise, while Tyagai [6, 7] investigated the cathodic reduction of iodine on platinum. The data obtained from these studies threw new light on the overall statistical features of the process (correlation between elementary acts of reaction and degree of nonequilibrium of the stationary state). There was agreement that studies of this type give the number of electrons γ participating in the slow step of a particular reaction, which is related to its exchange current. Further developments then included those by Cardon and Gomes [8], who studied oxidation reactions at the illuminated single-crystal zinc oxide electrode, and by Rangarajan [9], who proposed a stochastic theory of monolayer formation in electrocrystallization. Green and Yafuso [10, 11] studied the noise spectra associated with the ion transport through a membrane in an attempt to obtain information about the mechanism of ion exchange, and relaxation times that were apparently attributable to ion pairing in the membrane were determined.

Such microscopic noises are difficult to obtain as the measurement set-up is often too noisy to separate the often very low electrochemical noise of interest from the parasitic noise of the instrumentation. In addition, electrochemical noises at equilibrium of the Johnson type, or even of the Schottky type, give quantities which can be readily obtained by more conventional methods, for example, impedance measurements. As a consequence, investigators have more recently rather studied noises derived from more macroscopic random events arising from various phenomena, for example, pitting corrosion or gas-evolving electrodes, which give rise to electrical signals of higher amplitude and also have some practical interest. Some examples will be provided in the following sections.

11.1   Instrumentation

During the late 1970s, instrumentation was improved by the availability of operational amplifiers of increasingly better quality (high open loop gain, wide frequency bandwidth, low noise), enabling the design of low-noise amplifiers and control devices. This point was especially critical for the reliable measurement of the very low-amplitude noise generated by elementary processes.

When the fluctuations show current or voltage transients, data analysis may be performed in the time domain by investigating the shape, size and occurrence rate of the random events. It can be also performed by measuring the moments of the potential or current fluctuations (standard deviation, skewness, kurtosis) [12]. However, this approach is extremely limited for data interpretation. In the absence of current or voltage transients, the values of the moments are most likely close to zero, as for signals with a Gaussian distribution. Any deviation from zero indicates the existence of transients.

Data analysis may be also carried out in the frequency domain [6, 8, 11] by calculating the power spectral density of the fluctuations [13]. The interpretation of this quantity is often a challenge, especially when no transients are observable. When transients are observable the interpretation is sometimes easier, but not always so because different transient shapes may give the same power spectral density.

The random signal of low amplitude $S(t)$ of interest is often buried in a parasitic noise $N(t)$. To overcome this difficulty, two parallel and identical channels are used, with parasitic noises $N_1(t)$ and $N_2(t)$. The cross-power spectral density of the outputs of the two channels, $X(t)$ and $Y(t)$, $\Psi_{XY}(f)$, leads to the power spectral density of the studied fluctuations, $\Psi_{SS}(f)$, by using the Fast Fourier Transform(FFT).

$$X(t) = G(S(t) + N_1(t))$$
$$Y(t) = G(S(t) + N_2(t)) \tag{11.3}$$

hence

$$\Psi_{XY}(f) = G^2\Psi_{SS}(f) \tag{11.4}$$

as the cross spectra of $S(t)$ and $N_1(t)$, $S(t)$ and $N_2(t)$, and $N_1(t)$ and $N_2(t)$ are null because these quantities are not correlated. This approach is also valid for cross-correlation analysis of the noise by using a cross-correlator [14].

Figure 11.1 [15] shows an example of a potentiostatic arrangement used to measure the current noise, $i(t)$, around the steady-state current, I, through $S(t) = Ri(t)$, taking into account the potentiostat and amplifier noises [16]. Two parallel and identical channels measured the fluctuations $i(t)$ of the current flowing through an electrochemical cell by using two resistors, R, and two differential amplifiers with gain G. The current, I, is cancelled by applying an offset RI to the differential amplifiers. High-pass and antialiasing filters are added on each channel in a classical way to obtain a proper cross-spectrum of the electrochemical noise.

$$\Psi_{XY}(f) = G^2R^2\Psi_{II}(f) \tag{11.5}$$

The polarization potential E, the potentiostat, and the amplifiers are battery-powered to avoid parasitic noises coming from the mains (50 or 60 Hz). In order to minimize the parasitic noises, the potentiostat, differential amplifiers and the electrochemical cell can be placed in a shielding cage made from muferro steel (termed a Faraday cage). Some groups have proposed using wavelet techniques to improve the efficiency of spectral analysis, especially to detect the current or voltage transients more easily [17].

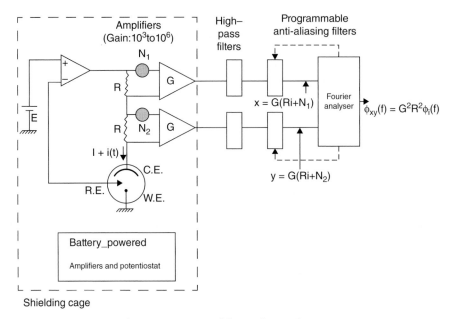

Figure 11.1 *Experimental arrangement used for analyzing the current noise in potentiostatic regime. N_1 and N_2 are the parasitic noises of the measurement channels. The figure shows the Fourier analyzer, amplifiers and filters. Reprinted from Ref. [89] with kind permission from Springer Science+Business Media.*

11.2   Applications

In an electrochemical system, all the chemical reactions or diffusion processes are stochastic by nature [18]. So, if the instrumentation is sufficiently efficient, the inherent fluctuations of these elementary phenomena due to their stochastic character become observable. Fluctuations are most readily observed when a stochastic event is followed by a large and essentially deterministic process, so that the electrochemical system itself amplifies the fluctuating signal. Examples of this type will be given below, and include gas-evolving electrodes, voltage-controlled transmembrane ion currents due to the insertion of pore-forming molecules, two-dimensional (2D) metal deposition, and corrosion. Fleischmann was one of the promoters of several applications of noise analysis in electrochemistry, and following these visionary ideas many groups have applied noise analysis to a wide variety of electrochemical problems.

11.2.1   Elementary Phenomena

Models of the stochastic behavior of the electrochemical interface have been given when a redox reaction limited by diffusion of the reacting species occurs on the electrode surface [19, 20]. The random fluctuations of the state variables (concentrations and voltage) are assumed to derive from Poisson elementary noise sources which are directly acting on the elementary fluxes – either reactive or diffusive – of the reacting species. Consequently, the evolution of the state variables is governed by the Langevin equations obtained from a linearization of the nonlinear electrochemical equations deduced from the heterogeneous electrochemical kinetics. The Langevin noise sources are derived from the Poisson noise. Finally, the power spectral density of the current and potential are derived and compared with the measured quantities. However, limitations appear due to the instrumentation noise.

Figure 11.2a shows the phase-sensitive detection of the current noise generated on a platinum electrode due to the limiting current diffusion in 1 M KCl of potassium ferri-ferrocyanide. On bare platinum (curve a) the increase of the noise below 10 Hz was due to fluctuations in the electrolyte velocity induced by natural convection, whereas on curve b a gel layer coating the electrode cancels this phenomenon. The thermal noise in current determined from the impedance $(4kT\text{Re}[1/Z(f)])$ and the parasitic current due to the potential control are also given. Above 1 kHz the parasitic regulation noise, which is inversely proportional to $|Z(f)|$, became preponderant. Figure 11.2b shows the calculated phase-sensitive detection of the total current noise $\Psi_{II}(f)$, the current thermal noise, and the noise generated by the kinetic and diffusional processes Ψ_{In} for the same conditions as for Figure 11.2a. The calculated and experimental power spectral densities $\Psi_{II}(f)$ are in rather good agreement when taking into account the extreme measurement difficulties. When a fluctuation–dissipation analysis of the noise of such an electrode was performed, the mass transfer step was shown to be correlated to the electric current polarizing the electrode [21].

More recently, a theoretical and experimental investigation has been carried out on the hydrogen evolution process, assuming a Volmer–Heyrovsky mechanism and following the technique proposed by Tyagai [7]. It was found that the apparent electron number values ranged between one and two, depending on the applied current [22].

11.2.2   Bioelectrochemistry

Fleischmann *et al.* studied conductance fluctuations induced by low concentrations of the polypeptide, alamethicin, in planar lipid bilayer membranes, using a computer-aided analysis [23]. In the time-recording of the conductance of the membrane (see Figure 11.3) there was a well-defined finite number of conductance levels, where the conductance was related to the size of the pore, which expanded by a defined increment as each additional molecule of alamethicin was added to the circumference. It was shown that the statistics of transition between nonadjacent states closely conform to a birth-and-death process. Based on the electrochemical rate constants, a complete set of rate parameters governing the steady-state distribution was derived, and it was shown that the electrochemical free energies of the conductance states varied quadratically with the state number for low-lying states. The Gibbs energy of activation of both the inclusion (birth process) and exclusion (death process) of alamethicin molecules was related to the Gibbs energy differences between the states. A detailed model based on the nucleation of a 2D pore accounts for these observations, provided that both birth and death take place via an intermediate expansion of the pore lumen. This model requires two energy parameters: the "edge" and "bulk" energies of the pore, together with a trigger rate of the initial process which is a sufficient description for the steady-state behavior of the voltage-controlled system.

As a few pores can be opened at the same time, the formalism required to derive the probabilities of the states of a single (elementary) pore from the observed probabilities of the compound states of the ensemble of pores was developed. This took into account

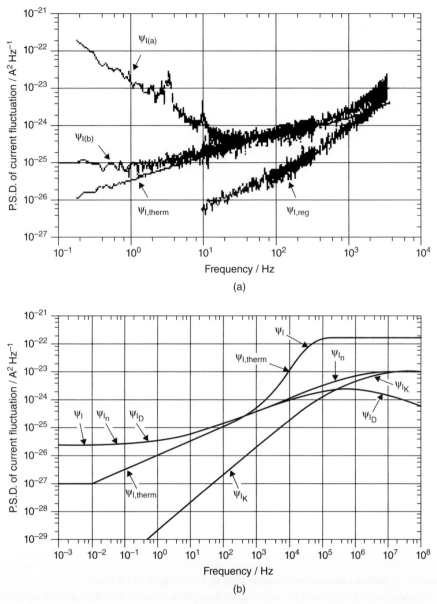

Figure 11.2 (a) Measured power spectral density of the voltage fluctuations Ψ_I, $\Psi_{I, therm}$, and $\Psi_{I, reg}$ for a platinum disk electrode (area 10^{-3} cm^2) in 1 M KCl solution with potassium ferri/ferrocyanide (10^{-2} M) at the potential + 115 mV versus SCE. Curve (a) = bare platinum; curve (b) = platinum covered with a gel layer; (b) Calculated phase-sensitive detection of Ψ_I, $\Psi_{I, therm}$, and Ψ_{In} in the same experimental conditions as in (a). Reprinted with permission from Ref. [20]. Copyright 1993, AIP Publishing LLC.

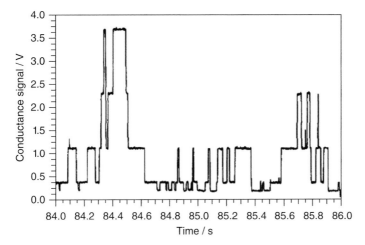

Figure 11.3 *Example of 2000 samples blocks of the conductance fluctuations obtained after 20-fold amplification of a 205 mV pA-1 sensitivity instrumentation (sampling frequency: 1000 Hz, 0.3 kHz filtering before sampling). Reprinted from Ref. [26] with permission from Elsevier.*

a Poisson process triggering a subsequent birth and death process [24, 25]. This model was applicable to all situations where the modeling was most readily cast in terms of the behavior of a single elementary system, whilst the experiment dealt with ensembles of such systems.

Following these studies, improvements in the methods led to experiments being performed that employed a patch–clamp technique and a membrane of only a few micrometers in diameter. In this case, only one pore was opened at any given time, at most, and a birth-and-death process could interpret the experimental results quite closely such that the kinetics of opening and closing of the pore can be evaluated [26].

11.2.3 Electrocrystallization

Metal deposition has been investigated by employing noise analysis for both two dimensional (2D) and three-dimensional (3D) electrocrystallizations [27]. For example, Budevski *et al.* studied 2D nucleation, namely the growth of silver on a perfect, single-crystal substrate, where the spatial and time statistics were distinguished under conditions in which elementary events could be separately observed (see Figure 11.4) [28].

A model which considers circular crystallites on a circular substrate at low nucleation rate would lead to the growth of a complete planar layer following the formation of a single nucleus, the growth of which would be limited only by the boundaries of the substrate. When the current–time transients were calculated, a statistical analysis of the transients alone was compared with experiments at low overpotentials. The moments of the transients alone consolidated the model and showed that nucleation was uniform over the substrate. The power spectral density of the whole experiment provided the steady-state nucleation rate and showed that, in a stationary state, nucleation could be adequately described as a Poisson process.

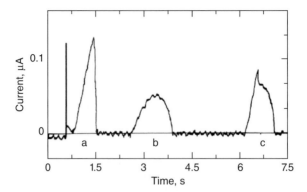

Figure 11.4   *Sections of data for the initial stages of Ag deposition on a single-crystal surface at 8 mV overpotential, nucleation at (point a) the center, (point b) the edge, and (point c) midway between the center and the edge of the circular electrode. Reprinted from Ref. [28] with permission from Elsevier.*

Electrochemical noise was also used to characterize 3D metal deposition, and a correlation of the characteristics of the noise with the deposit structure was developed. First, a phenomenological model based on a birth-and-death process of the crystallites, which explains the correlation between the noise power and the structural organization of the electrodeposits, was proposed [29]. The important ideas here were that crystallites are randomly nucleated, grow for a time and then stop growing, giving a noise power and spectrum that is quite different from that expected for a simple electrochemical reaction rate control of the deposit growth. It is, of course, an interesting question as to the mechanisms that might cause the growth of an individual crystallite to stop. These might include the possibility that the crystallite had been sterically crowded by others, that there was a local depletion of the reacting species, or that the growth has been stopped by the adsorption of an inhibitor. The statistics of the birth-and-death processes, as well as the size-dependent rate of growth of the crystallites, can – at least in principle – be deduced from the observable noise, provided that the mechanism is not too complex and would reflect and affect the form of the deposit. The problem of deducing the birth-and-death statistics, as well as the kinetics of growth, is strongly analogous to the problem of analyzing the birth, growth and death of corrosion pits (see below). When zinc electroplating was studied by means of electrochemical noise in conjunction with scanning electron microscopy observations, it was found that that the noise generated during the electroplating of a dendritic or a large conglomerate zinc deposit had a large potential oscillation amplitude and a positive potential drift, whereas the formation of a compact zinc deposit was characterized by small noise amplitude and little potential drift. A detailed analysis of the noise spectra led to a type of "fingerprint" that could be used to characterize the deposit morphology and relate it to the rate-limiting process, namely the surface reaction rate, diffusion, or mixed control [30]. More recently, another study was undertaken to investigate the influence of different additives (gelatin and thiourea) on the quality of copper electrorefining from a sulfuric acid electrolyte bath. The roughness of the cathodic deposit could be monitored and analyzed with noise, and then compared with various techniques such as profilometry, atomic force microscopy, X-ray diffraction and scanning electronic microscopy. The authors concluded that the homogeneity, smoothness and grain size of the deposit could indeed be characterized through noise [31].

11.2.4 Corrosion

In corrosion processes, many aspects of electrochemistry are integrated – specifically, the coupling of mass transport, local solution composition and reaction rate, the coupling of anodic and cathodic processes, and the effects of solution resistance. In a corrosion system at open circuit, the rates (total current) of the anodic and cathodic processes must balance. The anodic and cathodic processes may occur on spatially separated areas of the surface; the resistance to current flow through the electrolyte between these areas determines the difference in local electrode–solution potential difference between the anodic and cathodic areas. The coupling of current flow, mass transport (electromigration, diffusion, convection), local solution composition and local electric potential means that large gradients of solution composition and potential can develop on small spatial scales; these gradients can then trigger new electrode reactions. The result is characteristic fluctuations in measured electrode potential (at open circuit or under galvanostatic control) or current (under potentiostatic control).

The results of studies by Bertocci [32] and others have indicated that current fluctuations could be used as a means of studying the initiation stages of localized corrosion. Hladky and Dawson [33, 34] then showed that measurements of self-generated electrochemical potential fluctuations on electrodes undergoing either pitting or crevice corrosion confirmed that these forms of localized attack have quite distinct noise "signatures," and that both types of attack can be detected within seconds of their initiation. Hladky and Dawson's results indicated the possibility of a nonperturbative electrochemical corrosion monitoring technique capable of detecting pitting and crevice attack. These ideas have been generalized and it appears that, within the context of a particular environment and material, noise "signatures" can be recognized which indicate the onset of particular types of corrosion, whose magnitude is indicative of the corrosion rate [34]. For example, if a probe is constructed from two electrodes of the same material as the component whose corrosion is to be monitored, and these are coupled through a zero-resistance ammeter, then a fluctuating current may be observed. Moreover, its mean and variance, σ_I^2, is dependent on the nature of the corrosion reaction occurring on the probe electrodes, as well as the size and relative size of the two electrodes, Similarly, the electrode potential fluctuations of either one of the electrodes or of the two coupled together can be measured. The ratio σ_I/σ_E can be interpreted as a "noise conductance", the magnitude of which can be correlated with the corrosion rate on the probes [35, 36], and hence with the risk or rate of corrosion of the component being monitored. Other measurements aimed at deducing corrosion rates can be carried out on the same probe, including direct current (DC) polarization resistance, Tafel extrapolation, alternating current (AC) impedance and weight loss. Moreover, the comparison is instructive in building a picture of the electrochemical processes occurring *in-situ*.

A correct interpretation of the observed fluctuations requires an understanding of what processes are driving them. Generally speaking, for corrosion reactions this is not thermal fluctuations, so interpretation of the behavior in terms of Johnson noise in a simple resistor is not correct. The first approach to a rigorous analysis was made by Williams, Westcott and Fleischmann, in the case of the initiation of pitting corrosion of stainless steel [37–40].

This approach developed the idea of a deterministic evolution of current following a stochastic trigger, and borrowed from the theory of electrocrystallization. The key ideas – which subsequently were elaborated – were that the initiation of pitting corrosion could be separated into two processes. The "triggering" of a breakdown had the characteristics of a Poisson process and this resulted in an "unstable" or "metastable" pit: a small anode within which the local current density was very high and which was maintained by extreme local gradients of solution composition and electrode potential. "Maintenance" of the evolving pit was a deterministic process that could be described in terms of the evolution of surface area within the evolving defect, the electrical resistance around the mouth of the evolving pit (when the pit mouth acts as a microelectrode), and the stability of the local solution composition against variations of the transport processes (migration and diffusion) within and from the pit. If these transport fluctuations were such as to destroy the stability of the local solution environment, then the pit would stop growing and the current to that pit would fall abruptly. If, however, the pit survived beyond some critical depth (which is treated as a critical age) then the pit would become "stable." Studies of ensembles of current–time traces from many experiments, together with the theory and simulations, were used to validate these ideas [38, 39]. Subsequently, current transients due to individual metastable pits were measured [41–46] allowing a detailed understanding of the processes that were occurring.

Figure 11.5 [46] illustrates a train of current pulses resulting from the nucleation, growth and death of metastable pits in stainless steel, polarized in a dilute NaCl solution. The

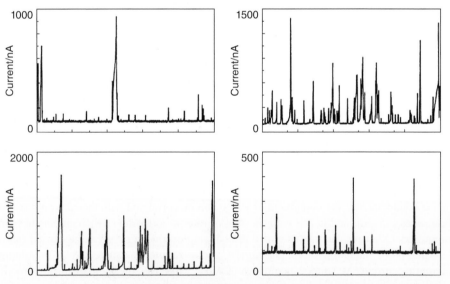

Figure 11.5 *Current pulses due to the initiation, temporary growth and termination of metastable pits. Variability between different specimens taken from the same steel. Each frame is 5000 s, taken from a measurement with a slow potential ramp (5×10^{-6} V s^{-1}). AISI type 304L stainless steel at 95–120 mV versus SCE in 0.028 mol dm^{-3} aqueous NaCl solution. From Ref. [46], J. Stewart, "Pit initiation on austenitic stainless steels," PhD Thesis, University of Southampton, 1990.*

frequency and amplitude of the current pulses varied very significantly from one specimen to another; these effects were subsequently understood when it was appreciated that pits nucleated around sulfide inclusions in the steel [46, 47], with the probability of nucleation being sensitive to the shape and composition of each particular inclusion. The general shape of the transients can be understood in terms of the evolution of the shape of the pit and the change in solution resistance for current flow to the pit mouth as the latter expands and the pit geometry changes [43, 44]. The detail of the transients is more complex, however (Figure 11.6), as the pit may develop facets, may re-nucleate inside the pit, and may develop

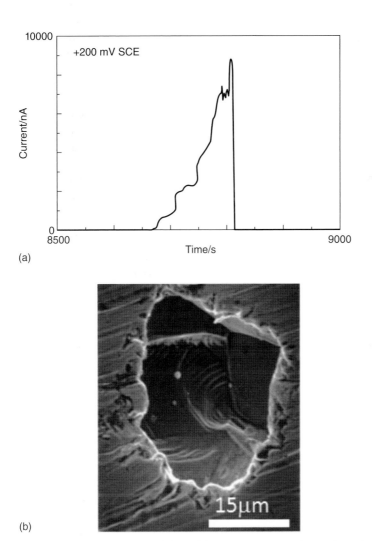

(a)

(b)

Figure 11.6 *(a) Micrograph of a metastable pit following cessation of growth, with the corresponding current transient (b), illustrating that the current rise is irregular and may be related to the development of facets within the pit. From Ref. [46], J. Stewart, "Pit initiation on austenitic stainless steels," PhD Thesis, University of Southampton, 1990.*

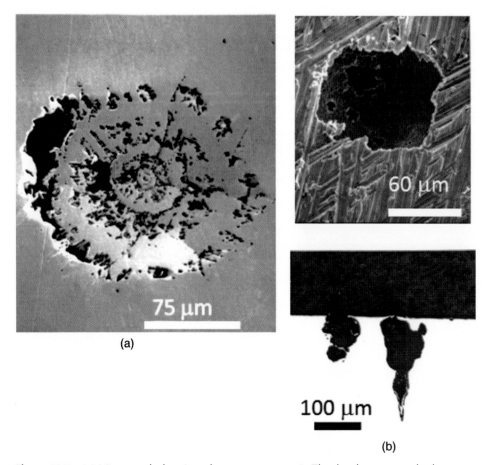

(a)

(b)

Figure 11.7   *(a) Micrograph showing a lacy cover over a pit. The development and subsequent detachment of such covers explains much of the detail in the current transients; (b) Micrograph illustrating re-nucleation within a metastable pit. From Ref. [46], J. Stewart, "Pit initiation on austenitic stainless steels," PhD Thesis, University of Southampton, 1990.*

a cover (Figure 11.7) that perforates during pit development and may eventually fall into the pit. If that occurs then the resistance for current flow to the pit mouth will fall abruptly such that the current jumps up in a spike and then falls, equally abruptly, as the solution inside the pit dissipates and the pitting stops [43, 46].

These measurements lead, in particular, to an understanding that there is a critical solution composition that must be maintained within the pit, and if this does not occur then the pit will die. As a consequence, there is an important effect of the kinetics of active dissolution of the alloy within this critical solution. Measurements of the characteristics of metastable pits, coupled with studies of the dissolution kinetics of steel within the aggressive solutions that characterize the local pit environment, have provided an explanation as to why certain alloy additions (specifically Mo) act to inhibit pitting corrosion by making the maintenance of metastable pits more difficult. Other important effects include those of salt films, which act to sustain the local solution composition, the effect of solution flow acting through an effect on the solution composition profile around the pit mouth and hence on the solution conductivity profile, and resistance to current flow to the pit, and the effect of the geometry of the pit and around the pit mouth. The behavior of metastable pits seems now fully understood (see Ref. [48]). Finally, the chemistry involved in the trigger events around the inclusions was resolved, primarily through *in-situ* imaging techniques using both photoelectrochemical imaging and scanning electrochemical microscopy [49]. Here, the critical factor appeared to be compositional changes in the inclusions at the inclusion–steel interface [48]. Subsequently, these ideas have been extended broadly, one particularly successful example being an application to the study of pitting of aluminum alloys [50–53]. The detection of metastable pitting events has been directly connected to the presence of intermetallic precipitates above a critical size, as measured using ultra-high-resolution transmission electron microscopy. However, if the precipitate particles are below a few nanometers in dimension then it appears that the oxide layer which confers stability on the aluminum can cover these particles without fracturing.

The approach of treating localized corrosion as a stochastic trigger, followed by a deterministic evolution, has also been effective in the study of stress–corrosion cracking(SCC). Early investigations identified current fluctuations associated with the transient (stop–start) propagation of cracks [54], whereby the current pulses correlated with acoustic emission as the crack "jumped." Techniques that had been developed previously to study pitting corrosion were then refined and applied to specific instances of SCC. One such instance was the thiosulfate- or tetrathionate-induced cracking of sensitized stainless steels [55–58], the studies of which revealed the mechanistic detail that can be derived from examining such trains of current pulses, random in time and varying in shape. In these studies, corrosion current pulses at the open-circuit potential were measured by coupling the specimen under load to a large cathode (a cylinder of annealed steel of the same type surrounding the specimen) through a zero-resistance ammeter. Current pulses were associated with the nucleation of microcracks and their movement across single-grain boundary facets. The cracks initiated as the consequence of a nonuniform deformation around grain boundaries, and most stopped after the penetration of, at most, a few grain-boundary facets. The idea of microstructural barriers to the propagation of short stress–corrosion cracks was developed, and such barriers became less important as the chemistry of the environment became more aggressive. A simple statistical model, based on a probability to jump across a barrier, was developed for crack advance, and in part for the statistics of failure, by intergranular stress–corrosion cracking(IGSCC). At a higher strain, fatal cracks were initiated from a pre-existing microcrack. Strain-induced martensite formation resulted in a decrease with increasing strain in both microcrack nucleation frequency and penetration. The same experimental methods and ideas were used successfully to describe the initiation of SCC of sensitized stainless steel in oxygenated water at 288 °C [58].

Erosion–corrosion is another example of a process with a stochastic trigger with a deterministic evolution. In this case, the trigger is the impact of a particle onto the surface, and the resultant exposure of unpassivated metal (for a passive system) or the removal of scale or a change in mass transport may cause a current pulse, the time-evolution of which can be modeled according to the metal–solution system being examined [59–64]. Unfortunately, most of the current induced by an impact is not captured by an external electrode, and this makes interpretation complex. Indeed, correlation with the results of

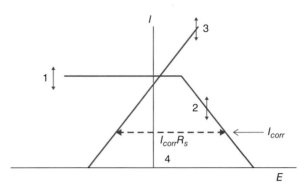

Figure 11.8   *Simplified schematic to illustrate possible sources of fluctuations in corrosion current, I_{corr} or corrosion potential measured at a distant reference electrode, for general corrosion with a diffusion-limited cathodic reaction such as oxygen reduction. Fluctuations leading to fluctuations in I_{corr} can be in: (1), the transport rate of the cathodic reagent, leading to changes in diffusion-limited current; (2) and (3), the relative areas of the anodic and cathodic processes, caused for example by detachment of surface scales or by changes in the electrode kinetics of these processes caused for example by the addition of corrosion inhibitors or change in surface concentration of such inhibitors; (4), in the solution resistance between cathodic and anodic areas, if these are spatially separated, caused for example by fluctuations in local electrolyte composition itself linked to the occurrence of the corrosion reaction.*

measurements in which the surface is momentarily scratched by a probe is generally necessary to develop a rigorous interpretation [62, 63]. The electrochemical noise signal can be correlated quantitatively with an acoustic noise measurement that characterizes the energy transferred in each impact, and from which the size of the induced defect can be deduced [59].

The extension of these ideas to provide a quantitative interpretation of the fluctuations associated with general corrosion is more difficult. Figure 11.8 provides a simplified scheme showing the possible sources of fluctuations in the corrosion current, I_{corr} or corrosion potential measured at a distant reference electrode, for general corrosion with a diffusion-limited cathodic reaction such as oxygen reduction. Fluctuations leading to variations in I_{corr} can occur in the transport rate of the cathodic reagent [65], leading to changes in the diffusion-limited current. Indeed, the study of such current fluctuations is an excellent way to characterize turbulent fluctuations in the boundary layer near an electrode [65–69] (see below). Fluctuations might also be caused by variations in the relative areas [70] of the anodic and cathodic processes, caused for example by the detachment of surface scales or by fluctuations in current distribution over the surface that may result from changes in electrode kinetics, mass transport, or solution resistance. Changes in the electrode kinetics, perhaps caused by the addition of corrosion inhibitors or a change in the surface concentration of such inhibitors, are other possible sources of fluctuations. Finally, fluctuations can be driven by variations in the solution resistance [71] between the cathodic and anodic areas if these are spatially separated, and caused by fluctuations in the local electrolyte composition which is itself linked to the occurrence of a corrosion reaction. Bubble evolution is also an obvious possible driver of fluctuations, either through changes in the exposed area as the bubble evolves, by stirring induced by the detachment of a bubble, or electrical resistance changes if a bubble is evolved within a zone of localized corrosion (e.g., a crevice or crack) and constricts the current path. The time scales for these different fluctuations can be very different. However, whereas fluctuations associated with the flow rate may encompass a very wide time range, those effects due to alterations within the corrosion scales, fluctuations in surface concentrations or bubble evolution can be rather slow. Thus, the spectrum of fluctuations may be different for each of these possible origins. The above-described analyses of specific cases demonstrate how the fluctuating signal may alter if there is a change in the phenomenon that is driving it, and also provide at least some nonempirical foundation for the application of electrochemical noise as a technique to infer changes in dominant corrosion processes or corrosion rates.

11.2.5   Other Systems

The pioneering studies initiated by Martin Fleischmann suggested many other applications of noise analysis to other research groups, including gaseous bubble evolution, hydrodynamics, battery state of charge [72–74], conductance of conjugated polymers [75, 76], and diagnostics of polymer electrolyte membrane fuel cells [77].

Noise analysis has been particularly fruitful in characterizing various aspects of hydrodynamics, as noted above for the specific case of corrosion processes. First of all, multiphase flows were investigated, either gas/water [78], solid/liquid [79, 80], oil/water [81] or oil/brine [82]. In these flows, fluctuations are due primarily either to fluctuations in transport rates to an electrode or to fluctuations in electrolyte resistance. If one phase preferentially wets the electrode, then there may be fluctuations due to variation in the effective electrode area. Each of these phenomena has a characteristic spectral signature. Turbulent flows close to a wall have been investigated by means of electrochemical noise by using electrochemical probes of various shapes, by measuring the power spectral density of the limiting diffusion current fluctuations [83–86].

The detail of bubble evolution from electrodes is a topic of practical importance. The size of a bubble at detachment is dependent on the gas and solution density, the cell geometry, hydrodynamics, interfacial tensions between the bubble, electrode and solution, and the roughness of the electrode. The way in which bubbles detach affects the resistance to current flow to the electrode, and rising bubbles effectively stir the electrolyte. Electrochemical noise measurements have been used to characterize the evolution of chlorine [87], oxygen [88], and hydrogen [89].

Hydrogen formation occurs through several interacting phenomena:

- Electrochemical reactions leading to molecular hydrogen which dissolves into the solution.
- The nucleation and growth of bubbles from the dissolved hydrogen on some active sites on, or close to, the electrode surface.
- The detachment of a bubble after some time, depending upon the physical conditions prevailing at the electrode–electrolyte interface such as surface tension and hydrodynamic conditions.

Under potentiostatic (galvanostatic) control, random fluctuations of the total current crossing a gas-evolving electrode (potential) can be observed. For low currents, individual

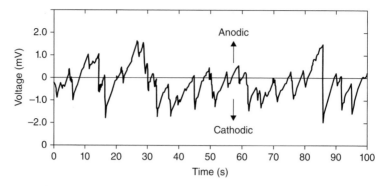

Figure 11.9   *Voltage fluctuations during iron dissolution in 2 N sulfuric acid medium. Voltage–time recording at a 3.7 mA cm^{-2} anodic current density in galvanostatic regime. Reprinted from Ref. [89] with kind permission from Springer Science+Business Media.*

events can be distinguished, as shown in Figure 11.9 [89], where the quasi-linear increases of potential are related to the growth of bubbles on the electrode surface; this induces an increase in electrolyte resistance, whereas the sharp potential changes are related to bubble detachment. However, when the current density is high the bubbles are so numerous that their occurrence cannot be counted and their individual effects cannot be separated on recording the current (or voltage) fluctuations. Hence, no analysis is possible in the time domain, and any identification of the characteristic parameters of the evolution regime of these electrolytically generated hydrogen bubbles (nucleation rate, life time, etc.) must be performed through the measured power spectral density (Figure 11.10).

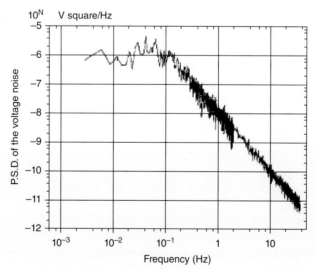

Figure 11.10   *Power spectral density of the fluctuations related to hydrogen evolution. (a) measurement in galvanostatic regime corresponding to Figure 11.9; (b) Measurement in potentiostatic regime. Reprinted from Ref. [89] with kind permission from Springer Science+Business Media.*

A simple stochastic model of the time series displayed by the electrolysis current during bubble evolution has been devised by considering the sum of elementary voltage transients, starting at the nucleation time and following a homogeneous Poisson law. The elementary transient is a linear increase of the overvoltage due to the screening effect of the bubble [90] before a sudden drop related to bubble detachment after a lifetime having an exponential probability distribution. This is in fact the same model as used initially to describe pitting corrosion [39]. The power spectral density of the total voltage fluctuations can be calculated using the generalized shot noise (or filtered Poisson process) theory [91]. More complicated models which take into account the control of bubble growth by diffusion, coalescence, and gas–oscillator phenomena [92] allow electrode surface wetting [93] or electrode roughness [94] to be investigated.

11.3 Conclusions

The study of electrochemical noise has proved to be a powerful source of insight into a wide range of electrochemical processes, as illustrated by the examples given in this chapter. These examples have encompassed the electrochemical signals resulting from fluctuations in transport or reaction rate which may act either directly on the reaction at the surface, or manifest though effects such as fluctuations in electrolyte resistance or triggered by variations in the exposed area. Fluctuations are most readily observed when a stochastic event is followed by a large and essentially deterministic process, so that the electrochemical system itself amplifies the fluctuating signal. The fields of corrosion and electrocrystallization provide numerous examples. Fluctuations are frequently examined empirically, through measurement of the spectral signature and qualitative comparison with some expected effects or with examples from previous experience. Such qualitative application of the technique has proven of value, for example, as a simple way to detect when a change has occurred in some electrochemical process; specific examples are the onset of corrosion or a change in corrosion type. On the other hand, in many cases a rigorous theory can be derived (those models that seem applicable are often common across a range of different phenomena), after which the method can provide a simple and powerful route to understanding such complex processes.

References

(1) Van der Ziel, A. (1959) *Fluctuation Phenomena in Semiconductors*, Butterworths, London.
(2) Nyquist, H. (1928) Thermal agitation of electric charge in conductors. *Physical Review*, **32**, 110–113.
(3) Schottky, W. (1918) Uber spontane stromschwankungen in verschiedenen elektrizitatsleitern. *Annals of Physics*, **57**, 541–567.
(4) Fleischmann, M. and Oldfield, J.W. (1970) Generation-recombination noise in weak electrolytes. *Journal of Electroanalytical Chemistry*, **27**, 207–218.
(5) Barker, G.C. (1969) Noise connected with electrode processes. *Journal of Electroanalytical Chemistry*, **21**, 127–136.
(6) Tyagai, V.A. and Lukyanchikova, N.B. (1968) Electrochemical noise of iodine reduction on a cadmium sulfide surface. *Surface Science*, **12**, 331–340.
(7) Tyagai, V.A. (1971) Faradaic noise of complex electrochemical reactions. *Electrochimica Acta*, **16**, 1647–1654.

(8) Gomes, W.P. and Cardon, F. (1971) Study of noise associated with oxidation reactions at the illuminated single crystal zinc oxide anode. *Journal of Solid State Chemistry*, **3**, 125–130.

(9) Rangarajan, S.K. (1973) Electrocrystallisation: a stochastic theory of monolayer formation. *Journal of Electroanalytical Chemistry*, **46**, 119–123.

(10) Green, M.E. and Yafuso, M. (1968) A study of the noise generated during ion transport across membranes. *The Journal of Physical Chemistry*, **72**, 4072–4078.

(11) Yafuso, M. and Green, M.E. (1971) Noise spectra associated with hydrochloric acid transport through some cation-exchange membranes. *The Journal of Physical Chemistry*, **75**, 654–662.

(12) Gabrielli, C., Keddam, M. and Raillon, L. (1979) Random signals: 3rd order correlation measurements. *Journal of Physics E: Scientific Instruments*, **12**, 632–636.

(13) Huet, F. (2005) Electrochemical noise technique, in *Analytical Methods in Corrosion Science and Engineering* (eds P. Marcus and F. Mansfeld), Corrosion Technology, Taylor & Francis, CRC Press, vol. **22**, pp. 507–570.

(14) Blanc, G., Epelboin, I., Gabrielli, C. and Keddam, M. (1975) Etude des comportements déterministe et stochastique de l'interface électrochimique à l'aide de corrélateurs numériques. *Journal of Electroanalytical Chemistry*, **62**, 59–94.

(15) Gabrielli, C., Huet, F. and Keddam, M. (1986) Investigation of electrochemical processes by an electrochemical noise analysis. Theoretical and experimental aspects in potentiostatic regime. *Electrochimica Acta*, **31**, 1025–1039.

(16) Epelboin, I., Gabrielli, C., Keddam, M. and Raillon, L. (1979) Measurement of the power spectral density of electrochemical noise: direct two-channel method. *Journal of Electroanalytical Chemistry*, **105**, 389–395.

(17) Aballe, A., Bethencourt, M., Botana, F.J. *et al.* (2001) Use of wavelets to study electrochemical noise transients. *Electrochimica Acta*, **46**, 2353–2361.

(18) Fleischmann, M., Labram, M., Gabrielli, C. and Sattar, A. (1980) The measurement and interpretation of stochastic effects in electrochemistry and bioelectrochemistry. *Surface Science*, **101**, 583–601.

(19) Gabrielli, C., Huet, F. and Keddam, M. (1993) Fluctuations in electrochemical systems. I. General theory on diffusion limited electrochemical reactions. *The Journal of Chemical Physics*, **99**, 7232–7239.

(20) Gabrielli, C., Huet, F. and Keddam, M. (1993) Fluctuations in electrochemical systems. II. Application to a diffusion limited redox process. *The Journal of Chemical Physics*, **99**, 7240–7252.

(21) Gravof, B.M., Kutnetsov, A.M. and Suntsov, A.N. (2002) Fluctuation-dissipation analysis of diffusion and discharge processes. *Journal of Electroanalytical Chemistry*, **533**, 19–23.

(22) Szenes, I., Meszaros, G. and Lengyel, B. (2007) Noise study of hydrogen evolution process on Cu and Ag microelectrodes in sulphuric acid medium. *Electrochimica Acta*, **52**, 4752–4759.

(23) Fleischmann, M., Gabrielli, C., Labram, M.T.G. *et al.* (1980) Alamethicin induced conductances in lipid bilayers: I. Data analysis and simple steady state model. *Journal of Membrane Biology*, **55**, 9–27.

(24) Fleischmann, M., Gabrielli, C., Labram, M.T.G. and Markvart, T. (1986) Alamethicin induced conductances in lipid bilayers. Part II. Decomposition of the observed (compound) process. *Journal of Electroanalytical Chemistry*, **214**, 427–440.

(25) Fleischmann, M., Gabrielli, C. and Labram, M.T.G. (1986) Alamethicin induced conductances in lipid bilayers. Part III. Derivation of the properties of the elementary process from the (compound) observed process. *Journal of Electroanalytical Chemistry*, **214**, 441–457.

(26) Clessienne, T., Gabrielli, C., Huet, F. *et al.* (1990) Analysis of the alamethicin induced single channel conductance fluctuations in lipid bilayers as a birth and death process. *Journal of Electroanalytical Chemistry*, **296**, 429–444.

(27) Bindra, P., Fleischmann, M., Oldfield, J.W. and Singleton, D. (1973) *Nucleation. Faraday Discussions of the Chemical Society*, **56**, 180–198.

(28) Budevski, E., Fleischmann, M., Gabrielli, C. and Labram, M. (1983) Statistical analysis of the 2-D nucleation and growth of silver deposition. *Electrochimica Acta*, **28**, 925–931.

(29) Gabrielli, C., Ksouri, M. and Wiart, R. (1978) Electrocrystallization noise: a phenomenological model. *Journal of Electroanalytical Chemistry*, **86**, 233–239.

(30) Zhang, Z., Leng, W.H., Cai, Q.Y. *et al.* (2005) Study of the zinc electroplating process using electrochemical noise technique. *Journal of Electroanalytical Chemistry*, **578**, 357–367.

(31) Safideh, F., Lafront, A.M., Ghali, E. and Houlachi, G. (2010) Monitoring the quality of copper deposition by statistical and frequency analyses of electrochemical noise. *Hydrometallurgy*, **100**, 87–94.

(32) Bertocci, U. and Kruger, J. (1980) Studies of passive film breakdown by detection and analysis of electrochemical noise. *Surface Science*, **101**, 608–618.

(33) Hladky, K. and Dawson, J.L. (1981) The measurement of localized corrosion using electrochemical noise. *Corrosion Science*, **21**, 317–322.

(34) Hladky, K. and Dawson, J.L. (1982) The measurement of corrosion using electrochemical 1/f noise. *Corrosion Science*, **22**, 231–237.

(35) Dawson, J.L. (1996) Electrochemical noise measurements, in *Electrochemical Noise Measurement for Corrosion Applications: ASTM Special Testing Publications 1277* (eds J.R. Kearns, J.R. Scully, P.R. Roberge *et al.*), ASTM International.

(36) Bertocci, U., Gabrielli, C., Huet, F. and Keddam, M. (1997) Noise resistance applied to corrosion measurements: I. Theoretical analysis. *Journal of the Electrochemical Society*, **144**, 31–37.

(37) Bertocci, U., Gabrielli, C., Huet, F. *et al.* (1997) Noise resistance applied to corrosion measurements: II. Experimental tests. *Journal of the Electrochemical Society*, **144**, 37–43.

(38) Williams, D.E., Westcott, C. and Fleischmann, M. (1984) Studies of the initiation of pitting corrosion on stainless steels. *Journal of Electroanalytical Chemistry*, **180**, 549–564.

(39) Williams, D.E., Westcott, C. and Fleischmann, M. (1985) Stochastic models of pitting corrosion of stainless steels: I. Modeling of the initiation and growth of pits at constant potential. *Journal of the Electrochemical Society*, **132**, 1796–1804.

(40) Williams, D.E., Westcott, C. and Fleischmann, M. (1985) Stochastic models of pitting corrosion of stainless steels: II. Measurement and interpretation of data at constant potential. *Journal of the Electrochemical Society*, **132**, 1804–1811.

(41) Williams, D.E., Westcott, C. and Fleischmann, M. (1983) A statistical approach to the study of localised corrosion, in *Passivity of metals and Semiconductors: Proceedings of the Fifth International Symposium on Passivity, Bombannes, France*, May 30–June 3, 1983 (ed. M. Froment), Elsevier, pp. 217–228.

(42) Frankel, G.S., Stockert, L., Hunkeler, F. and Boehni, H. (1987) Metastable pitting of stainless-steel. *Corrosion*, **43**, 429–436.

(43) Pistorius, P.C. and Burstein, G.T. (1992) Metastable pitting corrosion of stainless-steel and the transition to stability. *Philosophical Transactions of the Royal Society of London Series A - Mathematical Physical and Engineering Sciences*, **341**, 531–559.

(44) Williams, D.E., Stewart, J. and Balkwill, P.H. (1994) The nucleation, growth and stability of micropits in stainless steel. *Corrosion Science*, **36**, 1213–1235.

(45) Williams, D.E., Westcott, C. and Fleischmann, M. (1987) The effect of solution variables on pit initiation as measured by statistical methods, in *Corrosion Chemistry within Pits, Crevices and Cracks: Proceedings of a Conference Held at the National Physical Laboratory, Teddington, Middlesex*, October 1–3, 1984 (ed. A. Turnbull), HMSO, London, pp. 61–88.

(46) Stewart, J. (1990) Pit initiation on austenitic stainless steels. PhD thesis, University of Southampton.

(47) Stewart, J. and Williams, D.E. (1992) The initiation of pitting corrosion on austenitic stainless steel: on the role and importance of sulphide inclusions. *Corrosion Science*, **33**, 457–474.

(48) Williams, D.E., Kilburn, M.R., Cliff, J. and Waterhouse, G.I.N. (2010) Composition changes around sulphide inclusions in stainless steels, and implications for the initiation of pitting corrosion. *Corrosion Science*, **52**, 3702–3716.

(49) Williams, D.E., Mohiuddin, T.F. and Zhu, Y.Y. (1998) Elucidation of a trigger mechanism for pitting corrosion of stainless steels using submicron resolution scanning electrochemical and photoelectrochemical microscopy. *Journal of the Electrochemical Society*, **145**, 2664–2672.

(50) Gupta, R.K., Deschamps, A., Cavanaugh, M.K. *et al.* (2012) Relating the early evolution of microstructure with the electrochemical response and mechanical performance of a Cu-rich and Cu-lean 7xxx aluminum alloy. *Journal of the Electrochemical Society*, **159**, C492–C502.

(51) Gupta, R.K., Sukiman, N.L., Cavanaugh, M.K. *et al.* (2012) Metastable pitting characteristics of aluminium alloys measured using current transients during potentiostatic polarisation. *Electrochimica Acta*, **66**, 245–254.

(52) Gupta, R.K., Sukiman, N.L., Fleming, K.M. *et al.* (2012) Electrochemical behavior and localized corrosion associated with Mg_2Si particles in Al and Mg alloys. *ECS Electrochemistry Letters*, **1**, C1–C3.

(53) Ralston, K.D., Birbilis, N., Cavanaugh, M.K. *et al.* (2010) Role of nanostructure in pitting of Al-Cu-Mg alloys. *Electrochimica Acta*, **55**, 7834–7842.

(54) Newman, R.C., Sieradzki, K. and Woodward, J. (1987) Current fluctuations during transgranular stress corrosion cracking, in *Corrosion Chemistry within Pits, Crevices and Cracks: Proceedings of a Conference held at the National Physical Laboratory, Teddington, Middlesex*, October 1–3, 1984 (ed. A. Turnbull), HMSO, London, pp. 203–212.

(55) Balkwill, P., Stewart, J., Westcott, C. *et al.* (1989) Noise signals from corrosion processes - current fluctuations associated with the early stages of pitting and stress-corrosion cracking of stainless-steel, in *Instationary Processes and Dynamic Experimental Methods in Catalysis, Electrochemistry and Corrosion, Dechema Monographs*, vol. **120** (eds G. Sandstede, G. Kreysa), VCH, Weinheim, pp. 229–238.

(56) Breimesser, M., Ritter, S., Seifert, H.P. *et al.* (2012) Application of electrochemical noise to monitor stress corrosion cracking of stainless steel in tetrathionate solution under constant load. *Corrosion Science*, **63**, 129–139.

(57) Wells, D.B., Stewart, J., Davidson, R. *et al.* (1992) The mechanism of intergranular stress-corrosion cracking of sensitized austenitic stainless-steel in dilute thiosulfate solution. *Corrosion Science*, **33**, 39–71.

(58) Stewart, J., Wells, D.B., Scott, P.M. and Williams, D.E. (1992) Electrochemical noise measurements of stress-corrosion cracking of sensitized austenitic stainless-steel in high-purity oxygenated water at 288°C. *Corrosion Science*, **33**, 73–88.

(59) Oltra, R., Chapey, B. and Renaud, L. (1995) Abrasion corrosion studies of passive stainless-steels in acidic media - combination of acoustic-emission and electrochemical techniques. *Wear*, **186**, 533–541.

(60) Berradja, A., Deforge, D., Nogueira, R.P. *et al.* (2006) An electrochemical noise study of tribocorrosion processes of AISI 304 L in Cl^- and SO_4^{2-} media. *Journal of Physics D - Applied Physics*, **39**, 3184–3192.

(61) Wood, R.J.K., Wharton, J.A., Speyer, A.J. and Tan, K.S. (2002) Investigation of erosion–corrosion processes using electrochemical noise measurements. *Tribology International*, **35**, 631–641.

(62) Deforge, D., Huet, F., Nogueira, R.P. *et al.* (2006) Electrochemical noise analysis of tribocorrosion processes under steady-state friction regime. *Corrosion*, **62**, 514–521.

(63) Lu, B.T., Luo, J.L., Mohammadi, F. *et al.* (2008) Correlation between repassivation kinetics and corrosion rate over a passive surface in flowing slurry. *Electrochimica Acta*, **53**, 7022–7031.

(64) Lu, B.T., Mao, L.C. and Luo, J.L. (2010) Hydrodynamic effects on erosion-enhanced corrosion of stainless steel in aqueous slurries. *Electrochimica Acta*, **56**, 85–92.

(65) Deslouis, C., Gil, O., Tribollet, B. *et al.* (1992) Oxygen as a tracer for measurements of steady and turbulent flows. *Journal of Applied Electrochemistry*, **22**, 835–842.

(66) Adolphe, X., Danaila, L. and Martemianov, S. (2007) On the small-scale statistics of turbulent mixing in electrochemical systems. *Journal of Electroanalytical Chemistry*, **600**, 119–130.

(67) Bouazaze, H., Huet, F. and Nogueira, R.P. (2005) A new approach for monitoring corrosion and flow characteristics in oil/brine mixtures. *Electrochimica Acta*, **50**, 2081–2090.

(68) Elkin, V.V., Grafov, B.M., Nekrasov, L.N. *et al.* (2002) Turbulent electrochemical noise: a theoretical analysis in the frequency-potential coordinates. *Russian Journal of Electrochemistry*, **38**, 199–207.

(69) Nekrasov, L.N., Khomchenko, T.N., Alekseev, V.N. *et al.* (1999) Electrochemical cell with a mechanical stirrer and stationary working electrode: an instrument for studying turbulent

noise and steady-state polarization characteristics of electrode processes. *Russian Journal of Electrochemistry*, **35**, 966–972.

(70) Lowe, A.M., Eren, H. and Bailey, S.I. (2003) Electrochemical noise analysis: detection of electrode asymmetry. *Corrosion Science*, **45**, 941–955.

(71) Gabrielli, C., Huet, F. and Keddam, M. (1991) Real-time measurement of electrolyte resistance fluctuations. *Journal of the Electrochemical Society*, **138**, L82–L84.

(72) Huet, F., Nogueira, R.P., Lailler, P. and Torcheux, L. (2006) Investigation of the high frequency resistance of a lead-acid battery. *Journal of Power Sources*, **158**, 1012–1018.

(73) Martinet, S., Durand, R., Ozil, P. *et al.* (1999) Application of electrochemical noise analysis to the study of batteries: state-of-charge determination and overvoltage detection. *Journal of Power Sources*, **83**, 93–99.

(74) Baert, D.H.J. and Vervaet, A.A.K. (2003) Small bandwidth measurement of the noise voltage of batteries. *Journal of Power Sources*, **114**, 357–365.

(75) Parkhutik, V., Patil, R., Harima, Y. and Matveyeva, E. (2006) Electrical conduction mechanism in conjugated polymers studied using flicker noise spectroscopy. *Electrochimica Acta*, **51**, 2656–1661.

(76) Xue, W., Jiang, X. and Harima, Y. (2010) Influence of electrochemical doping on low frequency noise of conducting poly(3-methylthiophene) film. *Synthetic Metals*, **160**, 803–807.

(77) Legros, B., Thivel, P.X., Bultel, Y. and Nogueira, R.P. (2011) First results on PEMFC diagnosis by electrochemical noise. *Electrochemistry Communications*, **13**, 1514–1516.

(78) Gabrielli, C. and Huet, F. (1994) Fluctuation analysis in electrochemical engineering processes with two-phase flows. *Journal of Applied Electrochemistry*, **24**, 593–601.

(79) Gabrielli, C., Huet, F., Wiart, R. and Zoppas-Ferreira, J. (1994) Dynamic behaviour of an electrolyser with a two-phase solid-liquid electrolyte: I. Spectral analysis of potential fluctuation. *Journal of Applied Electrochemistry*, **24**, 1228–1234.

(80) Gabrielli, C., Huet, F., Wiart, R. and Zoppas-Ferreira, J. (1994) Dynamic behaviour of an electrolyser with a two-phase solid-liquid electrolyte: II. Investigation of elementary phenomena and electrode modelling. *Journal of Applied Electrochemistry*, **24**, 1235–1243.

(81) Huet, F. and Nogueira, R.P. (2003) Comparative analysis of potential, current, and electrolyte resistance fluctuations in two-phase oil/water mixtures. *Corrosion*, **59**, 747–755.

(82) Bouazaze, H., Huet, F. and Nogueira, R.P. (2007) Measurement of electrolyte resistance fluctuations generated by oil-brine mixtures in a flow-loop cell. *Corrosion*, **63**, 307–317.

(83) Van Shaw, P. and Hanratty, T.J. (1964) Fluctuations in the local rate of turbulent mass transfer to a pipe wall. *AIChE Journal*, **10**, 475.

(84) Campbell, J.A. and Hanratty, T.J. (1983) Turbulent velocity fluctuations that control mass transfer to a solid boundary. *AIChE Journal*, **29**, 215–221.

(85) Deslouis, C., Huet, F., Gil, O. and Tribollet, B. (1993) Spectral analysis of wall turbulence with a bicircular electrochemical probe. *Experiments in Fluids*, **16**, 97–104.

(86) Seguin, D., Montillet, A., Comiti, J. and Huet, F. (1998) Experimental characterization of flow regimes in various porous media. II: transition to turbulent regime. *Chemical Engineering Science*, **53**, 3897–3909.

(87) Hodgson, D.R. (1996) Application of electrochemical noise and in-situ microscopy to the study of bubble evolution on chlorine evolving anodes. *Electrochimica Acta*, **41**, 605–609.

(88) Huet, F., Musiani, M. and Nogueira, R.P. (2003) Electrochemical noise analysis of O_2 evolution on PbO_2 and PbO_2-matrix composites containing Co or Ru oxides. *Electrochimica Acta*, **48**, 3981–3989.

(89) Gabrielli, C., Huet, F. and Keddam, M. (1985) Characterization of electrolytic bubble evolution by spectral analysis. Application to a corroding electrode. *Journal of Applied Electrochemistry*, **15**, 503–508.

(90) Gabrielli, C., Huet, F., Keddam, M. *et al.* (1989) Potential drops due to an attached bubble on a gas-evolving electrode. *Journal of Applied Electrochemistry*, **19**, 617–629.

(91) Gabrielli, C., Huet, F., Keddam, M. and Sahar, A. (1989) Investigation of water electrolysis by spectral analysis. I. Influence of the current density. *Journal of Applied Electrochemistry*, **19**, 683–696.

(92) Gabrielli, C., Huet, F. and Nogueira, R.P. (2002) Electrochemical noise measurements of coalescence and gas-oscillator phenomena on gas-evolving electrodes. *Journal of the Electrochemical Society*, **149**, E71–E77.
(93) Bouazaze, H., Cattarin, S., Huet, F. *et al.* (2006) Electrochemical noise study of the effect of electrode surface wetting on the evolution of electrolytic hydrogen bubbles. *Journal of Electroanalytical Chemistry*, **597**, 60–68.
(94) Huet, F., Musiani, M. and Nogueira, R.P. (2004) Oxygen evolution on electrodes of different roughness: an electrochemical noise study. *Journal of Solid State Electrochemistry*, **8**, 786–793.

12

From Microelectrodes to Scanning Electrochemical Microscopy

Salvatore Daniele[1] and Guy Denuault[2]
[1]*University of Venice, Dipartimento Scienze Molecolari e Nanosistemi, Italy*
[2]*University of Southampton, Chemistry, UK*

This chapter considers advances in electroanalytical chemistry which resulted from the development of microelectrodes, an area of research where Martin Fleischmann made significant theoretical and experimental contributions. Briefly, a microelectrode is defined as an electrode with at least one dimension sufficiently small (typically less than 50 μm) that its amperometric properties are a function of this characteristic length [1, 2]. Historically, the development of microelectrodes resulted from the needs of biologists to perform measurements in real biological systems. For this, tiny electrodes were required to operate *in-situ*, typically *in vivo*, and to offer localized recordings without affecting the integrity of the biological tissues. Hence, as early as 1942 electrophysiologists were developing micrometer-sized electrodes to determine, amperometrically, the concentration of dissolved oxygen in animal tissues [3]. Subsequently, efforts moved on to the development of potentiometric microelectrodes [4, 5], and it was not until the 1970s that the advantages of amperometry at micrometer-sized electrodes began to be fully recognized, primarily for the ability to perform voltammetry *in vivo*, especially in studies of neurotransmitters [6, 7]. Although, previously, the electrochemical fraternity had frequently referred to microelectrodes, these were in fact millimeter-sized electrodes; consequently, in order to avoid confusion the truly micrometer-sized electrodes were denoted ultramicroelectrodes (UMEs) during the early 1980s [8–10].

Developments in Electrochemistry: Science Inspired by Martin Fleischmann, First Edition.
Edited by Derek Pletcher, Zhong-Qun Tian and David E. Williams.
© 2014 John Wiley & Sons, Ltd. Published 2014 by John Wiley & Sons, Ltd.

Martin Fleischmann had already toyed with the idea of submillimetric electrodes during the late 1960s, but this was for potentiometric applications [11]. Subsequently, during the early 1980s, he truly embraced the amperometric applications and over a ten-year period went on to publish over 40 theoretical and experimental articles on the subject. Martin Fleischmann's experimental reports were essentially focused on the use of microelectrodes in nonconventional media, where the amount of electrolyte was very small, for analyses of the mechanism and kinetics of coupled chemical reactions, and for the study of electrocrystallization and electrodeposition. Although his theoretical investigations were broadly focused on the development of analytical expressions for steady-state amperometry, chronoamperometry and even AC amperometry at microdisc electrodes [12–22], he also applied the same mathematical insight to microelectrodes with other geometries, such as the microring [12–16, 18–21, 23].

In this chapter, two areas are considered where the unique properties of microelectrodes have had a significant impact: (i) the use of microelectrodes and arrays of microelectrodes in electroanalytical studies (in foodstuffs, in concentrated industrial solutions, analysis with minimal sample preparation), especially in combination with pulsed amperometric techniques; and (ii) in scanning electrochemical microscopy (SECM; note that the acronym is used for both the instrument and the technique).

12.1 The Contribution of Microelectrodes to Electroanalytical Chemistry

12.1.1 Advantages of Microelectrodes in Electroanalysis

The advantages offered by microelectrodes in analytical chemistry have been evident since the mid-1980s/early 1990s, when it was shown that simple analytical assays of electroactive species could be performed directly (i.e., without prior preparation steps) in real samples having different physical forms spanning from simple liquid systems to colloid and semi-solid samples [24–35]. These possibilities derive from the unique properties of microelectrodes such as: (i) the low ohmic drop; (ii) the high faradaic to capacitive current ratio; (iii) the high rate of diffusion to the microelectrode in the steady state; (iv) the rapid achievement of steady-state currents; (v) the requirement for only two-electrode electrochemical cells; and (vi) a need for only small-volume samples essential for analysis [1, 2, 9, 10]. The advantages offered by microelectrodes have been exploited in many fields of electroanalysis, including environmental [36, 37], food [38], biomedical [39–41], and material science areas [42].

The majority of measurements for electroanalysis with microelectrodes are recorded under steady-state conditions by using either chronoamperometry (CA), linear sweep voltammetry (LSV) or cyclic voltammetry (CV) [1, 2, 9, 10]. Moreover, to solve problems related to the selectivity between species with similar redox potentials, pulsed techniques such as differential pulse voltammetry (DPV) [1, 7, 43–45] and square-wave voltammetry (SWV) [1, 45–49] have been employed. The use of the latter technique also minimizes the influence of oxygen in aerated natural samples [47]. In order to enhance sensitivity in these measurements, fast-scan voltammetry (FSV) [50] or the accumulation of analytes onto an electrode surface has also been performed, in conjunction with stripping analysis (SA) [51].

Table 12.1 *Steady state and quasi-steady state equations for diffusion controlled currents at microelectrodes of various geometries.*

Microelectrode geometry	Steady-state equation
Disc	$i = 4nFDc^*a$, where a = disc radius
Hemisphere	$i = 2\pi nFDc^*r$, where r = hemisphere radius
Sphere cap	$i = knFDc^*a$ ($4 < k < 2\pi \ln 2$), where a = radius of the base microdisk
Cylinder	$i = \dfrac{2nFADc^*}{r_0 \ln(4Dt/r_0^2)}$, where r_0 = cylinder radius
Band	$i = \dfrac{2\pi nFADc^*}{w \ln(64Dt/w^2)}$, where w = band width

i = steady-state current; n = number of electrons; F = Faraday constant; D = diffusion coefficient; c^* = bulk concentration. A = surface area

Fast-scan voltammetry has largely been developed for biological applications [50, 52], and employs scan rates up to kV s^{-1}. It has also been used for the detection of various anions and cations on submillisecond timescales [32, 53–56].

Stripping analysis is probably one field where microelectrodes find the largest number of applications [24–27, 29, 51, 54–56]. In particular, the enhanced mass transport to the microelectrode surface by diffusion can obviate the need for convective mass transport during the initial preconcentration step of the species to be analyzed on the electrode surface. Similarly, an enhanced diffusion rate leads to current responses being less affected by convective forces in flowing systems [51]. In fact, in quiescent solutions a steady-state current is established in a relatively short time for microelectrodes with disc, shrouded-hemisphere and sphere-cap geometries [1, 2, 9, 10]. In the case of microelectrodes which are not small enough in all of their dimensions (as with cylinders and bands), the current response attains only a quasi-steady state as the equations for their currents contain time-dependent terms, even at long times [1, 2, 9, 10] (see Table 12.1). Mass transport properties need also to be considered when optimizing analytical procedures, in order to achieve the best performance in terms of reproducibility and preconcentration efficiency for trace element analysis [51]. Natural convection, which may occur during relatively lengthy preconcentration step experiments, has no effect on the stripping responses at disc and sphere-cap microelectrodes [51], though some effects were observed with microwires [51].

Because the ohmic drop has a minimal influence on voltammetric responses, the addition of supporting electrolytes to the solutions is often unnecessary [1, 2, 9, 10]. This helps to avoids the contamination of samples by external chemicals when ultra-trace element analysis needs to be performed, and also leaves the chemical equilibria unaltered. Consequently, direct measurements of low-ionic strength samples or resistive media and speciation measurements can be performed in a straightforward manner, without any need for pretreatment [57, 58].

A lack of sufficient electrolyte in the media, however, makes the dependence of current on the concentration of electroactive species nonlinear, and interpreting the results requires that the migration of electroactive species/product must be considered [1]. Several reports and reviews have covered the theoretical problems related to the modeling of steady-state voltammograms at microelectrodes without or with only dilute supporting electrolyte [59]. Fundamental studies describing the combined effects of diffusion and migration at microelectrodes have provided a greater understanding and have also facilitated the prediction of amperometric experimental responses in complex systems, such as solutions of polyelectrolytes, large polymer molecules with one or more ionic groups per monomer unit, colloidal suspensions, and polymeric gels [59]. Moreover, migration coupled with homogeneous equilibrium and voltammetry in undiluted liquid organic substances has also been investigated from both theoretical and experimental points of view [59–64].

The further advantages of microelectrodes derive from their small size, such that micro-electrode systems are highly suited to assaying extremely small sample volumes [42, 65], and to detecting species on single biological cells [35] and biologically important analytes *in vivo* (e.g., neurotransmitters) [66, 67]. Microelectrodes are also suitable for miniaturization, allowing the development of portable analytical instruments for *in-situ* or on-site measurements of trace elements [58].

Microelectrode arrays [1, 2, 9, 10, 36–38, 41] are especially attractive for analysis, as the low current associated with a single microelectrode is amplified many times, sometimes by several orders of magnitude. This means that instrumental difficulties in detecting very low currents can be avoided while any benefits associated with the microelectrode configuration will be retained. In order to achieve a steady state (or pseudo steady-state, depending on the geometry of the single microelectrode), however, the dimensions of each electrode in the array must be considered, as well as its shape and the interelectrode spacing. Theoretical studies of arrays (or ensembles) of microelectrodes have been carried out to predict the current responses and voltammetric profiles of microelectrode arrays of discs, recessed discs, hemispherical, band and hemicylindrical, interdigitated configurations [28], and to establish the optimal electrode packing density [51, 68, 69]. Because of the complexity of these tasks, numerical simulations have often been used [68, 69].

12.1.2   Microelectrodes and Electrode Materials

Various geometries of bare solid microelectrodes fabricated from carbon, gold, platinum, silver and silver–copper alloys have been used to detect a variety of analytes in synthetic laboratory aqueous solutions, simulated real matrices (e.g., body fluids, saliva, sweat), and also "real" samples such as natural waters, sediments, wine, distillates, foodstuffs and pharmaceutical preparations [1, 2, 9, 10, 51]. Unfortunately, multielement analysis is often complicated by the formation of intermetallics and multiple stripping peaks (due to the formation of monolayers or multilayers, or of bulk metal deposits) [51]. In addition, background currents from processes such as hydrogen evolution, oxygen reduction and oxide formation/reduction often limit the application of microelectrodes in trace analysis [51]. Mercury microelectrodes have been widely used in stripping analysis [51, 58], and continue to be used especially for trace element analysis. For the latter applications mercury is the material of choice, based on its well-known properties such as a very smooth surface area and high hydrogen overvoltage [51], which also make mercury ideal for analyses in the negative potential region in aqueous media [51]. Stationary mercury microelectrodes are usually prepared by electrodepositing mercury onto metal microelectrode substrates [51] such that, depending on the nature and shape of the substrate, the resulting mercury deposit will form either a uniform mercury film or randomly dispersed small droplets. Carbon fibers, carbon microdiscs, platinum, gold, silver and iridium microdiscs and gold microwires have most frequently been employed as substrate materials [51]. The mercury deposit on inlaid microdiscs of materials that are wettable by mercury (i.e., Pt, Ir, Au, Ag) is usually in the shape of a sphere cap, with its basal plane coincident with the original inlaid disc [51]. Because of the well-known and controllable sphere-cap geometry, procedures that do not require calibration have been proposed for determining various metal ions as well as sulfide and other anions with mercury-coated platinum disc microelectrodes [51, 70, 71]. For *in-situ* applications, Nafion or gel-coated mercury microelectrodes are used [58] to protect the mercury deposit against physical or chemical damage caused by chemical or biological interferants. Gel-coated mercury microelectrodes are also useful for speciation investigations [58].

Concern about the toxicity of mercury has stimulated the development of mercury-free electrodes for voltammetric measurements. Subsequently, the use of mercury-coated metal microelectrodes in stripping analysis led to a great reduction in the consumption of metallic mercury [51], with such microelectrodes having been shown superior to other electrode systems for some *in-situ* applications [72]. Since 2000, bismuth film electrodes (BiFEs) have been used as an alternative to mercury electrodes [73]; these consist of a thin metallic bismuth film electrodeposited onto a solid electrode from solutions containing bismuth ions. Bismuth film *micro*electrodes (BiFμEs) have also been prepared (here, the substrate material of choice is carbon fiber [73]) and fabricated on carbon microdiscs, carbon paste, gold, copper and platinum microdisks [73]. Alternative electrode materials to address the environmental problems related to the use and disposal of liquid mercury and dilute amalgams include solid metal alloys and, indeed, microelectrodes of silver–copper alloys have been proposed for use in stripping analysis for detecting species of biological interest [51, 74]. Boron-doped diamond microelectrodes have also been used directly for analytical applications, or as substrates for the deposition of Pt nanoparticles that are then used to detect a variety of analytes down to parts-per-billion levels, by stripping voltammetry [75]. Solid microelectrodes have also been used to detect several metal ions by exploiting the underpotential deposition phenomenon (UPD) [76] which, when used in conjunction with stripping analysis, has advantages over bulk-metal deposition. In particular, because of the limited amount of material deposited during UPD, the necessary preconcentration step is often short-lived.

12.1.3 New Applications of Microelectrodes in Electroanalysis

12.1.3.1 Microelectrodes as "Electronic Tongues" in Food Analysis

The simultaneous multicomponent analysis in liquids is an important task in analytical chemistry. Recently, so-called "electronic tongues" have emerged as an excellent alternative to traditional techniques for the evaluation of food quality and processes [77, 78]. Electronic tongue systems are based on arrays of low-selectivity sensors that are simultaneously sensitive to several components in the measured sample (cross-sensitivity). The signals collected by these sensors are processed by means of pattern recognition tools [78] in order to generate prediction models that allow classification of the samples and the quantification of some of their physico-chemical properties. Several variants of electronic tongues have been developed, including those using voltammetric techniques [78, 79]. Usually, the evaluation approach is based on the recording of an entire voltammogram in untreated samples, after which the entire voltammetric spectrum is evaluated by pattern recognition and chemiometric procedures. Working microelectrodes based on both single microelectrodes, and arrays there of, have been used to characterize wines and other alcoholic beverages [80–82] and edible oils [83], all of which are low-polarity samples. In the latter case, in order to provide a suitable conductivity to the media, small amounts of room-temperature ionic liquids were added to the oils as supporting electrolytes. Electronic tongues based on microelectrodes have also been developed to improve fuel quality control [84] and, in the field of forensics, to discriminate gunshot residues [85], thus providing information about the type of gun and ammunition used in a crime.

12.1.3.2   *Microelectrodes for Detecting and Characterizing Metal Nanoparticles*

Nanoparticles possess unique properties as a result of their size and shape, which makes them especially valuable for specific functions in research and industry. Advances in nanotechnology continue to offer many benefits to the fields of electronics, medicine and energy production, as well as numerous consumer goods. However, the increasing number of commercially available products containing metal nanoparticles, and the release of the latter into global water and air systems, raises concerns with regards to the environment and public health. In fact, recent studies have reported a significant toxicity of metal nanoparticles to human cells and aquatic organisms [86]. Consequently, in order to assess the risk posed by an increased exposure to metal nanoparticles, and their environmental fate, suitable detection techniques have been developed, with electrochemical methods based on nanoparticle collision with microelectrodes having found increasing applications in this field [87–93]. Microelectrodes, because of their small surface area, limit the number of metal nanoparticles that can interact with the active electrode surface, thus simplifying the analysis of the electrochemical responses. Various detection strategies have been exploited to obtain information on the concentration, size and degree of aggregation of nanoparticles. For instance, Pt and IrOx nanoparticles have been detected electrochemically by their characteristic current–time transients for a particle-catalyzed reaction due to an indicator species (proton, hydrogen peroxide, water) present in solution [88–90]. Other detection strategies are based on the impact frequency of metal nanoparticles with the microelectrode, which is held at a constant potential where the oxidation process on the metal nanoparticle can occur. Collisions are observed as sharp peaks with durations of 2–10 ms. The oxidation charge involved in the peak is then used to obtain information on concentration, size and degree of aggregation. This approach has been used for the detection of Ag, Ni, and Au nanoparticles [91–93].

12.1.3.3   *High-Surface-Area Microelectrodes in Electroanalysis*

Microelectrodes with high real surface areas and well-defined periodic nanostructures have recently attracted much interest because of their potential applications in electrocatalysis and electroanalysis [94–96]. These electrode systems can be prepared, using templating techniques, from lyotropic liquid crystalline phases of nonionic surfactants [94, 95]. In particular, the normal topology hexagonal (H_1) liquid crystalline phase has been used as a template for the synthesis of mesoporous metal thin films via the electrochemical reduction of metal salts dissolved in the aqueous domain of the liquid crystalline phases [119, 120].

The pore diameters of the H_1-Pt films reflect those of the liquid crystal structure, and typically range between 2 and 5 nm. Depending on the thickness of the Pt films, which can be controlled by the deposition charge, the real surface areas of the electrodes may be up to two or three orders of magnitude greater than those of the bare electrodes. Nanoporous microelectrodes with a high surface area and an open pore network have also been prepared by a potential-modulated electrochemical alloying–dealloying procedure in ionic liquid [97]. All of the above types of microelectrode are advantageous since, in spite of the very large surface area, they retain the efficient mass transport characteristics of the bare microelectrodes. These electrode systems have been employed for the detection of hydrogen peroxide [98, 99], nitrite [97], oxygen [100, 101], formic acid [102], metal ions, such as Cu, Ag, Pb [103] and Bi [104] via UPD-anodic stripping voltammetry, and glucose [105]. In the latter case, the mesoporous platinum microelectrodes were shown to display a high performance towards the electro-oxidation of glucose, with low interference due to other compounds (e.g., ascorbic acid) which undergo electro-oxidation processes in the same potential ranges. Microsensors have also been created by electrodepositing mesoporous Pd films onto Pt microdiscs, electrochemically loading the films with hydrogen to form the $\alpha + \beta$ Pd hydride phase, and then switching to a potentiometric mode to monitor pH [95].

12.1.3.4 Application of Microelectrodes to Concentrated Solutions for Investigations in Real Industrial Liquors

Within the electroanalytical community, investigations of electrode processes by voltammetry are often performed in diluted solutions or in relatively simple (synthetic) media. In this way, experimental results can easily be compared with theory, where models exist. This dilution approach has often been extended to the study of industrial processes or the monitoring of compounds in industrial electrolytes; information, however, can be lost or distorted by the dilution process. Difficulties in carrying out voltammetric measurements in real and industrial samples derive from the fact that such media often involve highly concentrated solutions of electroactive species and, consequently, high current densities; under these conditions, conventional electrodes do not provide easy-to-interpret and reproducible results. The use of microelectrodes allows data to be obtained free from any significant IR drop. The good performance displayed by microelectrodes for such situations is briefly illustrated with the following four examples:

- The electroreduction of copper(II) ions has been studied directly in refinery electrolytes to obtain information on the rate of nucleation process at a series of gold microelectrodes [106].
- A number of nickel-based electrocatalysts have been investigated as potential oxygen evolution catalysts under conditions close to those met in modern, high-current density alkaline water electrolysers. The catalysts were deposited onto nickel or stainless steel microelectrodes as working electrodes [107].
- The reactions occurring at the cathode in the hydrodimerization of acrylonitrile to adiponitrile have been investigated in conditions close to those applied in the commercial electrolysis process [108].
- Using platinum microelectrodes, the content of acetic acid and free hydrogen ions, which could come from sample adulteration, was monitored by steady-state voltammetry

directly in diluted vinegar samples [109]. Because the approach is based on recording linear sweep voltammetry and followed by the mathematical analysis of the responses, the lack of significant effects due to IR drop made data analysis reliable.

12.2   Scanning Electrochemical Microscopy (SECM)

12.2.1   A Brief History of SECM

As a result of the very small steady-state diffusion layer thickness at their microelectrode, Davies and Brink [3] were able to make localized amperometric recordings in biological tissues with a spatial resolution of approximately 25 μm. Although localized potentiometric measurements had been developed since the 1940s for probing localized corrosion events [110–112], the concept of localized amperometric measurements remained dormant until the mid-1980s when Engstrom and coworkers reported the use of a microelectrode to probe the time- and distance-dependence of the concentration of a species involved in a redox process at a larger electrode [113–115]. A few years later, this approach was coined the substrate generation–tip collection (SG-TC) mode of SECM. Bard and coworkers were the first to describe the use of a microelectrode tip held by a three-dimensional (3D) micropositioning system as a scanning electrochemical microscope [116], and later coined the acronym SECM [117]. The SECM evolved as a byproduct of two maturing fields, namely scanning tunneling microscopy (STM) and electroanalytical measurements with microelectrodes. Having quickly moved from surfaces under high vacuum to samples in air and liquids, STM experiments were soon conducted with the sample surface under potentiostatic control with the help of a bipotentiostat. This provided the trigger to turn the tip of the STM into an electrochemical probe and record faradaic currents instead of tunneling currents [116]. From this moment, the tip was behaving as a microelectrode and no longer needed to be an atomically sharp cone. With the aid of micropositioners, faradaic information could be recorded in one dimension by moving the tip in a direction normal to the sample surface, thereby producing an approach curve or in two dimensions by rastering the tip a few micrometers above the sample surface so as to produce an image of the tip response from the combination of all the line scans. Alternatively, the SECM tip could be used to perform a highly localized reaction by using it as a working electrode, and the sample surface as the counterelectrode; this approach was coined "direct mode" SECM [117]. The unique properties of the SECM were recognized with the development of the feedback mode [118–120], where a redox mediator shuttles charges between the tip and the substrate under diffusion control. In this mode, the SECM is akin to a microelectrochemical radar as the tip remotely interrogates the sample–solution interface and senses the kinetics of the regeneration of the mediator by the sample. The rate constant for this interfacial process can be estimated by comparing an experimental approach curve with a family of theoretical approach curves obtained from simulations [121]. The ability to sense the local kinetics of the interfacial process can be exploited by rastering the tip over the sample so as to produce a map of the tip current. In principle, this should yield the distribution of electrochemical activity at the sample surface, but in reality the tip current results from a convolution between the surface kinetics and the rate of mediator diffusion. As the latter is heavily dependent on the tip–substrate distance, any variation in surface topography will affect the tip current and it will become impossible to decouple the surface kinetics from the topography. The main approaches followed to circumvent this problem are briefly reviewed in the next paragraph.

12.2.2 SECM with Other Techniques

In many cases, the electrochemical signal at the tip is not sufficient to assess the tip–substrate distance independently from the interfacial process. In consequence, a variety of instrumental concepts have been investigated to maintain the tip–substrate distance constant, irrespective of the topography, and to operate the SECM in constant-distance mode. In practice, another measurement is performed – either alternatively or simultaneously, and usually with a separate sensor – to determine the tip–substrate distance and continually to adjust the tip position along the z-axis while scanning along the x- and y-axes over the sample.

12.2.2.1 Picking Mode

The picking mode involves using a hydrodynamic enhancement of the amperometric response to control the distance [122]. This approach was attractive because it only required an electrochemical signal from the tip, but was slow because the desired diffusion-controlled feedback tip current could only be recorded after the tip had been retracted far away (200 μm) then moved quickly (50 μm s^{-1}) back towards the sample to reach a preset convective enhancement of the current and, after a rest time sufficient for the signal to decay to its steady state. Moreover, this sequence had to be repeated at every point along the scan.

12.2.2.2 Tip Position Modulation

Tip position modulation with an autoswitching controller [123] was successfully employed to record an SECM image under constant current, irrespective of the conducting or insulating properties of the sample surface. This was made possible because the tip current increases above its bulk value when approaching a conductor, but decreases below its bulk value when approaching an insulator. While this approach is effective when imaging conducting islands distributed within an insulating matrix (or insulating islands within a conducting matrix), the detection loses sensitivity when the contrast in conductivity is not marked or when the sample is complex, as would be the case with a porous surface.

12.2.2.3 Shear-Force SECM

Adapted from the technology implemented in scanning near field optical microscopy, SECM with shear-force detection involves vibrating the SECM tip with a piezoelectric actuator and recording variations in resonance frequency when the tip is within a few hundred nanometers from the sample surface [124, 125]. This is a fast method when operated in real time during the scan, but it is tricky to implement in practice. The tip must be long and narrow to be appropriately flexible but, more importantly, the control loop parameters need to be frequently adjusted as the resonance and shear force properties vary with the tip dimensions, solution viscosity, and sometimes with the elasticity of the sample surface. Moreover, the parameters need to be readjusted if the tip is removed for polishing.

Several groups have nevertheless mastered this technique and report high-resolution SECM images. Shear-force SECM instruments are currently sold by two research groups, but this technology has not yet been implemented by the main instrument manufacturers. However, this may change as Etienne *et al.* have recently reported an automated shear-force SECM where the tip is polished at regular intervals during the scan, without loss of the control loop parameters [126].

12.2.2.4   SECM with Atomic Force Microscopy

The performance of SECM experiments in association with atomic force microscopy (AFM) is typically carried out using a commercial instrument. The concept involves converting the AFM tip into an electrode to operate as the SECM tip, and using the force signal to control its position above the sample surface [127–131]. The tip design (see discussion in Section 12.2.3) is critical and its fabrication technically challenging; these points currently limit the take-up of this method.

12.2.2.5   Intermittent-Contact SECM

Intermittent-contact SECM (IC-SECM) involves intermittently tapping the tip against the sample surface and then retracting the tip a fixed distance before performing the SECM measurement [132]. Whereas, shear-force SECM and AFM-SECM are difficult to implement and have been applied by only a few groups worldwide, IC-SECM is technically less challenging and has recently been integrated into instruments manufactured by Uniscan and commercialized by Biologic.

12.2.2.6   SECM with Scanning Ion Conductance Microscopy

SECM associated with scanning ion conductance microscopy (SICM) requires a double tip, on one side of which is a conventional microdisc electrode and on the other side is a narrow pipette filled with electrolyte and an electrode that measures ionic conductance through the mouth of the pipette with respect to another electrode in the bulk solution. When the pipette mouth is within one pipette tip radius away from the sample surface, the conductance varies sufficiently to be used as a control signal to maintain the z-position of the tip during the scans, thereby affording constant-distance SECM operations [133, 134]. This methodology is fast and apparently less-challenging to implement than shear force SECM, but it requires the fabrication of double-barrel tips in which one channel is left empty and the other is filled with a conventional microdisc.

12.2.2.7   SECM with AC Impedance

SECM with AC impedance relies on measuring the conductance of the solution between the microdisc tip and the counterelectrode in the bulk of the solution. It was shown that the conductance depends on the tip–substrate distance, in the same way as the diffusion-controlled current, and this can be exploited to accurately position the tip [135]. This approach has the merit of being applicable in the absence of a redox mediator, and is convenient when the tip is operated as a passive probe [136]. This methodology has been exploited by Wipf and coworkers to construct an AC-SECM capable of harnessing AC and DC amperometric signals to record feedback images at constant distance with respect to the sample surface [137]. The AC signal is used to continually adjust the tip z-position, so as to maintain the distance with respect to the substrate, while the DC signal is the diffusion-controlled feedback current.

12.2.3 Tip Geometries and the Need for Numerical Modeling

Traditionally, SECM experiments have employed microdisc tips, but with thin insulating sheaths in order to approach the substrate as closely as possible. One important consequence of this is that diffusion round the corner of the insulating sheath contributes significantly to the tip current [138], and two parameters are now required to define the geometry of an SECM tip: a, the microdisc radius; and R_g, the radius of the glass, the latter being the most common insulator. R_g typically ranges between $5 \times a$ and $10 \times a$, and is a key parameter that needs to be determined experimentally as it affects the magnitude of the limiting current at all tip–substrate distances. In the bulk, the traditional expression for the limiting current at a microdisc, $i_{T,\infty} = 4nFDc^\infty a$, needs to be replaced by $i_{T,\infty} = \beta nFDc^\infty a$, where $i_{T,\infty}$ is the tip current at infinite tip–substrate distance and β is a parameter which reflects the contribution of diffusion round the corner of the insulating sheath. Close to the substrate the tip geometry also plays a major role, as thick insulating sheaths significantly hinder the diffusion of redox species in and out of the tip–substrate gap. Other microelectrode geometries have also been investigated and applied to SECM. For example, microring tips [139] have found a particular niche in the SG-TC mode of SECM for the detection of species generated under illumination. Typically, the tip consists of an optical fiber coated with a thin metallic film; the former illuminates the sample while the latter collects species generated under illumination. This approach has been used to screen arrays of photocatalysts [140–143]. Microring tips have also been used to study oxygen reduction catalysts in a tip generation–substrate collection–tip collection mode [144]. SECM with sphere-cap tips fabricated by electrodepositing mercury films on microdisc tips have been successfully applied and theoretically investigated [145]. The combination of SECM with AFM has led to a variety of tip geometries, including fully conducting AFM cantilevers [127, 146], pyramidal conducting tips [128], a protruding AFM tip on the edge of a microdisc electrode [147], AFM tips with integrated boron-doped diamond electrodes [148], AFM tips surrounded by framed shaped electrodes [149], and partially insulated conducting conical AFM tips [150–152].

Most SECM developments have been underpinned by numerical modeling of the tip response for the particular tip geometry concerned, and for a range of tip-substrate distances. In several cases, simulations have been used to produce approximate expressions to quantify the magnitude of the tip current and predict the shape of the approach curves for the geometry concerned [121, 138, 153–156]. While the simulations were initially "home-made" using either finite difference or finite element numerical algorithms, most are now performed with COMSOL Multiphysics, a commercial finite element solver with a library of discipline-specific modules, including several for electrochemical processes. These simulations are typically conducted in 2D domains assuming an axi-symmetric geometry, but a few have been conducted in 3D domains devoid of symmetry [153, 157], and it is even possible to simulate the SECM feedback image for a particular tip geometry and particular sample surface [158].

12.2.4 Applications of SECM

Scanning electrochemical microscopy has been exploited in a large number of applications that are too numerous to review at this point. Hence, only a very small selection of studies will be considered in this section, where the unique properties of SECM have been harnessed in the most elegant ways to reveal hitherto unobserved phenomena.

12.2.4.1 SECM as a Tool for Imaging Chemical Reactivity

One of the most striking examples of the application of the concept of scanning electro-chemical imaging with a redox mediator was the mapping of precursor sites for pitting on Ti oxide layers [159–162]. The study results showed clearly the ability of the technique to image a localized chemical process (in this case, a breakdown in passivation of the oxide layer) before the onset of corrosion. It also showed that the choice of redox mediator was key to obtaining the image, as the detection of precursor sites depended on the redox potential of the mediator being positive of the conduction band-edge of the Ti oxide layer. These studies also showed the unique capability of SECM to probe spatial- and potential-dependent electron-transfer rates at semiconductor electrodes. The SECM has since been employed to image a variety of interfacial reactions, including biological and chemical processes [163–168]. Recently, the imaging capability was demonstrated on a large scale with the use of flexible tip arrays [169, 170].

12.2.4.2 SECM as Tool to Provide Variable Mass Transfer Conditions

In many studies, the SECM is operated in a direction normal to the substrate, simply to vary the rate of mass transport. Far away, the tip operates as an independent microelectrode and its steady-state mass transfer coefficient, $k_{m,\infty}$, is determined by the tip geometry. k_m can be fine-tuned to values below $k_{m,\infty}$ when approaching an inert substrate (hindered diffusion), or to values above $k_{m,\infty}$ when approaching a conducting substrate (positive feedback). In effect, the SECM can be operated as an instrument providing a variable steady-state mass transfer coefficient, akin to the rotating disc electrode but without the inherent problems of hydrodynamically controlled systems. Several early studies made use of this property to investigate heterogeneous and homogeneous kinetics [171]. This property was recently exploited to unravel the effect of dissolved oxygen on the potentiometric response of Pd hydride microdiscs [172].

12.2.4.3 SECM as a Tool to Investigate Lateral Charge Transfer

The technique has been used in several studies to probe how charges propagate at the surface of the substrate, so as to sustain the charge-transfer process between the substrate and the redox mediator. Mandler and Unwin [173] investigated lateral charge propagation along ultrathin polyaniline layers (monolayer and multilayers) and demonstrated how the transient mode of the SECM could be used to obtain the true rate of charge transport within the molecular film. Subsequently, the same team developed an accurate SECM methodology to extract the conductivity of thin films from feedback measurements [174].

12.3   Conclusions

When Martin Fleischmann first began experimenting with microdisc electrodes he had to rely on the dexterity of highly skilled glassblowers to prepare them, and the experiments required great skill. Nowadays, the electrodes are commercially available in different materials and in sizes down to 5 μm radius, such that voltammetry at microelectrodes has become a routine technique. Smaller radii do remain technically challenging and still need to be homemade, however. Fleischmann also relied on his mathematical skills to develop the theories needed to predict the microelectrode properties. Nowadays, simple commercial simulation software (e.g., DigiSim or DigiElch) can be used to predict voltammograms, chronoamperograms and impedance results for the microdisc under combinations of heterogeneous and homogeneous kinetics. These packages allow simulations involving complex homogeneous reaction mechanisms to be easily computed in a matter of seconds on a personal computer, without the need to write program codes. Simulations can even be performed for complex 2D or 3D geometries using commercially available finite element solvers, although the software requires good mathematics and physics skills. So, even theoretical challenges arising from microelectrode arrays, recessed microelectrodes, or the combined effects of migration and diffusion in 2D and 3D, can be computed without programming knowledge.

By enabling measurements to be made in real solutions without the need for sample pretreatment, microelectrodes have become routine electroanalytical tools in some areas of research. However, a brief perusal of the electrochemical literature reveals that they are still not as widely exploited as might have been expected during the early 1980s. Nonetheless, this situation should evolve as the two main hurdles have been removed: (i) microelectrodes are now commercially available; and (ii) basic microelectrode theories are now included in all new electrochemistry textbooks. Microelectrodes have had a huge impact on interfacial science by enabling SECM, which is now a mature field with commercially available dedicated instrumentation. Indeed, SECM has allowed a number of fundamental studies to be conducted and is now being exploited for applied investigations in corrosion and materials science.

Overall, significant advances in experimental and theoretical electroanalytical chemistry have been made as a result of the unique properties of microelectrodes. Martin Fleischmann's contribution in this area has been of great importance to these developments.

References

(1) Montenegro, M.I., Queirós, M.A. and Daschbach, J.L. (1991) *Microelectrodes: Theory and Applications*, vol. **197**, Kluwer Academic Publishers, p. 497.

(2) Stulik, K., Amatore, C., Holub, K. *et al.* (2000) Microelectrodes. Definitions, characterization, and applications (Technical Report). *Pure and Applied Chemistry*, **72**, 1483–1492.

(3) Davies, P.W. and Brink, F. Jr (1942) Microelectrodes for measuring local oxygen tension in animal tissues. *Review of Scientific Instruments*, **13**, 524–533.

(4) Hinke, J.A.M. (1959) Glass microelectrodes for measuring intracellular activities of sodium and potassium. *Nature*, **184**, 1257–8.

(5) Walker, J.L. (1971) Ion specific liquid ion exchanger microelectrodes. *Analytical Chemistry*, **43**, A89–A93.

(6) Dayton, M.A., Brown, J.C., Stutts, K.J. and Wightman, R.M. (1980) Faradaic electrochemistry at micro-voltammetric electrodes. *Analytical Chemistry*, **52**, 946–950.

(7) Ewing, A.G., Dayton, M.A. and Wightman, R.M. (1981) Pulse voltammetry with microvoltammetric electrodes. *Analytical Chemistry*, **53**, 1842–1847.

(8) Bond, A.M., Fleischmann, M. and Robinson, J. (1984) The construction and behavior of ultramicroelectrodes - investigations of novel electrochemical systems. *Journal of the Electrochemical Society*, **131**, C109–C109.

(9) Deakin, M.R., Wipf, D. and Wightman, R.M. (1986) Ultramicroelectrodes. *Journal of the Electrochemical Society*, **133**, C135–C135.

(10) Fleischmann, M., Pons, S., Rolison, D.R. and Schmidt, P.P. (1987) *Ultramicroelectrodes*, Datatech Systems Inc., Morganton, NC, p. 363.

(11) Fleischmann, M. and Hiddleston, J.N. (1968) A palladium-hydrogen probe electrode for use as a microreference electrode. *Journal of Physics E: Scientific Instruments*, **1**, 667–668.

(12) Fleischmann, M. and Pons, S. (1987) The behavior of microdisk and microring electrodes. *Journal of Electroanalytical Chemistry*, **222**, 107–115.

(13) Fleischmann, M., Daschbach, J. and Pons, S. (1988) The behavior of microdisk and microring electrodes - mass-transport to the disk in the unsteady state - chronoamperometry. *Journal of Electroanalytical Chemistry*, **250**, 269–276.

(14) Fleischmann, M. and Pons, S. (1988) The behavior of microdisk and microring electrodes - mass-transport to the disk in the unsteady state - chronopotentiometry. *Journal of Electroanalytical Chemistry*, **250**, 257–267.

(15) Fleischmann, M. and Pons, S. (1988) The behavior of microdisk and microring electrodes - mass-transport to the disk in the unsteady state - the AC response. *Journal of Electroanalytical Chemistry*, **250**, 277–283.

(16) Fleischmann, M. and Pons, S. (1988) The behavior of microdisk and microring electrodes - mass-transport to the disk in the unsteady state - the effects of coupled chemical-reactions - the Ce mechanism. *Journal of Electroanalytical Chemistry*, **250**, 285–292.

(17) Abrantes, L.M., Fleischmann, M., Li, L.J. *et al.* (1989) the behavior of microdisk electrodes - chronopotentiometry and linear sweep voltammetric experiments. *Journal of Electroanalytical Chemistry*, **262**, 55–66.

(18) Daschbach, J., Pons, S. and Fleischmann, M. (1989) The behavior of microdisk and microring electrodes - application of Neumann Integral Theorem to the prediction of the steady-state response of microdisks - numerical illustrations. *Journal of Electroanalytical Chemistry*, **263**, 205–224.

(19) Fleischmann, M., Daschbach, J. and Pons, S. (1989) The behavior of microdisk and microring electrodes - application of Neumann integral theorem to the prediction of the steady-state response of microdisks. *Journal of Electroanalytical Chemistry*, **263**, 189–203.

(20) Fleischmann, M., Pletcher, D., Denuault, G. *et al.* (1989) The behavior of microdisk and microring electrodes - prediction of the chronoamperometric response of microdisks and of the steady-state for Ce and Ec catalytic reactions by application of Neumann Integral Theorem. *Journal of Electroanalytical Chemistry*, **263**, 225–236.

(21) Li, L.J., Hawkins, M., Pons, J.W. *et al.* (1989) The behavior of microdisk and microring electrodes - the chronoamperometric response at microdisk and microring electrodes. *Journal of Electroanalytical Chemistry*, **262**, 45–53.

(22) Pons, S., Daschbach, J. and Fleischmann, M. (1989) The use of Neumann Integral Theorem to describe the behavior of microdisk electrodes. *Abstracts of Papers - American Chemical Society*, **197**, 79.

(23) Fleischmann, M., Bandyopadhyay, S. and Pons, S. (1985) The behavior of microring electrodes. *Journal of Physical Chemistry*, **89**, 5537–5541.

(24) Wehmeyer, K.R. and Wightman, R.M. (1985) Cyclic voltammetry and anodic-stripping voltammetry with mercury ultramicroelectrodes. *Analytical Chemistry*, **57**, 1989–1993.

(25) Li, L.J., Fleischmann, M. and Peter, L.M. (1987) *In-situ* measurements of Pb-$^{2+}$ concentration in the lead-acid-battery using mercury ultramicroelectrodes. *Electrochimica Acta*, **32**, 1585–1587.

(26) Daniele, S., Baldo, M.A., Ugo, P. and Mazzocchin, G.A. (1989) Determination of heavy-metals in real samples by anodic-stripping voltammetry with mercury microelectrodes. 1. Application to wine. *Analytica Chimica Acta*, **219**, 9–18.

(27) Daniele, S., Baldo, M.A., Ugo, P. and Mazzocchin, G.A. (1989) Determination of heavy-metals in real samples by anodic-stripping voltammetry with mercury microelectrodes. 2. Application to rain and sea waters. *Analytica Chimica Acta*, **219**, 19–26.

(28) Daniele, S., Baldo, M.A., Ugo, P. and Mazzocchin, G.A. (1990) Voltammetric probe of milk samples by using a platinum microelectrode. *Analytica Chimica Acta*, **238**, 357–366.

(29) Wojciechowski, M. and Balcerzak, J. (1991) Square-wave anodic-stripping voltammetry of lead and cadmium at cylindrical graphite fiber microelectrodes with in-situ plated mercury films. *Analytica Chimica Acta*, **249**, 433–445.

(30) Stulik, K. (1989) Some aspects of flow electroanalysis. *Analyst*, **114**, 1519–1525.

(31) LaCourse, W.R. and Modi, S.J. (2005) Microelectrode applications of pulsed electrochemical detection. *Electroanalysis*, **17**, 1141–1152.

(32) Harman, A.R. and Baranski, A.S. (1990) Fast cathodic stripping analysis with ultramicroelectrodes. *Analytica Chimica Acta*, **239**, 35–44.

(33) Craston, D.H., Jones, C.P., Williams, D.E. and Elmurr, N. (1991) Microband electrodes fabricated by screen printing processes - applications in electroanalysis. *Talanta*, **38**, 17–26.

(34) Farrington, A.M., Jagota, N. and Slater, J.M. (1994) Simple solid wire microdisk electrodes for the determination of vitamin-C in fruit juices. *Analyst*, **119**, 233–238.

(35) Bixler, J.W. and Bond, A.M. (1986) Amperometric detection of picomole samples in a microdisk electrochemical flow-jet cell with dilute supporting electrolyte. *Analytical Chemistry*, **58**, 2859–2863.

(36) Davis, F. and Higson, S.P.J. (2013) Arrays of microelectrodes: technologies for environmental investigations. *Environmental Science: Processes & Impacts*, **15**, 1477–1489.

(37) Tan, F., Metters, J.P. and Banks, C.E. (2013) Electroanalytical applications of screen printed microelectrode arrays. *Sensors and Actuators B: Chemical*, **181**, 454–462.

(38) Peckova, K. and Barek, J. (2011) boron doped diamond microelectrodes and microelectrode arrays in organic electrochemistry. *Current Organic Chemistry*, **15**, 3014–3028.

(39) Marinesco, S. and Frey, O. (2013) Microelectrode designs for oxidase-based biosensors, in *Neuromethods* (eds S.Marinesco and N.Dale), vol. **80**, *Microelectrode Biosensors*, pp. 3–25.

(40) Sun, X.P., Luo, Y.L., Liao, F. *et al.* (2011) Novel nanotextured microelectrodes: electrodeposition-based fabrication and their application to ultrasensitive nucleic acid detection. *Electrochimica Acta*, **56**, 2832–2836.

(41) Ordeig, O., del Campo, J., Munoz, F.X. *et al.* (2007) Electroanalysis utilizing amperometric microdisk electrode arrays. *Electroanalysis*, **19**, 1973–1986.

(42) Li, C.M. and Hu, W.H. (2013) Electroanalysis in micro- and nano-scales. *Journal of Electroanalytical Chemistry*, **688**, 20–31.

(43) Molina, A., Laborda, E., Martinez-Ortiz, F. *et al.* (2011) Comparison between double pulse and multipulse differential techniques. *Journal of Electroanalytical Chemistry*, **659**, 12–24.

(44) Lovric, M. (1999) Differential pulse voltammetry on spherical microelectrodes. *Electroanalysis*, **11**, 1089–1093.

(45) daSilva, O.B. and Machado, S.A.S. (2012) Evaluation of the detection and quantification limits in electroanalysis using two popular methods: application in the case study of paraquat determination. *Analytical Methods-UK*, **4**, 2348–2354.

(46) Osteryoung, J.G. and O'Dea, J. (1987) Squarewave voltammetry, in *Electroanalytical Chemistry* (ed. A.J. Bard), Marcel Dekker, New York, p. 209.

(47) Wojciechowski, M., Go, W. and Osteryoung, J. (1985) Square-wave anodic-stripping analysis in the presence of dissolved-oxygen. *Analytical Chemistry*, **57**, 155–158.

(48) Bragato, C., Daniele, S. and Baldo, M.A. (2005) Low frequency square-wave voltammetry of weak acids at platinum microelectrodes. *Electroanalysis*, **17**, 1370–1378.

(49) Daniele, S., Bragato, C. and Baldo, M.A. (2002) Square wave voltammetry of strong acids at platinum microelectrodes. *Electrochemistry Communications*, **4**, 374–378.

(50) Bunin, M.A. and Wightman, R.M. (1998) Quantitative evaluation of 5-hydroxytryptamine (serotonin) neuronal release and uptake: an investigation of extrasynaptic transmission. *Journal of Neuroscience*, **18**, 4854–4860.

(51) Daniele, S., Baldo, M.A. and Bragato, C. (2008) Recent developments in stripping analysis on microelectrodes. *Current Analytical Chemistry*, **4**, 215–228.

(52) Takmakov, P., McKinney, C.J., Carelli, R.M. and Wightman, R.M. (2011) Instrumentation for fast-scan cyclic voltammetry combined with electrophysiology for behavioral experiments in freely moving animals. *Review of Scientific Instruments*, **82**, O74302.

(53) Wood, K.M. and Hashemi, P. (2013) Fast-scan cyclic voltammetry analysis of dynamic serotonin responses to acute escitalopram. *ACS Chemical Neuroscience*, **4**, 715–720.

(54) Alpuche-Aviles, M.A., Baur, J.E. and Wipf, D.O. (2008) Imaging of metal ion dissolution and electrodeposition by anodic stripping voltammetry-scanning electrochemical microscopy. *Analytical Chemistry*, **80**, 3612–3621.

(55) Munteanu, G., Munteanu, S. and Wipf, D.O. (2009) Rapid determination of zeptomole quantities of Pb^{2+} with the mercury monolayer carbon fiber electrode. *Journal of Electroanalytical Chemistry*, **632**, 177–183.

(56) Yang, Y.Y., Pathirathna, P., Siriwardhane, T. *et al.* (2013) Real-time subsecond voltammetric analysis of Pb in aqueous environmental samples. *Analytical Chemistry*, **85**, 7535–7541.

(57) Sigg, L., Black, F., Buffle, J. *et al.* (2006) Comparison of analytical techniques for dynamic trace metal speciation in natural freshwaters. *Environmental Science and Technology*, **40**, 1934–1941.

(58) Buffle, J. and Tercier-Waeber, M.L. (2005) Voltammetric environmental trace-metal analysis and speciation: from laboratory to in situ measurements. *Trends in Analytical Chemistry*, **24**, 172–191.

(59) Ciszkowska, M. and Stojek, Z. (1999) Voltammetry in solutions of low ionic strength. Electrochemical and analytical aspects. *Journal of Electroanalytical Chemistry*, **466**, 129–143.

(60) Daniele, S. and Mazzocchin, G.A. (1993) Stripping analysis at mercury microelectrodes in the absence of supporting electrolyte. *Analytica Chimica Acta*, **273**, 3–11.

(61) Pena, M.J., Fleischmann, M. and Garrard, N. (1987) Voltammetric measurements with microelectrodes in low-conductivity systems. *Journal of Electroanalytical Chemistry*, **220**, 31–40.

(62) Oldham, K.B. (1988) Theory of microelectrode voltammetry with little electrolyte. *Journal of Electroanalytical Chemistry*, **250**, 1–21.

(63) Oldham, K.B. (1997) Limiting currents for steady-state electrolysis of an equilibrium mixture, with and without supporting inert electrolyte. *Analytical Chemistry*, **69**, 446–453.

(64) Daniele, S., Baldo, M.A., Bragato, C. *et al.* (2002) Steady-state voltammetry of hydroxide ion oxidation in aqueous solutions containing ammonia. *Analytical Chemistry*, **74**, 3290–3296.

(65) Xu, X.L., Zhang, S., Chen, H. and Kong, J.L. (2009) Integration of electrochemistry in micro-total analysis systems for biochemical assays: recent developments. *Talanta*, **80**, 8–18.

(66) Zhang, D.A., Rand, E., Marsh, M. *et al.* (2013) carbon nanofiber electrode for neurochemical monitoring. *Molecular Neurobiology*, **48**, 380–385.

(67) Jaquins-Gerstl, A. and Michael, A.C. (2013) The advantage of microelectrode technologies for measurement in delicate biological environments such as brain tissue, in *Neuromethods* (eds S. Marinesco and N. Dale), vol. **80**, *Microelectrode Biosensors*, pp. 55–68.

(68) Prehn, R., Abad, L., Sanchez-Molas, D. *et al.* (2011) Microfabrication and characterization of cylinder micropillar array electrodes. *Journal of Electroanalytical Chemistry*, **662**, 361–370.

(69) Godino, N., Borrise, X., Munoz, F.X. *et al.* (2009) Mass transport to nanoelectrode arrays and limitations of the diffusion domain approach: theory and experiment. *Journal of Physical Chemistry C*, **113**, 11119–11125.

(70) Souto, R.M., Gonzalez-Garcia, Y., Battistel, D. and Daniele, S. (2012) *In-situ* Scanning Electrochemical Microscopy (SECM) detection of metal dissolution during zinc corrosion by means of mercury sphere-cap microelectrode tips. *Chemistry - A European Journal*, **18**, 230–236.

(71) Abdelsalam, M.E., Denuault, G. and Daniele, S. (2002) Calibrationless determination of cadmium, lead and copper in rain samples by stripping voltammetry at mercury microelectrodes - effect of natural convection on the deposition step. *Analytica Chimica Acta*, **452**, 65–75.

(72) Waite, T.J., Kraiya, C., Trouwborst, R.E. *et al.* (2006) An investigation into the suitability of bismuth as an alternative to gold-amalgam as a working electrode for the in situ determination of chemical redox species in the natural environment. *Electroanalysis*, **18**, 1167–1172.

(73) Svancara, I., Prior, C., Hocevar, S.B. and Wang, J. (2010) A decade with bismuth-based electrodes in electroanalysis. *Electroanalysis*, **22**, 1405–1420.

(74) Mikkelsen, O., Strasunskiene, K., Skogvold, S.M. and Schroder, K.H. (2008) Solid alloy electrodes in stripping voltammetry. *Current Analytical Chemistry*, **4**, 202–205.

(75) Jones, S.E.W. and Compton, R.G. (2008) Stripping analysis using boron-doped diamond electrodes. *Current Analytical Chemistry*, **4**, 170–176.

(76) Herzog, G. and Arrigan, D.W.M. (2005) Determination of trace metals by underpotential deposition-stripping voltammetry at solid electrodes. *Trends in Analytical Chemistry*, **24**, 208–217.

(77) Ciosek, P. and Wroblewski, W. (2007) Sensor arrays for liquid sensing - electronic tongue systems. *Analyst*, **132**, 963–978.

(78) Winquist, F., Wide, P. and Lundstrom, I. (1997) An electronic tongue based on voltammetry. *Analytica Chimica Acta*, **357**, 21–31.

(79) Twomey, K., de Eulate, E.A., Alderman, J. and Arrigan, D.W.M. (2009) Fabrication and characterization of a miniaturized planar voltammetric sensor array for use in an electronic tongue. *Sensors and Actuators B: Chemical*, **140**, 532–541.

(80) Novakowski, W., Bertotti, M. and Paixao, T.R.L.C. (2011) Use of copper and gold electrodes as sensitive elements for fabrication of an electronic tongue: discrimination of wines and whiskies. *Microchemical Journal*, **99**, 145–151.

(81) Ceto, X., Gutierrez, J.M., Gutierrez, M. *et al.* (2012) Determination of total polyphenol index in wines employing a voltammetric electronic tongue. *Analytica Chimica Acta*, **732**, 172–179.

(82) Gutierrez, M., Llobera, A., Ipatov, A. *et al.* (2011) Application of an E-tongue to the analysis of monovarietal and blends of white wines. *Sensors-Basel*, **11**, 4840–4857.

(83) Oliveri, P., Baldo, M.A., Daniele, S. and Forina, M. (2009) Development of a voltammetric electronic tongue for discrimination of edible oils. *Analytical and Bioanalytical Chemistry*, **395**, 1135–1143.

(84) Wiziack, N.K.L., Paterno, L.G., Fonseca, F.J. and Mattoso, L.H.C. (2011) A combined gas and liquid chemical sensor array for fuel adulteration detection. *AIP Conference Proceedings*, **1362**, 178–179.

(85) Salles, M.O., Bertotti, M. and Paixao, T.R.L.C. (2012) Use of a gold microelectrode for discrimination of gunshot residues. *Sensors and Actuators B: Chemical*, **166**, 848–852.

(86) Suresh, A.K., Pelletier, D.A. and Doktycz, M.J. (2013) Relating nanomaterial properties and microbial toxicity. *Nanoscale*, **5**, 463–474.

(87) Boika, A., Thorgaard, S.N. and Bard, A.J. (2013) Monitoring the electrophoretic migration and adsorption of single insulating nanoparticles at ultramicroelectrodes. *Journal of Physical Chemistry B*, **117**, 4371–4380.

(88) Xiao, X.Y. and Bard, A.J. (2007) Observing single nanoparticle collisions at an ultramicroelectrode by electrocatalytic amplification. *Journal of the American Chemical Society*, **129**, 9610–9612.

(89) Kwon, S.J., Fan, F.R.F. and Bard, A.J. (2010) Observing iridium oxide (IrOx) single nanoparticle collisions at ultramicroelectrodes. *Journal of the American Chemical Society*, **132**, 13165–13167.

(90) Kwon, S.J., Zhou, H.J., Fan, F.R.F. *et al.* (2011) Stochastic electrochemistry with electrocatalytic nanoparticles at inert ultramicroelectrodes-theory and experiments. *Physical Chemistry Chemical Physics*, **13**, 5394–5402.

(91) Stuart, E.J.E., Rees, N.V., Cullen, J.T. and Compton, R.G. (2013) Direct electrochemical detection and sizing of silver nanoparticles in seawater media. *Nanoscale*, **5**, 174–177.

(92) Stuart, E.J.E., Zhou, Y.G., Rees, N.V. and Compton, R.G. (2012) Determining unknown concentrations of nanoparticles: the particle-impact electrochemistry of nickel and silver. *RSC Advances*, **2**, 6879–6884.

(93) Zhou, Y.G., Rees, N.V., Pillay, J. *et al.* (2012) Gold nanoparticles show electroactivity: counting and sorting nanoparticles upon impact with electrodes. *Chemical Communications*, **48**, 224–226.

(94) Elliott, J.M., Birkin, P.R., Bartlett, P.N. and Attard, G.S. (1999) Platinum microelectrodes with unique high surface areas. *Langmuir*, **15**, 7411–7415.

(95) Imokawa, T., Williams, K.-J. and Denuault, G. (2006) Fabrication and characterization of nanostructured Pd hydride pH microelectrodes. *Analytical Chemistry*, **78**, 265–271.

(96) Szamocki, R., Velichko, A., Holzapfel, C. *et al.* (2007) Macroporous ultramicroelectrodes for improved electroanalytical measurements. *Analytical Chemistry*,**79**, 533–539.

(97) Jiang, J., Wang, X. and Zhang, L. (2013) Nanoporous gold microelectrode prepared from potential modulated electrochemical alloying–dealloying in ionic liquid. *Electrochimica Acta*,**111**, 114–119.

(98) Evans, S.A.G., Elliott, J.M., Andrews, L.M. *et al.* (2002) Detection of hydrogen peroxide at mesoporous platinum microelectrodes. *Analytical Chemistry*, **74**, 1322–1326.

(99) Mokrushina, A.V., Heim, M., Karyakina, E.E. *et al.* (2013) Enhanced hydrogen peroxide sensing based on Prussian Blue modified macroporous microelectrodes. *Electrochemistry Communications*, **29**, 78–80.

(100) Birkin, P.R., Elliott, J.M., Watson, Y.E. (2000) Electrochemical reduction of oxygen on mesoporous platinum microelectrodes. *Chemical Communications*, 1693–1694.

(101) Prien, R.D., Pascal, R.W., Attard, G.S. *et al.* (2001) Development and first Results of a new mesoporous Microelectrode DO- Sensor. Oceans 2001 MTS/IEEE: An Ocean Odyssey, Vols 1–4, Conference Proceedings, 5–8 November, pp. 1910–1914.

(102) Daniele, S., Bragato, C. and Battistel, D. (2012) Bismuth-coated mesoporous platinum microelectrodes as sensors for formic acid detection. *Electroanalysis*, **24**, 759–766.

(103) Sanchez, P.L. and Elliott, J.M. (2005) Underpotential deposition and anodic stripping voltammetry at mesoporous microelectrodes. *The Analyst*, **130**, 715–720.

(104) Battistel, D. and Daniele, S. (2013) Determination of trace bismuth by under-potential deposition-stripping voltammetry at mesoporous platinum microelectrodes: application to pharmaceutical products. *Journal of Solid State Electrochemistry*, **17**, 1509–1516.

(105) Daniele, S., Battistel, D., Bergamin, S. and Bragato, C. (2010) voltammetric determination of glucose at bismuth-modified mesoporous platinum microelectrodes. *Electroanalysis*, **22**, 1511–1518.

(106) Lukomska, A., Plewka, A. and Los, P. (2009) Electroreduction of cupric(II) ions at the ultramicroelectrodes from concentrated electrolytes - Comparison of industrial and laboratory prepared aqueous solutions of copper(II) ions in sulfuric acid electrolytes. *Journal of Electroanalytical Chemistry*, **633**, 92–98.

(107) Li, X.H., Walsh, F.C. and Pletcher, D. (2011) Nickel-based electrocatalysts for oxygen evolution in high current density, alkaline water electrolysers. *Physical Chemistry Chemical Physics*, **13**, 1162–1167.

(108) Watson, M., Pletcher, D. and Sopher, D.W. (2000) A microelectrode study of competing electrode reactions in the commercial process for the hydrodimerization of acrylonitrile to adiponitrile. *Journal of the Electrochemical Society*, **147**, 3751–3758.

(109) Daniele, S., Bragato, C. and Baldo, M.A. (2006) A steady-state voltammetric procedure for the determination of hydrogen ions and total acid concentration in mixtures of a strong and a weak monoprotic acid. *Electrochimica Acta*, **52**, 54–61.

(110) Evans, U.R. (1940) Report on corrosion research work at Cambridge University interrupted by the outbreak of war. *Journal of the Iron and Steel Institute*, **141**, 219–234.

(111) Isaacs, H.S. and Kissel, G. (1972) Surface preparation and pit propagation in stainless-steels. *Journal of the Electrochemical Society*, **119**, 1628–1632.

(112) Isaacs, H.S. and Vyas, B. (1981) Scanning reference electrode techniques in localized corrosion, in *Electrochemical Corrosion Testing*, vol. **ASTM STP 727** (ed. U.B.Florian Mansfeld), American Society for Testing and Materials, pp. 3–33.

(113) Engstrom, R.C., Weber, M., Wunder, D.J. *et al.* (1986) Measurements within the diffusion layer using a microelectrode probe. *Analytical Chemistry*, **58**, 844–848.

(114) Engstrom, R.C., Meaney, T., Tople, R. and Wightman, R.M. (1987) Spatiotemporal description of the diffusion layer with a microelectrode probe. *Analytical Chemistry*, **59**, 2005–2010.

(115) Engstrom, R.C., Wightman, R.M. and Kristensen, E.W. (1988) Diffusional distortion in the monitoring of dynamic events. *Analytical Chemistry*, **60**, 652–656.

(116) Liu, H.Y., Fan, F.R.F., Lin, C.W. and Bard, A.J. (1986) Scanning electrochemical and tunneling ultramicroelectrode microscope for high-resolution examination of electrode surfaces in solution. *Journal of the American Chemical Society*, **108**, 3838–3839.

(117) Wuu, Y.M., Fan, F.R.F. and Bard, A.J. (1989) High-resolution deposition of polyaniline on Pt with the scanning electrochemical microscope. *Journal of the Electrochemical Society*, **136**, 885–886.

(118) Bard, A.J., Fan, F.R.F., Kwak, J. and Lev, O. (1989) Scanning electrochemical microscopy - introduction and principles. *Analytical Chemistry*, **61**, 132–138.

(119) Kwak, J. and Bard, A.J. (1989) Scanning electrochemical microscopy - theory of the feedback mode. *Analytical Chemistry*, **61**, 1221–1227.

(120) Kwak, J. and Bard, A.J. (1989) Scanning electrochemical microscopy - apparatus and two-dimensional scans of conductive and insulating substrates. *Analytical Chemistry*, **61**, 1794–1799.

(121) Cornut, R. and Lefrou, C. (2008) New analytical approximation of feedback approach curves with a microdisk SECM tip and irreversible kinetic reaction at the substrate. *Journal of Electroanalytical Chemistry*, **621**, 178–184.

(122) Borgwarth, K., Ebling, D.G. and Heinze, J. (1994) Scanning electrochemical microscopy - a new scanning-mode based on convective effects. *Berichte der Bunsengesellschaft für Physikalische Chemie*, **98**, 1317–1321.

(123) Wipf, D.O., Bard, A.J. and Tallman, D.E. (1993) Scanning electrochemical microscopy. 21. Constant-current imaging with an autoswitching controller. *Analytical Chemistry*, **65**, 1373–1377.

(124) Ludwig, M., Kranz, C., Schuhmann, W. and Gaub, H.E. (1995) Topography feedback mechanism for the scanning electrochemical microscope based on hydrodynamic-forces between tip and sample. *Review of Scientific Instruments*, **66**, 2857–2860.

(125) Hengstenberg, A., Kranz, C. and Schuhmann, W. (2000) Facilitated tip-positioning and applications of non-electrode tips in scanning electrochemical microscopy using a sheer force based constant-distance mode. *Chemistry - A European Journal*, **6**, 1547–1554.

(126) Etienne, M., Layoussifi, B., Giornelli, T. and Jacquet, D. (2012) SECM-based automated equipment with a shearforce detection for the characterization of large and complex samples. *Electrochemistry Communications*, **15**, 70–73.

(127) Macpherson, J.V., Unwin, P.R., Hillier, A.C. and Bard, A.J. (1996) In-situ imaging of ionic crystal dissolution using an integrated electrochemical/AFM probe. *Journal of the American Chemical Society*, **118**, 6445–6452.

(128) Kranz, C., Friedbacher, G. and Mizaikoff, B. (2001) Integrating an ultramicroelectrode in an AFM cantilever: combined technology for enhanced information. *Analytical Chemistry*, **73**, 2491–2500.

(129) Macpherson, J.V. and Unwin, P.R. (2001) Noncontact electrochemical imaging with combined scanning electrochemical atomic force microscopy. *Analytical Chemistry*, **73**, 550–557.

(130) Kueng, A., Kranz, C., Lugstein, A. *et al.* (2003) Integrated AFM-SECM in tapping mode: simultaneous topographical and electrochemical imaging of enzyme activity. *Angewandte Chemie International Edition*, **42**, 3238–3240.

(131) Pobelov, I.V., Mohos, M., Yoshida, K. *et al.* (2013) Electrochemical current-sensing atomic force microscopy in conductive solutions. *Nanotechnology*, **24**, 115501.

(132) McKelvey, K., Edwards, M.A. and Unwin, P.R. (2010) Intermittent Contact-Scanning Electrochemical Microscopy (IC-SECM): a new approach for tip positioning and simultaneous imaging of interfacial topography and activity. *Analytical Chemistry*, **82**, 6334–6337.

(133) Takahashi, Y., Shevchuk, A.I., Novak, P. *et al.* (2010) Simultaneous noncontact topography and electrochemical imaging by SECM/SICM featuring ion current feedback regulation. *Journal of the American Chemical Society,* **132**, 10118–10126.

(134) Snowden, M.E., Güell, A.G., Lai, S.C.S. *et al.* (2012) Scanning electrochemical cell microscopy: theory and experiment for quantitative high resolution spatially-resolved voltammetry and simultaneous ion-conductance measurements. *Analytical Chemistry,* **84**, 2483–2491.

(135) Wei, C., Bard, A.J., Nagy, G. and Toth, K. (1995) Scanning electrochemical microscopy. 28. Ion-selective neutral carrier-based microelectrode potentiometry. *Analytical Chemistry,* **67**, 1346–1356.

(136) Horrocks, B.R., Schmidtke, D., Heller, A. and Bard, A.J. (1993) Scanning electrochemical microscopy. 24. Enzyme ultramicroelectrodes for the measurement of hydrogen-peroxide at surfaces. *Analytical Chemistry,* **65**, 3605–3614.

(137) Alpuche-Aviles, M.A. and Wipf, D.O. (2001) Impedance feedback control for scanning electrochemical microscopy. *Analytical Chemistry,* **73**, 4873–4881.

(138) Amphlett, J.L. and Denuault, G. (1998) Scanning electrochemical microscopy (SECM): an investigation of the effects of tip geometry on amperometric tip response. *Journal of Physical Chemistry B,* **102**, 9946–9951.

(139) Lee, Y., Amemiya, S. and Bard, A.J. (2001) Scanning electrochemical microscopy. 41. Theory and characterization of ring electrodes. *Analytical Chemistry,* **73**, 2261–2267.

(140) Casillas, N., James, P. and Smyrl, W.H. (1995) A novel-approach to combine scanning electrochemical microscopy and scanning photoelectrochemical microscopy. *Journal of the Electrochemical Society,* **142**, L16–L18.

(141) James, P., Casillas, N. and Smyrl, W.H. (1996) Simultaneous scanning electrochemical and photoelectrochemical microscopy by use of a metallized optical fiber. *Journal of the Electrochemical Society,* **143**, 3853–3865.

(142) Liu, W., Ye, H.C. and Bard, A.J. (2010) Screening of novel metal oxide photocatalysts by scanning electrochemical microscopy and research of their photoelectrochemical properties. *Journal of Physical Chemistry C,* **114**, 1201–1207.

(143) Ye, H., Lee, J., Jang, J.S. and Bard, A.J. (2010) Rapid screening of BiVO4-based photocatalysts by Scanning Electrochemical Microscopy (SECM) and studies of their photoelectrochemical properties. *Journal of Physical Chemistry C,* **114**, 13322–13328.

(144) Johnson, L. and Walsh, D.A. (2012) Tip generation-substrate collection-tip collection mode scanning electrochemical microscopy of oxygen reduction electrocatalysts. *Journal of Electroanalytical Chemistry,* **682**, 45–52.

(145) Lindsey, G., Abercrombie, S., Denuault, G. *et al.* (2007) Scanning electrochemical microscopy: approach curves for sphere-cap scanning electrochemical microscopy tips. *Analytical Chemistry,* **79**, 2952–2956.

(146) Holder, M.N., Gardner, C.E., Macpherson, J.V. and Unwin, P.R. (2005) Combined scanning electrochemical-atomic force microscopy (SECM-AFM): simulation and experiment for flux-generation at un-insulated metal-coated probes. *Journal of Electroanalytical Chemistry,* **585**, 8–18.

(147) Davoodi, A., Farzadi, A., Pan, J. *et al.* (2008) Developing an AFM-Based SECM system; Instrumental setup, SECM simulation, characterization, and calibration. *Journal of the Electrochemical Society,* **155**, C474–C485.

(148) Smirnov, W., Kriele, A., Hoffmann, R. *et al.* (2011) Diamond-modified AFM probes: from diamond nanowires to atomic force microscopy-integrated boron-doped diamond electrodes. *Analytical Chemistry,* **83**, 4936–4941.

(149) Sklyar, O., Kueng, A., Kranz, C. *et al.* (2005) Numerical simulation of scanning electrochemical microscopy experiments with frame-shaped integrated atomic force microscopy-SECM probes using the boundary element method. *Analytical Chemistry,* **77**, 764–771.

(150) Frederix, P., Bosshart, P.D., Akiyama, T. *et al.* (2008) Conductive supports for combined AFM-SECM on biological membranes. *Nanotechnology,* **19**, 384004.

(151) Avdic, A., Lugstein, A., Wu, M. *et al.* (2011) Fabrication of cone-shaped boron doped diamond and gold nanoelectrodes for AFM–SECM. *Nanotechnology*, **22**, 145306–145311.

(152) Leonhardt, K., Avdic, A., Lugstein, A. *et al.* (2013) Scanning electrochemical microscopy: diffusion controlled approach curves for conical AFM-SECM tips. *Electrochemistry Communications*, **27**, 29–33.

(153) Sklyar, O. and Wittstock, G. (2002) Numerical simulations of complex nonsymmetrical 3D systems for scanning electrochemical microscopy using the boundary element method. *Journal of Physical Chemistry B*, **106**, 7499–7508.

(154) Sklyar, O., Ufheil, J., Heinze, J. and Wittstock, G. (2003) Application of the boundary element method numerical simulations for characterization of heptode ultramicroelectrodes in SECM experiments. *Electrochimica Acta*, **49**, 117–128.

(155) Sklyar, O., Trauble, M., Zhao, C.A. and Wittstock, G. (2006) Modeling steady-state experiments with a scanning electrochemical microscope involving several independent diffusing species using the boundary element method. *Journal of Physical Chemistry B*, **110**, 15869–15877.

(156) Lefrou, C. and Cornut, R. (2010) Analytical expressions for quantitative Scanning Electrochemical Microscopy (SECM). *ChemPhysChem*, **11**, 547–556.

(157) Leonhardt, K., Avdic, A., Lugstein, A. *et al.* (2011) Atomic force microscopy-scanning electrochemical microscopy: influence of tip geometry and insulation defects on diffusion controlled currents at conical electrodes. *Analytical Chemistry*, **83**, 2971–2977.

(158) Leonhardt, K. (2012) Modelling of electrochemical processes at microelectrodes. PhD thesis, University of Southampton, Southampton.

(159) Casillas, N., Charlebois, S.J., Smyrl, W.H. and White, H.S. (1993) Scanning electrochemical microscopy of precursor sites for pitting corrosion on titanium. *Journal of the Electrochemical Society*, **140**, L142–L145.

(160) Casillas, N., Charlebois, S., Smyrl, W.H. and White, H.S. (1994) Pitting corrosion of titanium. *Journal of the Electrochemical Society*, **141**, 636–642.

(161) Basame, S.B. and White, H.S. (1995) Scanning electrochemical microscopy of native titanium-oxide films - mapping the potential dependence of spatially-localized electrochemical reactions. *Journal of Physical Chemistry*, **99**, 16430–16435.

(162) Basame, S.B. and White, H.S. (1998) Scanning electrochemical microscopy: measurement of the current density at microscopic redox-active sites on titanium. *Journal of Physical Chemistry B*, **102**, 9812–9819.

(163) Wittstock, G. and Schuhmann, W. (1997) Formation and imaging of microscopic enzymatically active spots on an alkanethiolate-covered gold electrode by scanning electrochemical microscopy. *Analytical Chemistry*, **69**, 5059–5066.

(164) Zhao, C.A. and Wittstock, G. (2004) Scanning electrochemical microscopy of quinoprotein glucose dehydrogenase. *Analytical Chemistry*, **76**, 3145–3154.

(165) Hengstenberg, A., Blochl, A., Dietzel, I.D. and Schuhmann, W. (2001) Spatially resolved detection of neurotransmitter secretion from individual cells by means of scanning electrochemical microscopy. *Angewendte Chemie, International Edition*, **40**, 905–908.

(166) Zhang, M., Becue, A., Prudent, M. *et al.* (2007) SECM imaging of MMD-enhanced latent fingermarks. *Chemical Communications*, 3948–3950.

(167) Zhang, M.Q. and Girault, H.H. (2007) Fingerprint imaging by scanning electrochemical microscopy. *Electrochemistry Communications*, **9**, 1778–1782.

(168) Zhang, M. and Girault, H.H. (2009) SECM for imaging and detection of latent fingerprints. *Analyst*, **134**, 25–30.

(169) Cortes-Salazar, F., Trauble, M., Li, F. *et al.* (2009) Soft stylus probes for scanning electrochemical microscopy. *Analytical Chemistry*, **81**, 6889–6896.

(170) Cortés-Salazar, F., Momotenko, D., Girault, H.H. *et al.* (2011) Seeing big with scanning electrochemical microscopy. *Analytical Chemistry*, **83**, 1493–1499.

(171) Bard, A.J. (2012) Scanning electrochemical microscopy, in *Scanning Electrochemical Microscopy*, 2nd edn (eds A.J. Bard and M.V. Mirkin), CRC Press, Boca Raton, p. 660.

(172) Serrapede, M., Denuault, G., Sosna, M. *et al.* (2013) Scanning electrochemical microscopy: using the potentiometric mode of SECM to study the mixed potential arising from two independent redox processes. *Analytical Chemistry*, **85**, 8341–8346.

(173) Mandler, D. and Unwin, P.R. (2003) Measurement of lateral charge propagation in polyaniline layers with the scanning electrochemical microscope. *Journal of Physical Chemistry B*, **107**, 407–410.

(174) Whitworth, A.L., Mandler, D. and Unwin, P.R. (2005) Theory of scanning electrochemical microscopy (SECM) as a probe of surface conductivity. *Physical Chemistry Chemical Physics*, **7**, 356–365.

13

Cold Fusion After A Quarter-Century: The Pd/D System

Melvin H. Miles[1] and Michael C.H. McKubre[2]
[1] *University of LaVerne, Department of Chemistry, USA*
[2] *SRI International, Energy Research, USA*

On March 23rd 1989, Fleischmann, Pons and Hawkins [1] publicly reported the results of an anomalous heat effect resulting from the intensive, electrochemical insertion of deuterium into palladium cathodes occurring over an extended period of time. This was already a well-studied system, and the SRI team, having worked on Pd/D_2O for more than a decade, was better positioned than most to judge this claim experimentally. If anomalous large heat production was indeed present in palladium cathodes loaded electrolytically with D, then the only possibility was that this occurred in the very high loading regime with the atomic loading ratio D/Pd greater than about 0.8, where the system had been infrequently and poorly studied. In the space available, the aim of this chapter is to focus on two details of Martin Fleischmann's final project: (i) the multithreshold materials constraints that prevented easy reproducibility of the Fleischmann–Pons heat effect (FPHE); and (ii) the brilliant, but largely not understood, implementation of the Fleischmann–Pons calorimeter, designed to take advantage of positive thermal feedback.

Some readers of this book will consider that this chapter on cold fusion represents Martin Fleischmann's greatest scientific failure; however, the authors believe that this may instead be one of the greatest contributions that Martin Fleischmann, along with Stanley Pons, made to science [1, 2]. Unfortunately, early attacks on this field were vigorous, even resorting to personal and unscientific criticisms [3–5], and progress in this field has been slow because of the consequential lack of funding and the difficult problem

Developments in Electrochemistry: Science Inspired by Martin Fleischmann, First Edition.
Edited by Derek Pletcher, Zhong-Qun Tian and David E. Williams.
© 2014 John Wiley & Sons, Ltd. Published 2014 by John Wiley & Sons, Ltd.

of reproducibility of the experimental results. Furthermore, the publication of research results on cold fusion soon became blocked by the editors of many scientific journals, and this discouraged academic entry into the field. The apparent irreproducibility is likely a materials problem, because results seem to depend on the palladium source and the metallurgical methods that were used in its preparation. Nevertheless, many research groups in various countries have now reported excess enthalpy effects for the Pd/D system [3–5]. The main nuclear reaction product that correlates with the excess enthalpy was found to be helium-4 [6–9]. Smaller amounts of other nuclear reaction products include tritium [10–12], neutrons [13, 14], radiation [3, 7, 15], and elemental transmutations [3–5]. There have been eighteen international conferences on cold fusion (ICCF) beginning in 1990 with alternating locations in the US or Canada, Europe and Asia.

Those of us who worked with Martin Fleischmann likely remember him for different aspects of his multifaceted personality and extreme scientific diversity. For the authors of this chapter, his major accomplishment was the discovery of anomalous heat effects in the electrochemical palladium deuterium system (Pd/D). Few would have had the vision to see such a possibility, the courage to pursue it, and the skill to test it. These anomalous effects were and are consistent in magnitude with excess enthalpy production by nuclear reactions. These are several orders of magnitude larger than can be explained by chemical reactions or lattice storage energy.

Initially called cold fusion,[1] this usage generated both confusion and hostility, and was in any case premature.[2] This field of endeavour, unveiled by Martin Fleischmann and Stanley Pons [1, 2], has broadened substantially in 25 years of research and is now known by several acronyms. The general classification of this field is Condensed Matter Nuclear Science (CMNS), and an electronic journal of that name is often used to publish papers in this field. This CMNS name highlights the observation, first suggested by Julian Schwinger [16], that nuclear reactions take place differently in a metal lattice than in free space. This chapter will be confined mostly to a discussion of the FPHE, a term which signifies anomalous excess heat production from electrochemical and gas-phase experiments in the Pd/D system. An attempt will be made in this chapter to demonstrate the consensus of the CMNS community that the FPHE is both anomalous and real, is nuclear in origin, has associated products that are quantitatively and temporally associated with the anomalous enthalpy, and results in products unique to nuclear reactions. As expected by Schwinger, the product distributions are not those of hot fusion and give rise to the possibility of practical energy production. Several dozen theoretical approaches are being actively developed, but none offers a clear and complete explanation for all observations. Martin Fleischmann often stated that a major reason for his study of the Pd/D system was to find experimental results that could only be explained by quantum electrodynamics (QED) and not by quantum mechanics. This led to his close relationship with Giuliano Preparata, who proposed an early theory for the FPHE that was based on QED [17] and correctly predicted helium-4 as the main fusion product [17].

---

[1] The term "cold fusion" was originally adopted to describe muon-catalyzed fusion.

[2] The original 1989 manuscript title contained a question mark "—cold fusion?". The word "fusion" was an unfortunate choice because high-energy and particle physicists reserve this historic term for well-known reactions that are not the same as those observed in the Pd/D system.

13.1   The Reproducibility Issue

While the lack of reproducibility in Pd/D experiments certainly makes the advancement of this science much more difficult, it does not mean that the FPHE is due to experimental errors. For example, the semiconductor field encountered a similar problem that persisted until the role of impurities was properly understood and the harmful impurities eliminated. Perhaps impurities in the palladium also play a critical role. The summary of the US Navy Laboratory (NAWCWD) at China Lake [18] shows that the FPHE was observed in 17 out of 28 experiments using palladium material from Johnson-Matthey. Also, palladium-boron (Pd-B) produced by the Naval Research Laboratory (NRL) gave excess enthalpy in seven out of eight experiments [18]; in contrast, palladium from other sources produced excess heat in only two of 24 experiments. Experiments at SRI also showed that the success rate varied greatly with the source of the palladium, with the most successful early set of experiments being performed with the most impure material [19].

Fleischmann and Pons had good success in the production of excess enthalpy using Johnson-Matthey palladium up until about the time of their 1989 announcement. However, at about this time Johnson-Matthey made a major change in their method of producing palladium, and this newer material was very poor for obtaining the FPHE. In private conversations, as well as in a later publication [20], Martin Fleischmann revealed that the good (or type A) Johnson-Matthey palladium was made under a cracked ammonia atmosphere [20]. The melting of the palladium under this $H_2 + N_2$ atmosphere likely resulted in H_2 reacting with any oxygen impurity in the palladium and removing it as H_2O vapor. However, this cracked ammonia atmosphere was no longer used when Johnson-Matthey changed their method of manufacturing palladium. The high success rate at China Lake for the NRL Pd-B material can also be explained by an oxygen removal mechanism. During arc-melting of the Pd-B material under argon, any oxygen in the palladium would react with the boron present to form B_2O_3, and this lower-density product would rise to the surface and skim off the top of the molten palladium.[3]

13.2   Palladium–Deuterium Loading

Directly related to the reproducibility issue is the requirement of high loading of the palladium with deuterium during electrolysis in heavy water. McKubre and others have shown that no excess power was produced with Pd wire cathodes until the D/Pd atomic ratio reached 0.85, or higher. Furthermore, the excess power production increases rapidly to large values above this 0.85 threshold. This effect of the D/Pd loading ratio on excess power production was reported simultaneously and independently by McKubre [21] and by Kunimatsu [22] at a conference in Japan in 1992. For bulk pure palladium wire cathodes, such as those used by Fleischmann and most early replicators, the problem is compounded by the multithreshold nature of the FPHE. Not only does initiation of the effect require D/Pd loadings rarely achieved before 1989, but these loadings must also be maintained

---

[3] Assuming that the palladium metal is a critical factor for the anomalous FPHE, it is not possible to make two palladium cathodes with exactly the same atomic arrangements, impurity levels, etc. Therefore, two "identical experiments" cannot be carried out.

for hundreds of hours in the presence of threshold current densities of 100 mA cm^{-2} or larger, well beyond the current density of maximum loading.[4] Very few bulk Pd samples are capable of sustaining high loading, because of the mechanical damage caused by mismatched lattice expansion and the formation of internal voids filled at equilibrium with D$_2$ at high gas pressure. Electrochemical damage of the interface and the build-up of surface impurities also tend to reduce the maximum loading achievable by a particular sample with time. Of equal importance, surface damage and poor interface conditioning and control, reduces the flux of deuterium through the interface. The magnitude (but not direction) of this flux is now known to be proportional to the magnitude of the excess heat effect as expressed in the following empirical equation [23].

$$P_X = M(x - x°)^2 \, (i - i°) \, |i_D| \quad \text{at } t > t° \tag{13.1}$$

where P_X is the excess power, $x = $ D/Pd, $x° \sim 0.875$, i is the electrochemical current density for the cathode, i_D is the absorption deuteron flux through the surface expressed as current density (2–20 mA cm^{-2}), and $t° > 20$ times the deuterium diffusional time constant in the cathode [23].

The failure to meet one or more of these threshold conditions provides an easy explanation for important early failures to reproduce the FPHE. Large significance was attached to early null heat results reported by a small number of groups at prestigious institutions. In light of the discussion above, it is useful to see whether these experiments – as well as other early experiments – were operated in a relevant regime. Perhaps the most cited early result reporting no anomalous effects was that of Lewis *et al.* [24] from Caltech, in which they stated that "D/Pd stoichiometries of 0.77, 0.79, and 0.80 obtained from these measurements were taken to be representative of the D/Pd stoichiometry for the charged cathodes used in this work." Also widely cited is the early null result of Albagli *et al.* [25] from MIT, who suggested that "... average loading ratios were found to be 0.75 ± 0.05 and 0.78 ± 0.05 for the D and H loaded cathodes, respectively." The Caltech and MIT reports of no excess heat effects are noted in Figure 13.1 in a histogram illustrating a number of early SRI and ENEA (Frascati) experiments producing positive excess power results as a function of maximum loading achieved.[5]

Even lower loading results were estimated by Fleming *et al.* [26] at Bell Labs in another report of no excess heat. In this case, the authors stated that "... the degree of deuterium incorporation was comparable to that for the open cells for the same time duration. The amount incorporated in longer electrolysis experiments was typically PdD$_x$ ($0.45 < x < 0.75$)." From what is known today (and as clearly shown in Figure 13.1), *none* of the cells in *any* of these cited studies would be expected to evidence any excess heat. In addition to insufficient loading, commonly the duration of the experiments were too short, the current density stimuli were too low and/or the deuterium flux was not measured. None of the criteria of Equation 13.1 was shown to be met, and at least two demonstrably were not met. In hindsight, it is evident that the authors were victims of "unknown unknowns", and

---

[4] The current density of maximum loading for D into Pd cathodes is between 15 and 25 mA cm^{-2} in 0.1–1.0 M LiOD solutions typically used in these experiments.

[5] Fleischmann and Pons were well aware of the significance of loading and the need to measure it, and they did so by means of the cathode overvoltage. Since it is now clear that the FPHE occurs at or near the cathode surface, this measurement is possibly more relevant than the average loading inferred from bulk resistivity measurements, but requires experienced interpretation.

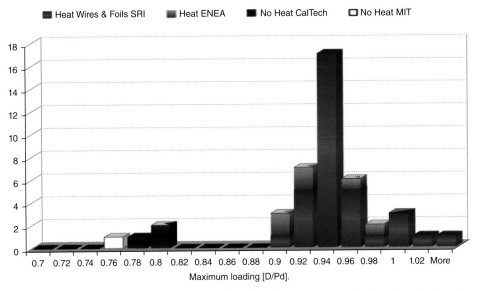

Figure 13.1   *Histogram illustrating the number of early experiments at SRI and ENEA, showing measurable excess power as a function of maximum cathode loading. Also illustrated are points for the MIT calorimetric experimental result, with a stated loading of 0.75 ± 0.05 (Ref. [25]); and for the Caltech null experimental result, with loading measurements quoted at 0.77, 0.78, and 0.80.*

perhaps "undue haste" – but this is understandable in the circumstances of 1989. What is important is that these experiments are recognized for what they are, not what they are not. They are important members of the experimental database that teaches us under what conditions one encounters the FPHE. They are not any part of a proof of nonexistence; absence of evidence is not evidence of absence.[6]

13.3   Electrochemical Calorimetry

There was never a significant problem with the isoperibolic Dewar calorimetric cell used by Fleischmann and Pons (F–P), but possible errors were often proposed by critics as an explanation for their excess enthalpy measurements [24,25]. However, improvements were made over time that increased the accuracy of the calorimetry and reduced the error to only ±0.1 mW [27]. It should be noted that significant calorimetry problems have been identified for several important groups that failed to find excess enthalpy production [28–30].

The Fleischmann–Pons studies of the Pd/D system required very accurate measurements over long time periods for the cell voltage (E), the cell current (I), the cell temperature (T), and the bath temperature (T_b). The limiting accuracy for the calorimetry is dictated by

---

[6] The NAWCWD group at China Lake (formerly NWC) reported no excess heat effects for various experiments until September of 1989, and they were listed with the unsuccessful Caltech, MIT, and other groups in the November, 1989 DOE ERAB Panel Report.

the temperature measurements which were measured to within ± 0.001 K by Fleischmann and Pons [27, 30] using carefully calibrated thermistors. Fleischmann considered the cell to be a "well-stirred tank," and the calorimetric error limit could be estimated from the error in the temperature measurement. For example, if the temperatures being measured differ by about 10 K for the cell and bath temperatures, then the ± 0.001 K error in the temperature measurements gives a minimum calorimetric error of $\pm 0.001/10$ or $\pm 0.01\%$. For an input power to the cell of 1000 mW, the goal for the minimum error in the excess power measurements should be ± 0.1 mW. This was Fleischmann's thinking, and this was the calorimetric accuracy that he eventually achieved [27]. The MIT calorimetry [25] measured temperatures only to within ± 0.1 K, and therefore their minimum calorimetric error in excess power measurements would be a much larger value of ± 10 mW [29, 30]. A number of other errors in the MIT calorimetry increased their error to about ± 50 mW [25, 29]. Nevertheless, it is likely that no excess power was actually present in the MIT experiments because they did not reach the required threshold for D/Pd loading.

A stirring problem was the first proposed error source for the Fleischmann–Pons Dewar calorimetry, and this was claimed to be the authoritative explanation for the reported excess enthalpy [4, 24]. However, to obtain a "well-stirred tank," Fleischmann and Pons used Dewar cells of small diameters. The three F–P Dewar cells used extensively by one of the present authors (M.H.M.) at the New Hydrogen Energy (NHE) laboratory in Sapporo, Japan, had an inner diameter of only 2.5 cm. The height of these cells was 25.0 cm, with the top 8.0 cm silvered to give a well-defined surface area for heat transport by radiation [31]. Because of the Dewar vacuum, the outside diameter of these cells was 4.2 cm. The two thermistors in each cell were well-separated in locations, and both gave similar results. For cells operating at currents of 50 mA or higher, it could be observed visually that these Dewar cells were indeed "well-stirred" by the electrolysis gases. Fleischmann presented a video to an overflow audience at the May Electrochemical Society 1989 meeting in Los Angeles, showing that a drop of phenolphthalein added to a working electrolysis cell containing a $D_2O + LiOD$ solution gave an intense red coloration that was thoroughly mixed in a matter of seconds [4]. These visual observations inside the calorimeter were a major advantage for the Fleischmann–Pons Dewar cells. In private conversations, Fleischmann stated that more than 10 thermistors had been used simultaneously in various locations in his Dewar cell, and they all provided the same calorimetric results. A major problem for stirring, however, was described for the short and fat calorimetric cell used by Caltech, that even required the use of a magnetic stirring bar [24]. It stands to reason that Fleischmann and Pons made great strides in perfecting their calorimetric cells and modeling equations over several years of research prior to their 1989 announcement. During the short-lived frenetic activity following the 1989 announcement, it would be quite unexpected that any other research group could match the calorimetric accuracy of Fleischmann and Pons.

13.4   Isoperibolic Calorimetric Equations and Modeling

Fleischmann had great skills in mathematics and the modeling of systems used throughout his research programs. This was also true in cold fusion, but unfortunately most scientists could not, or did not, follow his footsteps into the mathematical forest that led to a new and accurate calorimetric system for the study of electrochemical reactions. Even today, only a few scientists fully understand Fleischmann's equations, modeling, integration and data analysis methods that provided a calorimetric accuracy of ±0.01%.

The isoperibolic Dewar calorimetry and equations developed by Fleischmann and Pons remain relevant today because this accurate electrochemical calorimetry could be applied to the investigation of the thermal behavior of a wide range of electrochemical reactions, especially irreversible processes. Thus, the understanding and adoption of this electrochemical calorimeter concept remains important both for the Pd/D system and elsewhere in science. The detailed discussions of Fleischmann's calorimetry are found mainly in US Navy reports [31–33] because the length and subject matter prohibited publication in scientific journals.

The modeling equation developed by Fleischmann and Pons [2] for isoperibolic calorimetry can be represented by Equation (13.2) [30, 34, 35]

$$P_{calor} = P_{EI} + P_X + P_H + P_C + P_R + P_{gas} + P_W \tag{13.2}$$

where $P_C = -k_C (T - T_b)$, $P_R = -k_R (T^4 - T_b^4)$, and T and T_b are the cell and bath temperatures, respectively.

This equation represents the First Law of Thermodynamics as applied to electrochemistry in terms of power (watts), where the thermodynamic system is the calorimetric cell. This equation can occupy nearly a full page if each term were fully expressed mathematically [2, 31, 33]. Only a very limited discussion can be given here. The net power that flows into and out of the calorimetric system (P_{calor}) is determined by the electrochemical power (P_{EI}), excess power (P_X), the internal heater power (P_H), the heat conduction power (P_C), the heat radiation power (P_R), the power carried away by the gases evolved (P_{gas}), and the power due to pressure–volume work (P_W). Each power term is a function of time (t) such as $P_{calor} = C_p M \, dT/dt$; thus, Equation (13.2) is a nonlinear, inhomogeneous differential equation that can be used directly or numerically integrated for greater accuracy. Equation (13.2), originally developed by Fleischmann and Pons, has not been challenged, yet many groups reporting calorimetric results have ignored important terms [28–30].

13.5 Calorimetric Approximations

Although Fleischmann preferred to use the full calorimetric equation [Equation (13.2)] in order to measure the excess power as accurately as possible, he also provided useful approximations that greatly simplify the mathematics. A lower bound heat transfer coefficient, k'_R or k'_C, should first be calculated by assuming $P_X = 0$. For the Dewar calorimeter with heat transfer mainly by radiation, Equation (13.2) becomes

$$P_{calor} = P_{EI} + 0 + P_H + P_C - k'_R f(T) + P_{gas} + P_W \tag{13.3}$$

where $f(T) = T^4 - T_b^4$, and k'_R is smaller than the true coefficient, k_R, when $P_X > 0$. The simple subtraction, Equations (13.2) and (13.3), yields [31, 34]

$$P_X = (k_R - k'_R) f(T) \tag{13.4}$$

Similarly, for a calorimeter with heat transfer mainly by conduction

$$P_X = (k_C - k'_C) \Delta T \tag{13.5}$$

where $\Delta T = T - T_b$. All other terms are subtracted out without making any other assumptions. For control systems such as Pt/D_2O with no significant excess power, the lower bound heat transfer coefficient should always be nearly equal to the true heat transfer coefficient. For an active Pd/D_2O system, the lower bound heat transfer coefficient will vary with the amount of excess power and will always be less than the true heat transfer coefficient. Controls are critical for accurate calibrations of the calorimetric systems. It is not possible to calibrate a system when an unknown excess power is present, and this has been a source of confusion about the accuracy of this isoperibolic calorimetry [20, 31].

13.6   Numerical Integration of Calorimetric Data

It is well known from chemical reaction kinetics that it is much more accurate to use the integrated equations rather than the differential rate equations. This also applies to Equation (13.2) for electrochemical calorimetric studies. Fleischmann stated that many scientists did not understand the difference between differential and integral calorimetric coefficients and the disadvantage of differentiating "noisy" data as compared to integrating such data [32]. However, even today, no research group, except for Fleischmann and Pons, has exploited the advantages of integration methods applied to Equation (13.2).

The ± 0.1 mW accuracy of the Fleischmann-Pons Dewar calorimetry requires temperature measurements accurate to ± 0.001 K, the use of several averaging methods for the data, the casting of the calorimetric equation into a straight-line form, and the numerical integration of the calorimetric differential equation [27, 30–34]. The trapezium rule, Simpson's rule and the mid-point rule have all been used to carry out the numerical integrations, but the trapezium rule is generally used [31–33]. Fleischmann's straight-line method begins with the calorimetric Equation (13.2) expressed as

$$C_p M dT/dt = -k_R f(T) + P_{net} + P_X \tag{13.6}$$

for Dewar cells where $P_R \gg P_C$ and where $P_{net} = P_{EI} + P_H + P_{gas} + P_W$, and $f(T) = T^4 - T_b^4$. Rearrangement of this differential equations yields

$$(P_{net} + P_X)dt/f(T)dt = C_p M dT/f(T)dt + k_R \tag{13.7}$$

Let $y = (P_{net} + P_X)dt / f(T)dt$ and $x = dT/f(T)dt$, where y has units of WK^{-4} and x has units of $K^{-3}s^{-1}$. This gives a straight-line form for the differential equation

$$y = C_p M x + k_R \tag{13.8}$$

The substitution $x = x'/C_p M'$, where $C_p M'$ is an estimated value for $C_p M$, gives y, x' and k_R the same units (WK^{-4}), thus

$$y = C_p M x'/C_p M' + k_R \tag{13.9}$$

Multiplying Equation (13.8) by 10^9 removes the large exponents for each term, and numerical integration over a selected time period yields

$$10^9 [y] = 10^9 C_p M[x']/CpM' + 10^9 k_R \tag{13.10}$$

where $[y]$ and $[x']$ represent numerically integrated values for these variables. At $[x'] = 0$, $10^9 k_R$ equals the intercept value. It should be noted that the integral of $f(T)dt$ can be

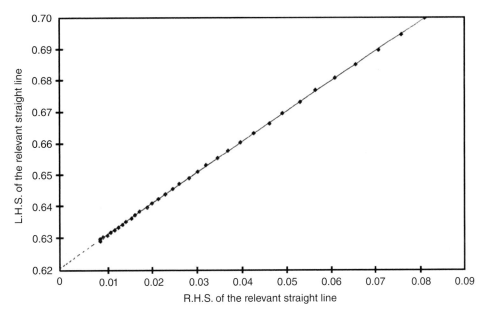

Figure 13.2   *Straight line from backward integration of calorimetric data.*

evaluated separately between the selected time of t and t_2. Also, the integral for dT is simply $T - T_2$ [31].

An example of the straight line obtained from the backward integration of the calorimetric data for a platinum/D_2O control experiment is shown in Figure 13.2. This line is in the form of Equation (13.10), and Figure 13.2 shows the left-hand side (LHS) versus the right-hand side (RHS). The line intercept at $x = 0$ yields $k_R = 0.62085 \times 10^{-9}$ WK^{-4}, and the slope of this line yields $C_pM = 341.1$ JK^{-1} [27, 34]. This cell was run at a current of $I = 0.2000$ A for 16 days using two-day cycles and refilling the cell with D_2O at the beginning of each cycle [27]. Many other examples of these methods are given elsewhere [27, 30–34]. However, the mathematical steps provided here of exactly how this was achieved is not reported in any of Fleischmann's publications. The calculation of various types of radiative heat transfer coefficients has also been reported by Fleischmann [27, 31–34]; these include differential, integral, lower bound, and true coefficients investigated over different time periods [27, 31–34].

An advantage of the straight-line method shown in Figure 13.2 is that possible errors from C_pM are eliminated at $[x'] = 0$. The slope of the straight line is C_pM/C_pM', and is near unity when a good estimate of C_pM' is used. The actual heat capacity value is given by $C_pM = (\text{slope})(C_pM')$. Various integration methods have been reported by Fleischmann [27, 31–34], but backward integrations from the midpoint of the two-day cycle (t_2), through the heating pulse, to near the beginning of the cycle (t), gives the best results. Furthermore, the k_R value obtained is for the mid-point of the two-day cycle and is not affected by any errors in the value for C_pM. Fleischmann's straight-line method assumes that the excess power (P_X) is zero or constant over the selected time period. It is not possible to accurately calibrate any calorimetric system when the excess power is changing, and thus control experiments such as Pd/H$_2$O or Pt/D$_2$O are essential [20, 31–34].

13.7   Examples of Fleischmann's Calorimetric Applications

The most detailed discussions of Fleischmann's application of his calorimetric methods are contained in three US Navy reports [31–33] that include a simulation of his calorimetry [32]. Two of these reports [31, 33] involved a study conducted by one of the present authors (M.H.M.) at NHE in Japan using a Fleischmann–Pons Dewar calorimeter and a Pd-B cathode (prepared by Dr M.A. Imam at the NRL). The results of this Pd-B data analysis by Fleischmann are shown in Figure 13.3 in terms of the excess enthalpy for each day. The average excess power can be readily calculated; for example, 40 000 J day^{-1}/86 400 s day^{-1} = 0.463 W. The dimensions of this Pd-B rod was 0.471 cm diameter and 2.01 cm length (A = 3.15 cm^2; V = 0.350 cm^3). The actual experimental measurements for key portions of this experiment are given in the NRL report [31], and the entire experimental data set from this Pd-B study has been made available for analysis by other groups. As this was a larger Dewar cell, both k_R and C_pM are larger than in Figure 13.2.

Many features of this experiment are similar to results for other Pd/D experiments, such as "positive feedback" (an increase in the rate of excess enthalpy generation with temperature) and several examples (Days 26 and 69) of "Heat-after-Death" (a persistence of an excess power effect when the current is lowered or turned off). The cell contents

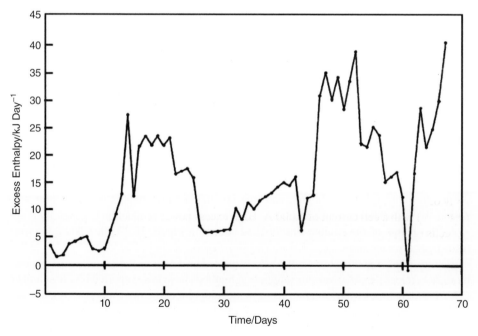

Figure 13.3 *Excess enthalpy generated per day using the radiative heat transfer coefficient $k_R = 0.85065 \times 10^{-9}$ WK^{-4} and $C_pM = 450$ JK^{-1} [31].*

evaporated to dryness on Day 68 (not shown in Figure 13.3), and the rate of excess enthalpy generation rose to nearly 10 W [31, 35, 36]. The visual observation of the furious boiling and swirling of the electrolyte was centered at the Pd-B cathode at this time, and this observation suggested that the cathode was the hot spot in the cell [31].

The experimental data following the evaporation to dryness (Days 68, 69) showed a continuation of excess power that gradually decreased, as would be expected for a process controlled by a diffusional relaxation time for deuterium inside the palladium. The cooling of this cell was also slower than expected, and there was at least one period (Day 69) during which the cell contents reheated with no applied electrochemical or heater power [31, 33]. Illustrations of these effects are shown in Ref. [31] (Figures A22, A23, and A24).

There was a striking difference in this Pd-B study that has not been observed with pure palladium cathodes. This was the very early development of the excess power and "positive feedback" effect. This excess power was likely present even on Day 1 because the measured enthalpy for the exothermic charging of the Pd-B material with deuterium was significantly larger than expected [31, 33]. These effects became more obvious on the first application of the heater power (0.2500 W on Day 3) [31, 33, 35, 36]. This can be seen simply by the fact that the cell temperature baseline does not relax back to the expected baseline following the end of the heater application (see Figure 1 of Ref. [35]). This temperature baseline for the cell is a simple method for detecting any significant changes in the excess power [31, 33]. For example, it has often been observed that the cell temperature for an active Pd/D cell will increase despite a decreasing input power to the cell. One of Fleischmann's goals in measuring small effects was to study conditions for the initiation of the larger FPHE. Although it is not known why the excess power appears early for the Pd-B material, this material was found to have two distinct phases of the same cubic structure [31, 35], and perhaps one phase reaches the required loading much quicker than the other phase. It was also observed that the deuterium de-loading process was much slower for the Pd-B material so that the palladium bulk acts less effectively as a deuterium sink. Two US patents have been granted for this unusual Pd-B material [37].

The small negative enthalpy for Day 61 (-436 J or -5.0 mW) in Figure 13.2 was troublesome for Fleischmann because it violated the Second Law of Thermodynamics. This was later explained, however, by the neglected $P_W = -RT\,(0.75\,I/F)$ term that became significantly more negative on Day 61 (-22.3 mW) due to the large cell current (1.000 A) and high cell temperature (345.4 K) on that day. However, it is strange that the excess power suddenly decreased to near zero on Day 61 and then quickly recovered to yield 17 200 J of excess enthalpy or 0.199 W of average excess power for the following day (Day 62). The peak excess enthalpy production prior to the boil-off was 41.0 kJ on Day 67 (0.475 W) with a cell current of 1.000 A. High excess power production is generally found for cells operating above a threshold temperature of about 60 °C [32, 33]. This condition was reached for only the later portions of this experiment (Days 46–52, 55, and 60–68).

A second detailed example of Fleischmann's calorimetric application was given for a control study using a platinum cathode in place of palladium in 0.1 M LiOD + D_2O [27, 34]. The small excess power measured (1.1 ± 0.1 mW) for a cell input power of 0.8 W was attributed to the electrochemical reduction of the oxygen generated in the cell [27, 34]. This result supports other studies that show that recombination of D_2 and O_2 is not an important factor at the high currents used in these electrochemical cells [31, 38]. An attempt was made to publish these results in a major journal in 2007, but despite recommendations for publication by two of the three reviewers, the journal editor concluded: "Since *The Journal of Physical Chemistry* has a broad audience, I feel this work is not appropriate for the journal. I have decided to reject your paper". One would think that a new and accurate calorimetric method for electrochemical processes would be a valid topic for a journal that covers physical chemistry. This letter from the editor, dated September 4, 2007, likely marked Fleischmann's final attempt at publishing his electrochemical calorimetry in a major journal. The reported result of 1.1 mW, nevertheless, is in good agreement with both theoretical calculations and experimental measurements for recombination in an electrochemical cell operating at 0.2000 A for D_2O electrolysis [38].

13.8   Reported Reaction Products for the Pd/D System

13.8.1   Helium-4

Fleischmann and Pons were actually the first to observe the production of helium-4 in the Pd/D system [4]. However, due to the extensive criticism of their 1989 announcement, they did not want " . . . to open another front" for attacks on their work, and so their measurements of helium-4 were never officially reported. The first reported experiments correlating the calorimetric excess enthalpy and helium-4 production were conducted by Miles in 1990 at the Naval Weapons Center (now NAWCWD) in China Lake, California, and the helium measurements were performed under the supervision of Bush at the University of Texas [6–8]. The presence of helium-4 was observed in eight out of nine effluent gas samples collected during the presence of excess heat [7, 8]. No helium-4 was observed for six out of the six samples of effluent gas for a Pd/H_2O control study. Measurements were also conducted for helium-3 in these studies, but none was detected [6]. In summary, for all experiments conducted by Miles at NAWCWD, 12 out of 12 produced no excess helium-4 when no excess heat was measured, and 18 out of 21 experiments gave a correlation between the measurements of excess heat and helium-4 [8, 18]. Three of the experiments that produced helium-4 were conducted under "double-blind" rules [8, 18]. An exact statistical treatment for all experiments shows that the probability is only one in 750 000 that the China Lake set of heat and helium-4 measurements could be this well correlated due to random experimental errors [18]. Furthermore, the rate of helium-4 production was always in the appropriate range of 10^{10} to 10^{12} atoms per second per watt of excess power for D + D or other likely nuclear reactions [8, 18].

An early confirmation for the correlation of excess heat and helium production in the Pd/D system was reported by McKubre and coworkers at SRI [9]. Several different experiments, using three different calorimetric methods, gave a strong time correlation between the rates of heat and helium production [9]. The production of helium-4 was also observed in sealed cells containing a Pd-C catalyst and D_2 gas where the helium increase exceeded the amount of helium-4 in room air [39]. There are a number of other reports of helium-4 production in Pd/D systems [3], including those of Gozzi in Italy [40], DeNinno in Italy [40], and Arata in Japan [41].

13.8.2   Tritium

There exist many reliable reports of tritium production in cold fusion experiments for the Pd/D system [3–5]. However, the amounts of tritium production are always too small to explain excess heat events. Three of the earliest reports of tritium production were by Bockris of Texas A&M [10], Storms of Los Alamos National Laboratory [11], and extensively by various workers operating independently at the Bhabha Atomic Research Center (BARC) in India [42]. Extensive experiments at the SPAWAR US Navy laboratory over many years on the Pd/D system have produced various nuclear products, including tritium [12]. A summary of tritium production lists more than 60 other reports of tritium production in the Pd/D system [3]. An unusual result is that the tritium to neutron ratio in cold fusion experiments is at least 10^6 [3], also confirming Schwinger's early declaration that "... the circumstances of cold fusion are not those of hot fusion" [16].

13.8.3   Neutrons, X-Rays, and Transmutations

Despite the many experiments designed to detect neutrons, their production in the Pd/D system has been difficult to prove [3,4]. This problem is explained by the small neutron production rate of 5–50 s^{-1} W^{-1} in active Pd/D cells reported by Pons and Fleischmann in 1992 [13]. Recent studies at the US Navy SPAWAR laboratory have reported significant evidence for neutron production using CR-39 integrating detectors [14]. These experiments used codeposition methods, where both Pd and D are deposited onto a substrate from a $PdCl_2$ + LiCl + D_2O solution [12]. The Navy SPAWAR CR-39 results also gave a low neutron production rate of 2.5 cm^{-2} s^{-1}, and it would be difficult to measure this low neutron flux using real-time detectors (P.A. Mosier-Boss, personal communication; see also Ref. [14]).

There are many reports of X-ray production in the Pd/D system [3], including studies at China Lake [7], SPAWAR [15], and Mitsubishi [43]. There are also reports of anomalous high counts from Geiger–Mueller (GM) detectors placed near Pd/D_2O electrolysis experiments. For example, many periods of high counts up to 73 σ were measured at China Lake during Pd/D_2O electrolysis experiments, while the counts were always normal during control studies using Pd/H_2O or when no electrolysis experiments were running [44].

Transmutation continues to be an active area of study for the Pd/D system [3], but because of possible contamination and molecular ion interferences this area is even more controversial than the excess heat found in cold fusion experiments. Among the leading contributors to transmutation results is the team at Mitsubishi Heavy Industries in Japan, led by Iwamura [45]. Transmutation results generally report the detection of new elements such as copper or zinc that were not initially present in the system [3]. Occasionally, these are reported with non-natural isotope distributions, but technical challenges make these results less certain than the heat, helium, and tritium results.

13.9   Present Status of Cold Fusion

The final field of study left to the world by Fleischmann entered its twenty-fifth year in 2014. If it proves to be not only right but also practical, this will likely be the greatest contribution that Fleischmann has made to science and the world. The FPHE, the production of enthalpy at levels consistent with nuclear but not chemical reaction, from palladium charged with deuterium by electrolysis or gas-phase methods, is the sort of invention that only a man of Fleischmann's knowledge, genius, confidence and courage was capable of making. Those of us in the field accept that the FPHE is real and novel, but remains both poorly understood and unbounded. Despite the considerable amount of work done – with several thousand total publications and over 60 person-years experimentation at SRI alone – we have no idea about its limitations or its potential.

Yet, we are willing to make an optimistic projection, against which history may judge us. In 2002, the field that Fleischmann started was redesignated as Condensed Matter Nuclear Science, in recognition of the apparent new fact that not just one but many processes of nuclear physics take place in condensed matter by different pathways, with different rates and with different product distributions than similar processes in free space. Fleischmann recognized this possibility early and set out to test it. Schwinger, Preparata, Hagelstein and others also perceived this possibility. What if it is true? What if we can wield the power of nuclear physics on a tabletop and not just in facilities the size of football stadiums and the cost of small cities? What technologies and products might ensue?

The pathway to the future is likely to be very different from the past 25 years spent to corroborate the existence of the FPHE, to determine the conditions under which it occurs, and then learning to control and overcome the important materials constraints. For a variety of reasons the future of Fleischmann's dream must be practical, and therefore the heat effects must be cheaper, easier and of much larger scale and gain. This rules out palladium unless much higher power densities can be attained, electrochemistry (for anything more than initial loading), and ambient temperature calorimetry (where the heat produced has little value), thus effectively disposing of more than 90% of the work to the present. Future experimentation is likely to be directed towards small-dimension materials including metals other than palladium in high-temperature, gas-phase systems of deuterium or natural hydrogen, perhaps doped with extra D_2. In fact, such investigations have already begun [46].

Acknowledgments

One of the authors (M.H.M.) is grateful for an anonymous fund from the Denver Foundation administered by Dixie State University, as well as an adjunct faculty position at the University of LaVerne.

References

(1) Fleischmann, M., Pons, S. and Hawkins, M. (1989) Electrochemically induced nuclear fusion of deuterium. *Journal of Electroanalytical Chemistry*, **261**, 301–308; errata, **263**, 187 (1989).

(2) Fleischmann, M., Pons, S., Anderson, M.W. *et al.* (1990) Calorimetry of the palladium-deuterium-heavy water system. *Journal of Electroanalytical Chemistry*, **287**, 293–348.

(3) Storms, E. (2007) *The Science of Low Energy Nuclear Reaction*, World Scientific, New Jersey.

(4) Beaudette, C.G. (2002) *Excess Heat, Why Cold Fusion Research Prevailed*, 2nd edn, Oak Grove Press, South Bristol, Maine.

(5) Krivit, S.B. and Winocur, N. (2004) *The Rebirth of Cold Fusion*, Pacific Oak Press.

(6) Bush, B.F., Lagowski, J.J., Miles, M.H. and Ostrom, G.S. (1991) Helium production during the electrolysis of D_2O in cold fusion experiments. *Journal of Electroanalytical Chemistry*, **304**, 271–278.

(7) Miles, M.H., Hollins, R.A., Bush, B.F. *et al.* (1993) Correlation of excess enthalpy and helium production during D_2O and H_2O electrolysis using palladium cathodes. *Journal of Electroanalytical Chemistry*, **346**, 99–117.

(8) Miles, M.H. (2006) Correlation of excess enthalpy and helium-4 production: a review, in *Condensed Matter Nuclear Science* (eds P. Hagelstein and S. Chubb), World Scientific, New Jersey, pp. 123–131.

(9) McKubre, M., Tanzella, F., Tripodi, P. and Hagelstein, P. (2000) The emergence of a coherent explanation for anomalies observed in D/Pd and H/Pd systems: Evidence for ^{4}He and ^{3}H production, in *Proceedings of the 8th International Conference on Cold Fusion* (ed. F. Scaramuzzi), Italian Physical Society, Bologna, Italy, pp. 3–10.

(10) Packham, N.J.C., Wolf, K.L., Wass, J.C. *et al.* (1989) Production of tritium from D$_2$O electrolysis at a palladium cathode. *Journal of Electroanalytical Chemistry*, **270**, 451–458.

(11) Storms, E. and Talcott, C.L. (1990) Electrolytic tritium production. *Fusion Technology*, **17**, 680–695.

(12) Szpak, S., Mosier-Boss, P.A., Boss, R.D. and Smith, J.J. (1998) On the behavior of the Pd/D system: evidence for tritium production. *Fusion Technology*, **33**, 38–50.

(13) Pons, S. and Fleischmann, M. (1992) Concerning the detection of neutrons and gamma-rays from cells containing palladium cathodes polarized in heavy water. *Il Nuovo Cimento*, **105**, 763–772.

(14) Mosier-Boss, P.A., Dea, J.Y., Forsley, L.P.G. *et al.* (2010) Comparison of Pd/D co-deposition and DT neutron generated triple tracks observed in CR-39 detectors. *The European Physical Journal - Applied Physics*, **51**, 20901 (10 pp).

(15) Szpak, S., Mosier-Boss, P.A. and Smith, J.J. (1996) On the behavior of the cathodically polarized Pd/D system: search for emanating radiation. *Physics Letters A*, **210**, 382–390.

(16) Schwinger, J. (1991) Nuclear energy in an atomic lattice. *Progress of Theoretical Physics*, **85**, 711–712.

(17) Preparata, G. (1995) *QED Coherence in Matter*, World Scientific, Singapore.

(18) Miles, M.H., Bush, B.F. and Johnson, K.B. (1996) *Anomalous Effects in Deuterated Systems*, NAWCWPNS TP 8302, September 1996.

(19) McKubre, M.C.H., Crouch-Baker, S., Rocha-Filho, R.C. *et al.* (1994) Isothermal flow calorimetric investigations of the D/Pd and H/Pd systems. *Journal of Electroanalytical Chemistry*, **368**, 55–66.

(20) Fleischmann, M. (1998) *Cold fusion; past, present and future. ICCF-7 Proceedings*, ENECO, Inc., Salt Lake City, Utah, pp. 119–132.

(21) McKubre, M.C.H., Crouch-Baker, S., Riley, A.M. *et al.* (1993) Excess power observations in electrochemical studies of the D/Pd system; the influence of loading, in *Frontier of Cold Fusion* (ed. H. Ikegami), Universal Academy Press, Tokyo, pp. 5–19.

(22) Kunimatsu, K., Hasegawa, N., Kubota, A. *et al.* (1993) Deuterium loading ratio and excess heat generation during electrolysis of heavy water by a palladium cathode in a closed cell using a partially immersed fuel cell anode, in *Frontiers of Cold Fusion* (ed. H. Ikegami), Universal Academy Press, Tokyo, pp. 31–45.

(23) McKubre, M.C.H., Crouch-Baker, S., Hauser, A.K. *et al.* (1995) Concerning reproducibility of excess power production. Proceedings of the 5th International Conference on Cold Fusion, IMRA, France (ed. S. Pons), pp. 17–33.

(24) Lewis, N.S., Barnes, C.A., Heben, M.J. *et al.* (1989) Searches for low-temperature nuclear fusion of deuterium in palladium. *Nature*, **340**, 525–530.

(25) Albagli, D., Ballinger, R., Cammarata, V. *et al.* (1990) Measurements and analysis of neutron and gamma-ray emission rates, other fusion products, and power in electrochemical cells having Pd cathodes. *Journal of Fusion Energy*, **9**, 133–147.

(26) Fleming, J.W., Law, H.H., Sapjeta, J. *et al.* (1990) Calorimetric studies of electrochemical incorporation of hydrogen isotopes into palladium. *Journal of Fusion Energy*, **9**, 517–524.

(27) Fleischmann, M. and Miles, M.H. (2006) The instrument function of isoperibolic calorimeters: excess enthalpy generation due to the parasitic reduction of oxygen, in *Condensed Matter Nuclear Science* (eds P. Hagelstein and S. Chubb), World Scientific, New Jersey, pp. 247–268.

(28) Miles, M.H., Bush, B.F. and Stilwell, D.E. (1994) Calorimetric principles and problems in measurements of excess power during Pd-D$_2$O electrolysis. *The Journal of Physical Chemistry*, **98**, 1947–1952.

(29) Miles, M.H. and Hagelstein, P.L. (2012) New analysis of MIT calorimetric errors. *The Journal of Condensed Matter Nuclear Science*, **8**, 132–138.

(30) Miles, M.H. (2014) Examples of isoperibolic calorimetry in the cold fusion controversy. *The Journal of Condensed Matter Nuclear Science*, in press.

(31) Miles, M.H., Fleischmann, M. and Imam, M.A. (2001) Calorimetric Analysis of a Heavy Water Electrolysis Experiment Using a Pd-B Alloy Cathode. Naval Research Laboratory Report NRL/MR/6320-018526, March 26, 2001, pp. 155.

(32) Fleischmann, M. (2002) *Thermal and Nuclear Aspects of the Pd/D$_2$O System, Volume 2: Simulation of the Electrochemical Cell (ICARUS) Calorimetry* (eds S. Szpak and P.A. Mosier-Boss), Technical Report 1862, SSC San Diego, February 2002, pp. 180.

(33) Szpak, S., Mosier-Boss, P.A., Miles, M.H. *et al.* (February 2002) Analysis of experiment MC-21: a case study, in *Thermal and Nuclear Aspects of the Pd/D$_2$O System, Vol. 1: A Decade of Research at U.S. Navy Laboratories* (eds S. Szpak and P.A. Mosier-Boss), SSC San Diego, pp. 31–89.

(34) Miles, M.H. and Fleischmann, M. (2008) Accuracy of isoperibolic calorimetry used in a cold fusion control experiment, in *Low-Energy Nuclear Reactions Sourcebook* (eds J. Marwan and S.B. Krivit), American Chemical Society, Washington, DC, pp. 153–171.

(35) Miles, M.H., Imam, M.A. and Fleischmann, M. (2003) Calorimetric analysis of a heavy water electrolysis experiment using a Pd-B alloy cathode, in *Batteries and Supercapacitors* (eds G.A. Nazri, E. Takeuchi, R. Koetz and B. Scrosati), Proceedings Volume 2001–21, The Electrochemical Society, Inc., New Jersey, pp. 795–806.

(36) Miles, M.H., Imam, M.A. and Fleischmann, M. (2000) "Case studies" of two experiments carried out with the ICARUS systems, in *Italian Physical Society Conference Proceedings, ICCF8*, Editrice Compositori, Bologna, Italy, vol. 70 (ed. F. Scaramuzzi), pp. 105–119.

(37) Miles, M.H. and Imam, M.A. (2004) Palladium-Boron Alloys for Excess Enthalpy Production, U.S. Patent No. 5,904,990, May 18, 1999 and 6,764,561, June 20, 2004.

(38) Will, F.G. (1997) Hydrogen + oxygen recombination and related heat generation in undivided electrolysis cells. *Journal of Electroanalytical Chemistry*, **426**, 177–184.

(39) Gozzi, D., Cellucci, F., Cignini, P.L. *et al.* (1997) X-ray, heat excess and ^{4}He in the D/Pd system. *Journal of Electroanalytical Chemistry*, **435**, 113–136.

(40) DeNinno, A., Frattolillo, A., Rizzo, A. and Del Giudice, E. (2006) ^{4}He Detection in a cold fusion experiment, in *Condensed Matter Nuclear Science* (eds P. Hagelstein and S. Chubb), World Scientific, New Jersey, pp. 133–137.

(41) Arata, Y. and Zhang, Y.C. (1999) Observation of anomalous heat release and helium-4 production from highly deuterated fine particles. *Japanese Journal of Applied Physics: Part 2*, **38**(7A), L774.

(42) Iyengar, P.K. and Srinivasan, M. (1990) Overview of BARC studies in cold fusion. The First Annual Conference on Cold Fusion: Conference Proceedings (ICCF1), National Cold Fusion Institute, Salt Lake City, March 28–31, 1990, pp. 62–81.

(43) Iwamura, Y., Itoh, T., Gotoh, N. and Toyoda, I. (1998) Detection of anomalous elements, X-ray, and excess heat in a D$_2$-Pd system and its interpretation by the electron-induced nuclear reaction model. *Fusion Technology*, **33**, 476–492.

(44) Miles, M.H. and Bush, B.F. (1998) Radiation measurements at China Lake: Real or Artifacts? Proceedings ICCF-7 Vancouver, ENECO, Salt Lake City, April 19–24, 1998, pp. 236–240.

(45) Iwamura, Y., Itoh, I., Sakano, M. *et al.* (2006) Low energy nuclear transmutation in condensed matter induced by D$_2$ gas permeation through Pd complexes: correlation between deuterium flux and nuclear products, in *Condensed Matter Nuclear Science* (eds P. Hagelstein and S. Chubb), World Scientific, New Jersey, pp. 435–446.

(46) McKubre, M.C.H., Bao, J., Tanzella, F.L. and Hagelstein, P.L. (2014) Calorimetric studies of the destructive stimulation of palladium and nickel fine wires. *Journal of Condensed Matter Nuclear Science*, in press.

14

In-Situ X-Ray Diffraction of Electrode Surface Structure

Andrea E. Russell[1], Stephen W.T. Price[2] and Stephen J. Thompson[1]
[1]University of Southampton, Chemistry, UK
[2]Diamond Light Source Ltd, Harwell, UK

Motivated by the recent advances in the understanding of the structure at the metal/gas interface being achieved in ultra-high vacuum (UHV), Martin Fleischmann sought to bring that level of detail to the exploration of the electrode/solution interface. During the 1980s, most laboratory-based structural techniques available to UHV surface scientists, such as low-energy electron diffraction (LEED), electron energy loss spectroscopy (EELS), and X-ray photoelectron spectroscopy (XPS), relied on the interaction of electron beams with the metal surface or the detection of ejected electrons, and so were not readily employed at the metal/solution interface. Techniques relying on the transmission or reflection of photons, such as infrared, Raman and UV-visible spectroscopies, were proving to be of use in identifying molecules adsorbed at electrode surfaces, their orientation, and bonding modes, but did not provide the level of structural detail available to the UHV surface scientists. Although neutron diffraction and surface-extended X-ray absorption fine structure (SEXAFS) techniques were becoming available, these required the use of reactors and synchrotrons, respectively, and their availability was not yet widespread. Realizing that X-ray diffractometers were readily available and, in the Southampton tradition of combining anything at hand with electrochemical measurements (although the true *in-situ* electrochemical measurements would come a bit later), Fleischmann, Hendra and Robinson, reported the "… first observation of X-ray diffraction from a two-dimensional adsorbate" at the metal/solution interface [1], and so initiated another line of research that continues to this day, as will be presented in this chapter.

Developments in Electrochemistry: Science Inspired by Martin Fleischmann, First Edition.
Edited by Derek Pletcher, Zhong-Qun Tian and David E. Williams.
© 2014 John Wiley & Sons, Ltd. Published 2014 by John Wiley & Sons, Ltd.

14.1   Early Work

In that first study, Fleischmann *et al.* [1] relied on a high-surface area (20 m^2 g^{-1}) electrode material, namely Papyex sheets, which are a graphitic material consisting of 0.15 mm-thick sheets with half the basal planes of the graphite flakes of which the material is composed being parallel to the surface of the sheet. Diffractograms were obtained for Papyex sheets in KI solutions with and without a monolayer of preadsorbed I$_2$ with the incident X-rays normal to (in-plane diffraction) and parallel (out-of-plane diffraction) to the surface of the Papyex sheets. An additional peak was found in the in-plane diffractograms in the presence of the adsorbed I$_2$ monolayer at $2\theta = 19.6°$, and this was attributed to the {10} diffraction of a two-dimensional (2D) layer of hexagonally close-packed adsorbed I$_2$, with a nearest-neighbor spacing of 5.3 Å. The width of the diffraction peak indicated an island size of 110 ± 20 Å, which was in excellent agreement with size of the graphite flakes making up the Papyex sheets.

The results of the first *in-situ* electrochemical study were published three years later, and briefly described the change in the structure of the water layer at the surface of a silver electrode, prepared as a 100 nm-thick film on a Mylar window, upon the underpotential deposition of a lead monolayer. The silver surface was hydrophilic, which led to a more structured water layer at the interface than for the hydrophobic lead surface [2]. A more complete report followed in which cell designs for both transmission and reflection modes of collection of the X-ray diffraction were described [3]. The working electrode for the transmission mode was a thin metal film evaporated onto an X-ray-transparent window, in this case Mylar. A second Mylar window was fixed over the end of a hollow syringe barrel to trap a thin layer of the electrolyte across the surface of the working electrode, with the reference being brought into close contact with this layer of electrolyte via a Luggin capillary (see Figure 14.1a). The working electrode for the reflection mode was a disc of metal sealed into the end of a syringe barrel, which was then pushed against a spacer placed against the Mylar window to define the thickness of the electrolyte layer (Figure 14.1b).

The reflection mode was used to investigate the deposition of thin (~150 nm) films of α-nickel hydroxide onto a smooth nickel electrode from a nickel nitrate solution, and its subsequent transformation into a crystalline β-nickel hydroxide phase upon immersion in concentrated KOH solution [3]. An example of the *in-situ* diffractograms obtained is shown in Figure 14.2a, and is characterized by a broad peak at $2\theta = 30°$, attributed to scattering from the water in the cell, and two narrower peaks at 2θ angles greater than 40° attributed to the {111} and {200} reflections from the nickel substrate. To obtain sufficient sensitivity to observe the nickel hydroxide overlayer, difference diffractograms were accumulated as the potential was modulated between +0.4 and +0.1 V versus SCE at 5×10^{-3} Hz for 6 h in 1 mol dm^{-3} KOH solution to prevent further aging of the deposit. Figure 14.2b shows the resulting difference diffractogram and a sketch in which the various peaks are assigned. The results show that the dominant species at +0.4 V is β-nickel oxyhydroxide, whilst that at +0.1 V is β-nickel hydroxide, in agreement with previous *ex-situ* studies [4,5].

The first *in-situ* X-ray diffraction (XRD) investigations of phase transitions of adsorbed monolayers and multilayers [6] and reconstruction of a metal surface [7, 8] were also reported by Fleischmann and Mao. The phase transitions were reported for the underpotential deposition (upd) and overpotential deposition (opd) of thallium onto a roughened silver electrode surfaces (similar to those used in surface-enhanced Raman spectroscopy (SERS) using the reflection mode of collection), and for upd of lead onto gold and silver

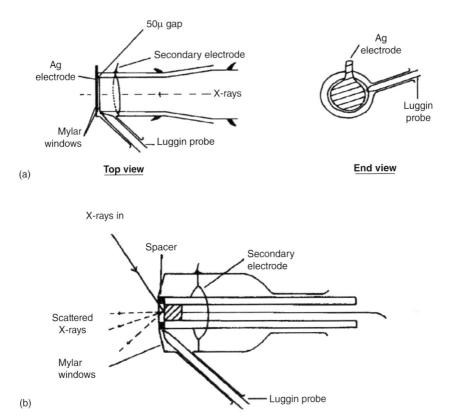

Figure 14.1 *Cell design for (a) transmission mode and (b) reflection mode X-ray diffraction experiments. Reprinted from Ref. [3]. Copyright (1986) with permission from Elsevier.*

thin film electrodes, using the transmission collection mode. Although Fleischmann and Mao were not using single-crystal surfaces, they were still able to show that the first upd layer of Tl on Ag was commensurate with the {111} facets of the Ag surface, whilst the second (opd) layer was not. Surface reconstruction was observed for Pt electrodes using the reflection mode at potentials where hydrogen atoms are weakly adsorbed in 1 mol dm^{-3} H_2SO_4 solution. Initially, the differences between the diffractograms at +0.2 V and −0.2 V were interpreted as the formation of an overlayer of Pt atoms, with little relation to the underlying bulk metal lattice of the polycrystalline substrate [7]. However, upon further consideration this was later interpreted as the formation of {110} facets brought about by extensive cycling between the Pt-oxide and Pt-H regions [9] and the relaxation/expansion of the surface layer of these {110} facets [8].

   That Fleischmann and his coworkers could make such progress towards introducing *in-situ* XRD is testament to their determination to succeed, and the long hours they were willing to work while collecting the data. Most of the studies that followed – as will be introduced briefly in the remainder of this chapter – relied on the increasing availability of synchrotron radiation, as a tunable and higher intensity X-ray source, enabling much more rapid data collection. The detectors also continued to be improved, and this too contributed to the advances described below.

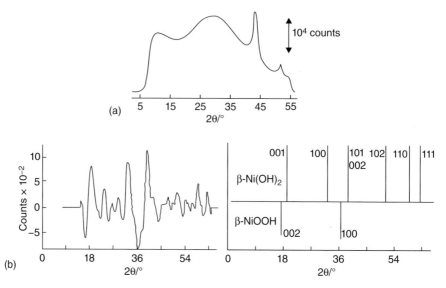

Figure 14.2 *(a) In-situ X-ray diffractogram for the aged Ni(OH)₂ deposit on a Ni substrate electrode at +0.1 V versus SCE in 1 mol dm⁻³ KOH solution following 3 h of data collection time, alternating between +0.4 and +0.1 V at 5 × 10⁻³ Hz. The data were obtained using Cu Kα radiation and a position-sensitive (one-dimensional) detector; (b) Difference diffractograms (data obtained at +0.1 V minus that at +0.4 V) obtained after 6 h of modulation. The sketch plot shows the assignment of the various peaks in the difference diffractogram. Reprinted from Ref. [3]. Copyright (1986) with permission from Elsevier.*

14.2   Synchrotron-Based *In-Situ* XRD

As noted in the first article published by Fleischmann and colleagues on *in-situ* XRD [1], synchrotron radiation sources were beginning to be available for the *in-situ* study of materials and surfaces. However, at that time they would not have been able to predict that synchrotron-based surface XRD would come to dominate *in-situ* XRD studies of electrode surfaces, and especially the study of single-crystal electrode surfaces that have contributed so much to fundamental studies of the electrochemical interface.

The high intensity of the X-ray beam from a synchrotron source enables surface XRD (scattering) studies. These differ from diffraction studies of bulk materials in that the X-ray scattering from the 2D (single-crystal) surface includes both the conventional Bragg peaks from the bulk of the crystal and additional features known as crystal truncation rods (CTRs). The latter run perpendicular to the surface and pass through the Bragg points, as depicted schematically in Figure 14.3 for scattering from a {111} crystal surface [10].

The CTRs are classed as specular when the scattering vector Q is normal to the surface, and nonspecular when there is a component of Q parallel to the surface. The specular CTRs provide information about the structure of the electrode and the electrolyte perpendicular to the surface, whilst the nonspecular CTRs only provide information about the in-plane structure of the electrode. The collection of experimental data involves collecting the scattering intensity along a CTR to yield a so-called "rocking curve" [11], as shown in Figure 14.4

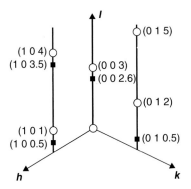

Figure 14.3   *Schematic representation of the reciprocal space for a {111} crystal, showing the bulk Bragg reflections (open circles) and the crystal truncation rods (thick lines). The squares indicate where the data were collected in the study reported in Ref. [10]. Reprinted from Ref. [10]. Copyright (1999) IOP Publishing. Reproduced by permission of IOP Publishing. All rights reserved.*

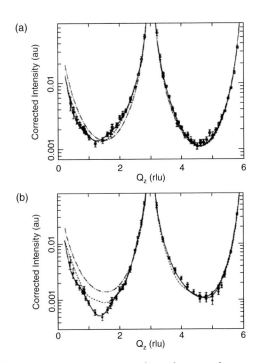

Figure 14.4   *Example of a rocking curve scan along the specular crystal truncation rod of a Ag{111} surface at −0.23 V versus Ag/AgCl (a) and +0.52 V (b), with the best fits shown as the solid lines. The differences in the profiles observed have been attributed to differences in the distribution of water molecules at the interface as a function of the applied potential. Reprinted from Ref. [11]. Copyright (1995) with permission from Elsevier.*

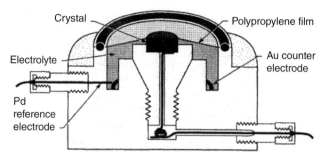

Figure 14.5 *Sectional view of the electrochemical cell commonly employed for SXS measurements at synchrotron beamlines. Note that additional feedthroughs for the solution inlet and outlet and the counter electrode connection are not shown. Reprinted with permission from Ref. [15]. Copyright (1993) by The American Physical Society.*

[11]. These rocking curves are then modeled to yield the structure of the interface, which includes identification of the scattering elements. As the details of the data analysis are beyond the scope of this brief review, readers are referred elsewhere for further information [12–14].

Most *in-situ* electrochemical surface X-ray scattering (SXS) studies use a cell design similar to that shown in Figure 14.5 [15], which is easily mounted on the standard Huber goniometer found on most beamlines. A key feature of this cell design is the polypropylene film window, which can be inflated with solution during electrochemical measurements (such as cyclic voltammetry) and then deflated during collection of the SXS data so as to minimize scattering by the solution. The cell can be placed inside an enclosure that is overpressured with nitrogen to ensure that the solution remains oxygen-free during the long (several hours) data collection periods [10].

14.3   Studies Inspired by Martin Fleischmann's Work

In-situ XRD or SXS studies of single-crystal electrode surfaces have now been used to characterize a large range of electrochemical systems, from fundamental studies of ordering within the electrochemical double layer to studies of the electrodeposition of metal overlayers [16]. Examples of the application of *in-situ* SXS that re-visit and further develop the research areas initially explored by Fleischmann and coworkers, are described in the following sections.

14.3.1   Structure of Water at the Interface

An understanding of the structure of water at the electrochemical interface is of fundamental importance in electrochemistry, as water is such a commonly used solvent. Previously, SXS had been used to study the effects of the applied potential on the orientation and spatial distribution of water molecules at Ag electrode surfaces [11, 17]. Models of the electrochemical interface predict that water molecules form an ice-like structure at the

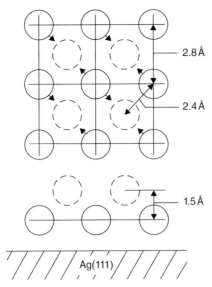

Figure 14.6   *Schematic of the structure of water at the Ag{111} surface in 0.1 mol dm^{-3} NaF at +0.52 V versus Ag/AgCl, top view (top) and side view (bottom). The water molecules lie in two bcc {001} planes; the solid circles represent water molecules in the plane closest to the Ag surface, and the dashed circles represent the second layer. The arrows indicate possible hydrogen bonding between water molecules. Reprinted from Ref. [11]. Copyright (1995) with permission from Elsevier.*

interface with the oxygen of the first layer of water molecules oriented towards the metal surface (oxygen-down) at potentials positive of the potential of zero charge (pzc), and away from the surface (oxygen-up) at potentials negative of the pzc [18]. As noted by Toney *et al.* [17], this would result in a decrease in the spacing between the oxygen atoms and the electrode surface at positive potentials, and this is indeed what they observed. By comparing the SXS data of a Ag{111} surface measured at the nonspecular and specular CTRs at +0.52 and −0.23 V versus Ag/AgCl in 0.1 mol dm^{-3} NaF (see Figure 14.4 above), it could be shown that water molecules near the Ag{111} surface were arranged into several potential dependent layers, as depicted in Figure 14.6. The spacing between the Ag surface and the first plane of O atoms from the water molecules was found to be 1 Å smaller at +0.52 V than at −0.23 V, and the areal density (2.6 × 10^{15} cm^{-2}) of this layer of water molecules was greater than that expected from simulations of the density of bulk water. These differences were attributed to effects of the strong electric field (~10^7 V cm^{-1}) present at the electrode/electrolyte interface. The observed increase in the areal density of water at the Ag surface was consistent with the earlier interpretation by Fleischmann and coworkers that the Ag surface is hydrophilic [2]. It would also be interesting to know if the areal density of water is lower for Pb surfaces, as indicated in Fleischmann's earlier study, where Pb was found to be less hydrophilic; however, that aspect of the study has not yet been repeated.

14.3.2   Adsorption of Ions

The adsorption and ordering of ions at the electrode/electrolyte interface to form the electrochemical double layer is a fundamental concept of electrochemistry. *In-situ* SXS has been used to study both anion adsorption [19] and cation adsorption [20].

Magnussen *et al.* studied the adsorption of bromide adlayers on Au{111} electrodes in 0.1 mol dm^{-3} HClO$_4$ + 0.001 to 0.1 mol dm^{-3} NaBr. In this case, Br$^-$ ions were selected as a model system, in contrast to the more usually studied I$^-$, as Br$^-$ ions are less strongly adsorbed but are still classed as specifically adsorbed ions [19]. At potentials more positive than a critical potential, which would depend on the NaBr concentration, it was found that Br$^-$ ions would form a hexagonal adlayer rotated relative to the $\sqrt{3}$ direction of the Au{111} substrate, with the extent of the rotation and adlayer density being dependent on the potential and NaBr concentration. The Br$^-$–Br$^-$ spacing within the layer was found to vary continuously, from 4.24 Å at the critical potential to 4.03 Å at a potential 0.3 V more positive.

In a more recent study, Lucas *et al.* investigated the interaction of water and K$^+$ ions at the Ag{111} surface in 0.1 mol dm^{-3} KOH solutions as a function of the applied potential [20]. At potentials between −0.8 and −0.4 V versus SCE, the cyclic voltammograms showed a broad reversible feature that had been attributed to the adsorption of OH$^-$ anions, which precedes the formation of the Ag$_2$O surface oxide at potentials positive to −0.1 V. The corresponding X-ray voltammograms (XRV), which measured the scattered X-ray intensity at a structure-sensitive reciprocal lattice position, showed very little change from the Ag{111} surface in the plane of the surface over the −1.0 to −0.2 V potential range explored, as evidenced by measurements at positions corresponding to nonspecular CTRs. In contrast, significant variations were observed in the XRVs measured at a position corresponding to a specular CTR, indicating a substantial change in the structure in the electrolyte near the electrode surface. Rocking curves for the CTRs were recorded at −0.2 and −1.0 V to further investigate the structure (Figure 14.7a–d). A schematic of the interfacial structure is also shown in the lower part of Figure 14.7. Here, at −0.2 V the data were best modeled by a 1.1% inward relaxation of the Ag{111} lattice spacing with an adsorbed layer of OH with a coverage of 0.5 and a layer of hydrated K$^+$ ions (coverage = 0.22 ± 0.05) situated 3.6 ± 0.2 Å from the Ag surface. Interestingly, at −1.0 V the inward surface relaxation of the Ag{111} surface was only 0.7%, and the layer of K$^+$ ions was situated further from the Ag surface at a distance of 4.1 ± 0.3 Å. The shorter Ag–K$^+$ distance observed at the more positive potential was attributed to an increased stabilization of the K$^+$ cations by the coadsorbed OH$^-$ anions.

14.3.3   Oxide/Hydroxide Formation

Oxide film formation is important in a number of areas in electrochemistry, including corrosion, where the oxide film frequently offers protection to the surface, and in electrocatalysis, where oxide formation may limit the rate of the desired reaction or even inactivate the electrocatalyst all together. In 2003, Scherer *et al.* reported a study of passive oxide formation on the Ni surface [21]. In contrast to Fleischmann's earlier study in KOH electrolyte on polycrystalline Ni [3], this more recent investigation was on Ni{111} in air and 0.05 mol dm^{-3} H$_2$SO$_4$ solutions. The SXS measurements were collected *in-situ*, with the window of the cell deflated after the electrochemical reduction or passivation had been completed with

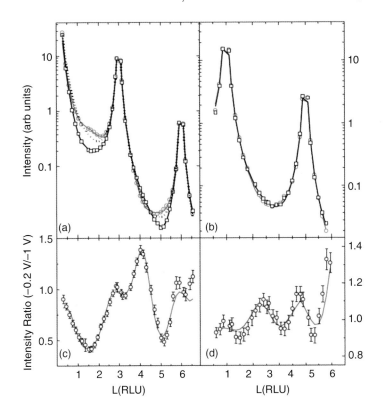

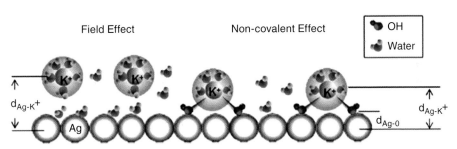

Figure 14.7   *Crystal truncation rod (CTR) data of the Ag{111}/0.1 mol dm^{-3} KOH interface measured at −1.0 V (circles) and −0.2 V (squares) versus SCE (a) the specular CTR, {0, 0, L} and (b) the nonspecular CTR, {1, 0, L}. The CTR data measured at −0.2 V normalized to the data measured at −1.0 V (c) {0, 0, L} and (d) {1, 0, L}. The solid lines are fits to the data according to the structural model shown in the schematic: −1.0 V left side of figure, −0.2 V right side of figure. The dashed line in (a) is a calculation of the specular CTR without inclusion of any ordering in the electrolyte. Reprinted from Ref. [20]. Copyright (2011) with permission from Elsevier.*

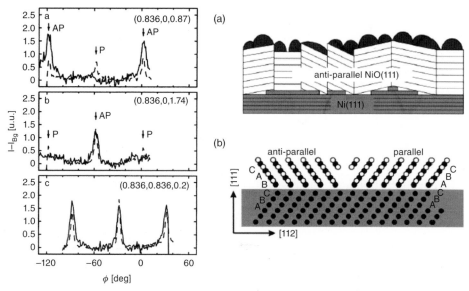

Figure 14.8 *Left: (a–c) Rocking curve scans through three low-order CTRs (indices indicated in the figures) for passive oxide formation on Ni{111} at 0.5 V in 0.05 mol dm⁻³ H₂SO₄ (solid lines) with the data collected for an air-formed oxide (dashed lines). Expected positions of the antiparallel (AP) and parallel (P) NiO phases are indicated by the arrows. Right: (a) Structural model of the film showing the underlying Ni{111} substrate, the NiO{111} crystallites, and an outer Ni hydroxide phase; vertical lines indicate the grain boundaries in the NiO{111} phase; (b) Illustration of the difference in stacking of an antiparallel and parallel {111} layer on to a {111} substrate. Reprinted from Ref. [21]. Copyright (2003) with permission from Elsevier.*

the window inflated. The rocking curves were recorded using the specular CTRs and the data fitted by Gaussians, as shown in Figure 14.8 (left). At +0.5 V versus Ag/AgCl, the diffraction pattern corresponded to crystalline fcc-NiO with the {111} face slightly tilted and with the layers of the fcc structure antiparallel to the underlying Ni{111} surface, as depicted schematically in Figure 14.8 (right). The average diameter of the NiO{111} crystallites was <100 Å, which suggested that a large number of nuclei had formed during formation of the passive layer at +0.5 V. When the potential for passivation was increased (to +0.7 or +0.9 V), the average crystallite size was found to decrease further, indicating the presence of increasing numbers of nuclei at the higher potentials. The thickness of the NiO layer was determined by analysis of the rocking curves, and found to vary approximately linearly with the passivation potential. The detailed information obtained from this careful study of the single-crystal Ni{111} surface, illustrated how far *in-situ* XRD (scattering) studies had developed in less than 20 years.

14.3.4   Underpotential Deposition (upd) of Monolayers

The modification of a metal electrode surface by the controlled deposition of a monolayer of a second metal at potentials positive to the bulk deposition potential is known as "underpotential deposition" (upd) [22]. This area of research has proved to be especially well suited to *in-situ* SXS studies, as scattering from the metal monolayer is more easily detected than scattering from lighter elements. When revisiting the areas of Fleishmann's research, only the applications of *in-situ* SXS to upd of Tl [23], Ag [24–27] and Pb [28,29] will be reviewed at this point.

In 1992, Toney *et al.* investigated the upd of Tl on Ag [23], by examining the {111} surface and comparing their results to those obtained by Fleishmann and Mao on roughened Ag [6]. Thus, data were obtained for both monolayers and bilayers of Tl on Ag{111}. The Tl monolayer structure was found to be incommensurate with the underlying Ag{111}, having a hexagonal 2D structure that was compressed relative to bulk Tl by 1.4–3.0% and rotated from the Ag{011} direction by 4–5°. Thus, it was concluded that the monolayer structure was determined by interactions within the Tl layer, rather than by interactions with the Ag surface. The bilayer was also found to have a hexagonal structure, but had only a 1% compression compared to bulk Tl. On comparing this structure to the interpretation of the Tl deposition on roughened Ag previously reported [6], Toney and coworkers commented that the original interpretation of the enhanced intensity of the Ag{111} peak as evidence of a commensurate Tl adlayer was incorrect. Instead, as the Ag{111} *d* spacing (2.36 Å) was comparable to the spacings obtained between the two Tl layers in the bilayer (~2.8 Å) and between the bottom Tl layer and the Ag surface (~3 Å), it was speculated that the increased intensity was more accurately attributed either to bulk Tl deposition or a Tl bilayer.

In contrast to the other studies that followed on from Fleischmann's work (as described in this chapter), Rayment's group uniquely employed a difference technique similar to Fleischmann's original methodology, namely surface differential diffraction (SDD), in combination with laboratory-based X-ray diffractometry, to study the overlayer structure formed by upd Ag on Au{111} [24]. The SDD method, as the name implies, involved an analysis of the change in diffraction profile accompanying a potential change or adsorption to extract the components of the diffraction originating from the substrate, the adsorbate, and interference between these two components. Use of this technique provided an improved sensitivity and enabled the exploration of subtle changes in the structure of the overlayer during adsorption, with a time resolution on the order of a few milliseconds. The Au{111} surface was prepared as a 300 Å-thick film of Au deposited on to a freshly cleaved mica substrate by vacuum evaporation, and the X-rays were incident on the reverse (through the mica) side of the Au electrode. The Ag upd process was then investigated in 0.002 mol dm^{-3} Ag$_2$SO$_4$ in 0.5 mol dm^{-3} H$_2$SO$_4$. The SDD patterns and calculated adlayer spacings, measured as a function of the potential during a potential sweep, are shown in Figure 14.9 [24]. The results showed that Ag atoms in the first upd adlayer were situated predominantly in the bridge and atop sites with an adlayer spacing of 2.68 ± 0.05 Å. The spacing was found to increase between the first and second layers (to 2.75 Å), and then to decrease upon formation of the third layer (to 2.20 Å) and was 2.35 Å for bulk Ag deposition, corresponding to a transition to threefold hollow sites. The shorter spacing of 2.20 Å could have been accounted for by either an incomplete discharge of the Ag$^+$ ions or a reconstruction of the Au{111} surface, such as a contraction of the spacing between the top two layers of the Au surface. The location of the Ag atoms in the first upd layer at bridge and atop sites in a (1 × 1) structure and at threefold sites in a (3 × 3) structure for bulk deposition of Ag on Au was confirmed with SXS measurements, using synchrotron radiation as the X-ray source. The CTR data and the models for the two conditions are shown in Figure 14.10.

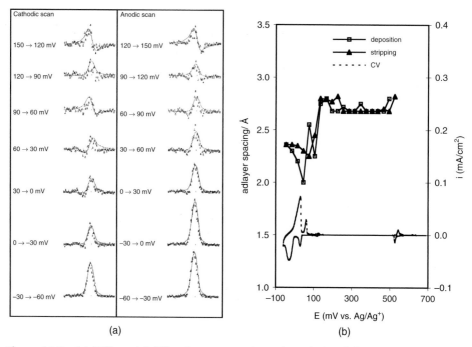

Figure 14.9 (a) Differential diffraction patterns (triangles) obtained during cyclovoltammetric measurements of Au{111} surface in 0.002 mol dm^{-3} Ag$_2$SO$_4$ and 0.5 mol dm^{-3} H$_2$SO$_4$ solution at a scan rate of 2 mV s^{-1}. Patterns were fitted with simulated profiles (solid line) for cathodic and anodic scans. For clarity, only a few potential ranges are shown; (b) Adlayer spacings, deduced from the fitting of (a), plotted as a function of potential. Reprinted with permission from Ref. [24]. Copyright (2002), The Electrochemical Society.

In a series of reports, Kondo *et al.* also investigated the Ag upd on Au system [25–27]. In one report, made during the same year as that of Rayment and coworkers (described above), the Ag atoms were found to be situated in threefold hollow (ccp) sites for both monolayer and bilayer deposition on Au{111} [25]. This difference in interpretation may be due to the higher quality of the Au{111} surface used by Kondo *et al.*, which was a bulk Au single-crystal electrode. The fits to the specular and nonspecular CTR rocking curves, as shown in Figure 14.11, demonstrate a much greater agreement for the ccp site compared to those for the hcp and atop sites. In a subsequent communication, Kondo *et al.* showed that the bilayer was more stable than the monolayer, and that the monolayer would convert to the bilayer when potential control was disconnected [26].

Toney *et al.* also studied the upd of Pb on Ag{111} [28, 29] and Au{111} [29] surfaces using *in-situ* SXS and explored the effect of the substrate, Ag or Au, and the anions in the supporting electrolyte (e.g., perchlorate, weakly adsorbing, or acetate, strongly adsorbing) on the structure of the Pb overlayer. In all cases, the Pb overlayer was found to be incommensurate with the underlying Ag{111} and Au{111} structures, being rotated by several degrees along the {01$\bar{1}$} direction. These results were in agreement with those reported previously by Fleischmann and colleagues for Pb upd on roughened Ag [2] and Au film

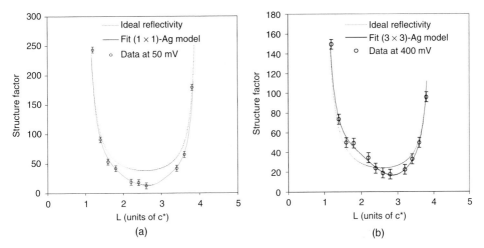

Figure 14.10 *Nonspecular rocking curves across the {1, 0, L}) CTRs from the Au{111} sur-face in 0.002 mol dm⁻³ Ag₂SO₄ and 0.5 mol dm⁻³ H₂SO₄ solution at (a) 50 mV, first upd adlayer and (b) 400 mV, bulk Ag deposition. The solid lines correspond to (a) (1 × 1) and (b) (3 × 3) models, and the dotted lines correspond to the ideally truncated surface. Reprinted with permission from Ref. [24]. Copyright (2002), The Electrochemical Society.*

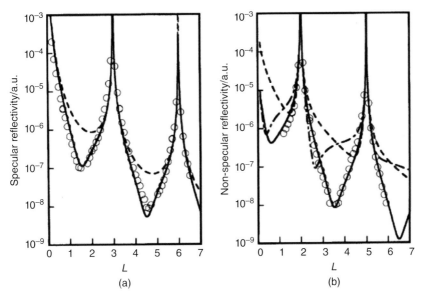

Figure 14.11 *(a) Specular and (b) nonspecular rod profiles measured at 50 mV. (a) The solid and dotted lines are the calculated curves for bare Au{111} and one monolayer Ag on Au{111}, respectively; (b) Solid, dotted, and dot-dashed lines are calculated for the Ag adsorption on the ccp, hcp, and atop sites, respectively, of Au{111}. Reprinted from Ref. [25]. Copyright (2009) with permission from Elsevier.*

electrodes [6], where incommensurate Pb overlayers on the {111} faces were proposed. For both surfaces [29], the structures of the overlayers were found to be dominated by adatom–adatom interactions, as the Pb atoms are much larger than the underlying Ag or Au atoms. The Pb–Pb near-neighbor spacing for deposition on Ag was approximately 0.01 Å less than that on Au, with both being smaller than that for bulk Pb. The effect of the substrate was modeled within the framework of effective medium theory [29], which accounted for the observation that the near-neighbor distance was greater for deposition on the relatively more electron-rich Au surface than on Ag. That the difference was not very large supporting the interpretation of there being a stronger Pb–Ag bond.

The upd of Pb on Pt has also been studied by Adzic *et al.*, using *in-situ* SXS at the {111}, {110}, and {100} single-crystal faces in 0.1 mol dm^{-3} HClO$_4$ [30]. In this case, it was found that Pb formed an ordered (3 × $\sqrt{3}$) structure with a Pb coverage of 0.67, whereas no ordered structures were found for Pb on the {110} and {100} faces. This system was further investigated at Pt{100} [31, 32] and Pt{111} [33] by Lucas *et al.*, who also examined the displacement of the Pb upd layer by CO. In agreement with Adzic *et al.*, Lucas and colleagues reported the (3 × $\sqrt{3}$) structure on Pt{111} but also found an ordered c(2 × 2) structure on the Pt{100} surface (as shown in Figure 14.12). An analysis of the SXS for Pb deposition on Pt{100} showed that, at full monolayer coverage corresponding to a coverage of 0.63, the surface layer of Pt atoms was displaced upwards from its bulk lattice position by 0.03 ± 0.02 Å due to interaction with the Pb [31].

It is interesting to compare the observations of Lucas *et al.*, that Pb deposition has an effect on the Pt interlayer spacing [31], with Fleischmann's observation of loss of order in the low-angle region for Pb deposition onto roughened Ag [2], and the lack of such an

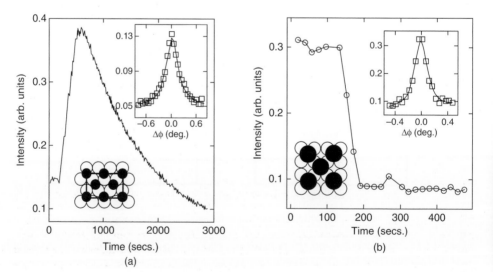

Figure 14.12 *Rocking curves obtained at CTRs corresponding to (a) the (3 × $\sqrt{3}$) Pb overlayer structure on Pt{111} [33] and (b) the c(2 × 2) structure on Pt{100}. The schematics show the overlayer structures [32]. Reprinted from Refs [32, 33]. Copyright (2000) with permission from Elsevier.*

effect for Pb deposition onto Au [6]. Fleischmann attributed the loss of order for the Ag surface to a change in the O–O distance for water at the interface, and the lack of an effect for Au to differences in the double-layer properties of the two metals. Taken together with Toney's comments on the relative strengths of the Pb–Ag and Pb–Au bonds [29], it may be that the difference in diffraction reported by Fleischmann for Ag and Au could be attributed to differences in the relative expansion of the top layers of the underlying metal substrate.

14.3.5 Reconstructions of Single-Crystal Surfaces

The study of single-crystal electrode surfaces has allowed the effects of the applied potential and adsorbates (ions and covalently bound species) on the structure of the metal surface to be determined. In particular, reconstructions of metal electrode surfaces, in which the top layer of atoms adopts a different ordered structure from that of the underlying crystal structure, have been studied extensively [34, 35]. As with upd studies, a number of surface reconstructions have been investigated using SXS [16]. In keeping with the theme of this chapter, only those reports which revisited the reconstruction of the Pt surface under the influence of adsorbed hydrogen, as reported previously by Fleischmann [8], have been included in the following sections.

The effects of hydrogen adsorption on the low-index {111} [35], {100} [36] and {110} [37, 38] planes of Pt have been investigated by Ross and coworkers, using synchrotron-based *in-situ* SXS. In full agreement with Fleischmann's earlier studies [8], Ross *et al.* found that the adsorption of hydrogen is accompanied by an expansion of the top layer of Pt atoms. For the {111} and {100} surfaces, which do not reconstruct, the expansion in 0.1 mol dm^{-3} KOH was 0.05 Å for Pt{100} and 0.03 Å for Pt{111}, whereas in 0.1 mol dm^{-3} HClO$_4$ it was 0.03 Å for Pt{100} and 0.015 Å for Pt{111} [35, 36]. However, for Pt{110}, which exhibits a (1 × 2) reconstruction, the expansion was more dramatic, being as much as 25% or 0.35 ± 0.11 Å at 0.1 V versus a Pt reference in 0.1 mol dm^{-3} NaOH [37].

Hoshi *et al.* have used *in-situ* SXS to study the effects of hydrogen adsorption on the surface structure of stepped Pt {311} [39] and {331} and {511} [40] surfaces in 0.1 mol dm^{-3} HClO$_4$. The {311} surface undergoes a missing row reconstruction to form the (1 × 2) surface. Interestingly, the adsorption of hydrogen under electrochemical conditions did not result in an expansion of the topmost Pt layer for this surface, which indicated that the reconstructed surface was more rigid than the lower-index planes [39]. The period of the corrugation of the stepped Pt surface and the effects of hydrogen adsorption increased in the order of {311} < {331} < {511}: 0% for the {311} surface, 13% for the {331} and 37% for the {511} [40].

14.3.6 High-Surface-Area Electrode Structures

Although the application of *in-situ* XRD to studies of electrode surfaces and structures continues to be investigated by the electrochemistry group at Southampton, the studies are currently more focused on examining the practical high-surface-area electrode materials that are used as electrocatalysts in fuel cells and as electrode materials in batteries, rather than on the structure of the adsorbed layers described above.

Carbon-supported metal nanoparticles are often employed as electrocatalysts in low-temperature, proton-exchange membrane, fuel cells. In the Southampton group, *in-situ* XRD has been used as a probe for both the composition and particle size as a function

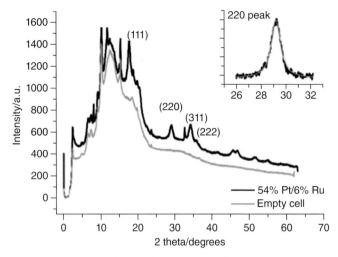

Figure 14.13 In-situ XRD pattern for a carbon-supported 54 wt% Pt/6 wt% Ru catalyst electrode at 0.0 V versus RHE in 1 mol dm^{-3} H$_2$SO$_4$ and the empty cell. The inset shows an example of the peak fit for the {220} reflection used to determine the particle size. Reproduced from Ref. [42]. Copyright (2009) with permission of the PCCP Owner Societies.

of the applied potential. The data have been obtained using a transmission geometry, with synchrotron radiation as the X-ray source [41]. For example, a series of PtRu/C catalysts used both for enhanced CO tolerance for reformatted fuel, H$_2$/O$_2$ fuel cells and enhanced methanol oxidation activity for direct methanol fuel cells (DMFCs) were studied with *in-situ* XRD. Studies of varying Pt:Ru ratio catalysts showed that the electrochemical environment and the electrode potential [measurements were made at 0.0 V versus RHE where hydrogen is adsorbed (see Figure 14.13) and 1.05 V versus RHE at the onset of oxide formation] had little effect on the crystal structure and the crystallite size of these catalysts [42].

In-situ XRD with synchrotron radiation has also been used to study the lattice expansion accompanying hydride formation at negative potentials for a Pd/C catalyst in 1 mol dm^{-3} H$_2$SO$_4$ [43]. Whilst this particular catalyst is of little interest in the acidic environment of a polymer electrolyte membrane (PEM) fuel cell, because Pd is known to dissolve under the conditions present during oxygen reduction at the cathode, it has been of interest as the core in a core–shell catalyst with a Pt shell [44, 45]. Consequently, *in-situ* XRD has been used (among other techniques) to study the stability of the core under electrochemical conditions [46]. Ultimately, the *in-situ* XRD measurements were found to be of great value in establishing how the electrochemical testing procedure might contribute to the accelerated degradation of core–shell catalysts.

Most recently at Southampton, Owen and Hector and their research groups have used *in-situ* XRD to study phase changes during the discharge of an electrode composed of LiFePO$_4$, carbon, and a poly(tetrafluoroethylene) (PTFE) binder in a lithium half-cell (M.R. Roberts, unpublished results). These measurements also relied on synchrotron radiation as the X-ray source. In this case, by constructing a cell in which the incident angle of the X-rays could be changed it was possible to vary the depth of penetration of the X-rays and, thereby, to examine the Li concentration gradient within the material as a function of the extent of discharge.

14.4 Conclusions

The application of *in-situ* XRD/SXS to studies of the electrode/electrolyte interface, as inspired by the research of Martin Fleischmann, has helped to shape the present understanding of the structure of the electrolyte and ions at the interface, the structure of the metal electrode itself, and the effects of adsorbed species and oxides on this structure, and the structure of overlayers. Future investigations using SXS are likely to involve the examination of increasingly complex surface structures, including those of alloys [47], while *in-situ* XRD is likely to be of increased value in the study of practical high-surface-area materials [42, 43, 46]. The completion of this chapter has highlighted areas where, with the relatively limited laboratory-based resources available to them, Fleischmann and coworkers largely anticipated the results that have been obtained at later stages, using more sophisticated synchrotron-based methods. This clearly, is a testament to Martin Fleischmann's forward-thinking approach to research.

References

(1) Fleishmann, M., Hendra, P.J. and Robinson, J. (1980) X-ray diffraction from adsorbed iodine on graphite. *Nature*, **288**, 152–154.
(2) Fleischmann, M., Graves, P., Hill, I. *et al.* (1983) Raman spectroscopic and X-ray diffraction studies of electrode-solution interfaces. *Journal of Electroanalytical Chemistry*, **150**, 33–42.
(3) Fleishmann, M., Oliver, A. and Robinson, J. (1986) In situ X-ray diffraction studies of electrode solution interfaces. *Electrochimica Acta*, **31**, 899–906.
(4) Briggs, G.W.D. and Wynne-Jones, W.F.K. (1962) The nickel hydroxide electrode; the effects of ageing – I. X-ray diffraction study of the electrode process. *Electrochimica Acta*, **7**, 241–248.
(5) Bode, H., Dehmelt, K. and Witte, J. (1969) Zur Kenntnis der Nickelhydroxidelektrode. II. Über die Oxydationsprodukte von Nickel(II)-hydroxiden. *Zeitschrift für Anorganische und Allgemeine Chemie*, **366**, 1–21.
(6) Fleishmann, M. and Mao, B.W. (1988) In-situ X-ray diffraction investigations of the upd of Tl and Pb on Ag and Au electrodes. *Journal of Electroanalytical Chemistry*, **247**, 297–309.
(7) Fleishmann, M. and Mao, B.W. (1987) In situ X-ray diffraction studies of platinum. *Journal of Electroanalytical Chemistry*, **229**, 125–139.
(8) Fleishmann, M. and Mao, B.W. (1988) In situ X-ray diffraction measurements of the surface structure of Pt in the presence of "weakly" adsorbed H. *Journal of Electroanalytical Chemistry*, **247**, 311–313.
(9) Yamamoto, K., Kolb, D.M., Kotz, R. and Lehmpfuhl, G. (1979) Hydrogen adsorption and oxide formation on platinum single crystal electrodes. *Journal of Electroanalytical Chemistry*, **96**, 233–239.
(10) Lucas, C.A. (1999) Atomic structure at the electrochemical interface. *Journal of Physics D: Applied Physics*, **32**, A198–A201.
(11) Toney, M.F., Howard, J.N., Richer, J. *et al.* (1995) Distribution of water molecules at Ag{111}/electrolyte interface as studied with surface X-ray scattering. *Surface Science*, **335**, 326–332.
(12) Robinson, I.K. (1983) Direct determination of the Au{110} reconstructed surface by X-ray diffraction. *Physical Review Letters*, **50**, 1145–1148.

(13) Feidenhans'l, R. (1989) Surface structure determination by X-ray diffraction. *Surface Science Reports*, **10**, 105–188.

(14) Robinson, I.K. (1991) Surface crystallography, in *Handbook on Synchrotron Radiation*, vol. **3** (eds G.S. Brown and D.E. Moncton), North-Holland, Amsterdam, pp. 221–266

(15) Tidswell, I.M., Markovic, N.M., Lucas, C.A. and Ross, P.N. (1993) In situ X-ray scattering study of the Au{001} reconstruction in alkaline and acidic electrolytes. *Physical Review B*, **47**, 16542–16553.

(16) For a more comprehensive review than that presented here please see: Lucas, C.A. and Markovic, N.M. (2006) *In situ* X-ray diffraction studies of the electrode/solution interface, in *Advances in Electrochemical Science and Engineering*, vol. **9** (eds R.C. Alkaire, D.M. Kolb, J. Lipkowski and P.N. Ross), Wiley-VCH, Weinheim, pp. 1–45.

(17) Toney, M.F., Howard, J.N., Richer, J. *et al.* (1994) Voltage-dependent ordering of water molecules at an electrode–electrolyte interface. *Nature*, **368**, 444–446.

(18) See, Theil, P.A. (1987) The interaction of water with solid surfaces: fundamental aspects. *Surface Science Reports*, **7**, 211–385 for an excellent review and references in [11, 15].

(19) Magnussen, O.M., Ocko, B.M., Wang, J.X. and Adzic, R.R. (1996) In-situ X-ray diffraction and STM studies of bromide adsorption on Au(111) electrodes. *Journal of Physical Chemistry*, **100**, 5500–5008.

(20) Lucas, C.A., Thompson, P., Gründer, Y. and Markovic, N.M. (2011) The structure of the electrochemical double layer: Ag{111} in alkaline electrolyte. *Electrochemistry Communications*, **13**, 1205–1208.

(21) Scherer, J., Ocko, B.M. and Magnussen, O.M. (2003) Structure, dissolution, and passivation of Ni{111} electrodes in sulfuric acid solution: an in situ STM, X-ray scattering, and electrochemical study. *Electrochimica Acta*, **48**, 1169–1191.

(22) See this review article for other examples: Herrero, E., Buller, L.J. and Abruña, H.D. (2001) Under potential deposition at single crystal surfaces of Au, Pt, and Ag and other materials. *Chemical Reviews*, **101**, 1897–1930.

(23) Toney, M.F., Gordon, J.G., Samant, M.G. *et al.* (1992) Underpotentially deposited thallium on silver {111} by *in situ* surface X-ray scattering. *Physical Review B*, **45**, 9362–9374.

(24) Lee, D., Ndieyira, J.W. and Rayment, T. (2002) In situ time-resolved structural study of an electrode process by surface differential diffraction, in *Spectroscopic tools for the analysis of electrochemical systems* (ed. J. McBreen), Electrochemical Society Proceedings, vol. **99** (15), pp. 111–121.

(25) Kondo, T., Morita, J., Okamura, M. *et al.* (2009) In situ structural study on underpotential deposition of Ag on Au{111} electrode using surface X-ray scattering technique. *Journal of Electroanalytical Chemistry*, **532**, 201–205.

(26) Kondo, T., Takakusagi, S. and Uosaki, K. (2009) Stability of underpotentially deposited Ag layers on a Au{111} surface studied by surface X-ray scattering. *Electrochemistry Communications*, **11**, 804–807.

(27) Kondo, T., Tamura, K., Takakusagi, S. *et al.* (2009) Partial stripping of Ag atoms from silver bilayer on a Au{111} surface accompanied with the reductive desorption of hexanethiol SAM. *Journal of Solid State Electrochemistry*, **13**, 1141–1145.

(28) Melroy, O.R., Toney, M.F., Borges, G.L. *et al.* (1988) Two-dimensional compressibility of electrochemically adsorbed lead on silver {111}. *Physical Review B*, **38**, 10962–10965.

(29) Toney, M.F., Gordon, J.G., Samant, M.G. *et al.* (1995) In-situ atomic structure of underpotentially deposited monolayers of Pb and Tl on Au{111} and Ag{111}: a surface X-ray scattering study. *Journal of Physical Chemistry*, **99**, 4733–4744.

(30) Adzic, R.R., Wang, J., Vitus, C.M. and Ocko, B.M. (1993) The electrodeposition of Pb monolayers on low-index Pt surfaces – an X-ray scattering and scanning-tunneling-microscopy study. *Surface Science*, **293**, L876–L883.

(31) Lucas, C.A., Markovic, N.M. and Ross, P.N. (1997) Underpotential deposition of lead on to Pt{001}: interface structure and the influence of adsorbed bromide. *Langmuir*, **13**, 5517–5520.

(32) Lucas, C.A., Markovic, N.M., Grgur, B.N. and Ross, P.N. (2000) Structural effects during CO adsorption on Pt-bimetallic surfaces I. The Pt{100} electrode. *Surface Science*, **448**, 65–76.

(33) Lucas, C.A., Markovic, N.M. and Ross, P.N. (2000) Structural effects during CO adsorption on Pt-bimetallic surfaces II. The Pt{111} electrode. *Surface Science*, **448**, 77–86.

(34) Kolb, D.M. (1996) Reconstruction phenomena at metal-electrolyte interfaces. *Progress in Surface Science*, **51**, 109–173.

(35) Tidswell, I.M., Markovic, N.M. and Ross, P.N. (1994) Potential dependent surface structure of the Pt{111}|electrolyte interface. *Journal of Electroanalytical Chemistry*, **376**, 119–126.

(36) Tidswell, I.M., Markovic, N.M. and Ross, P.N. (1993) Potential dependent surface structure of the Pt{001}/electrolyte interface. *Physical Review Letters*, **71**, 1601–1604.

(37) Lucas, C.A., Markovic, N.M. and Ross, P.N. (1996) Surface structure and relaxation at the Pt{110}/electrolyte interface. *Physical Review Letters*, **77**, 4922–4925.

(38) Markovic, N.M., Grgur, B.N., Lucas, C.A. and Ross, P.N. (1997) Surface electrochemistry of CO on Pt{110}-(1 × 2) and Pt{110}-(1 × 1) surfaces. *Surface Science*, **384**, L805–L814.

(39) Nakahara, A., Nakamura, M., Sumitani, K. *et al.* (2007) In situ surface X-ray scattering of stepped surfaces of platinum: Pt{311}. *Langmuir*, **23**, 10879–10822.

(40) Hoshi, N., Nakamura, M., Sakata, O. *et al.* (2011) Surface X-ray scattering of stepped surfaces of platinum in an electrochemical environment: Pt(331) = 3(111)-(111) and Pt(511) = 3(100)-(111). *Langmuir*, **27**, 4236–4242.

(41) Maniguet, S., Mathew, R.J. and Russell, A.E. (2000) Carbon monoxide oxidation on supported Pt fuel electrocatalysts. *Journal of Physical Chemistry B*, **104**, 1998–2004.

(42) Wiltshire, R.J.K., King, C.R., Rose, A. *et al.* (2009) Effects of composition on the structure and activity of PtRu/C catalysts. *Physical Chemistry Chemical Physics*, **11**, 2305–2313.

(43) Rose, A., Maniguet, S., Mathew, R.J. *et al.* (2003) Hydride phase formation in carbon supported palladium nanoparticle electrodes investigated using in situ EXAFS and XRD. *Physical Chemistry Chemical Physics*, **5**, 3220–3225.

(44) Wells, P.P., Crabb, E.M., King, C.R. *et al.* (2009) Preparation, structure, and stability of Pt and Pd monolayer modified Pd and Pt electrocatalysts. *Physical Chemistry Chemical Physics*, **11**, 5773–5781.

(45) Tessier, B.C., Russell, A.E., Theobald, B. and Thompsett, D. (2009) PtML/Pd/C core-shell electrocatalysts for the ORR in PEMFCs. *ECS Transactions*, **16**, 1–11.

(46) Wise, A.M. (2012) Characterisation of bimetallic alloy and core-shell electrocatalysts. PhD thesis, University of Southampton.

(47) Lucas, C.A., Cromack, M., Gallagher, M.E. *et al.* (2008) From ultra-high vacuum to the electrochemical interface: X-ray scattering studies of model electrocatalysts. *Faraday Discussions*, **140**, 41–58.

15

Tribocorrosion

Robert J.K. Wood
University of Southampton, Engineering Sciences, UK

Martin Fleischmann always sought to build bridges between fundamental electrochemistry and engineering, including mechanical engineering. During the late 1980s, he led a project to develop a liquid Van de Graaff generator which used microelectrode arrays and line electrodes to form a cloud of charged water droplets that could potentially be used to coat and help disable ballistic missiles in space. Martin was passionate about the research into single ion electrochemistry, and always positive at project meetings. Although most of his efforts were invested in cold fusion research in the United States, he still found the time and energy to develop the underpinning theory to help interpret the experimental results on microelectrode arrays and micro-line electrodes. Although my career has centered on tribocorrosion, the Fleischmann approach to and enthusiasm for research has remained with me.

15.1 Introduction and Definitions

The term "tribo" has its origins in the Greek word "tribos" (rubbing), which led to the sciences of lubrication, friction and wear being called "tribology," while the term "corrosion" has its origins in the medieval Latin word "corrodere," which mean to gnaw through, with "cor" meaning intensive force and "rodere" to gnaw [1]. Therefore, the term tribocorrosion refers to the surface degradation mechanisms when mechanical wear and chemical/electrochemical processes interact with each other. The subject marries two personal passions of the author, inspired by three professors who taught or supervised me: Freddie Barwell (tribology); Lionel Shrier (corrosion); and Martin Fleischmann (electrochemistry).

Developments in Electrochemistry: Science Inspired by Martin Fleischmann, First Edition.
Edited by Derek Pletcher, Zhong-Qun Tian and David E. Williams.
© 2014 John Wiley & Sons, Ltd. Published 2014 by John Wiley & Sons, Ltd.

15.1.1   Tribocorrosion

The subject of tribocorrosion includes the interaction of corrosion and erosion (solids, liquid flow and droplet impingements or cavitation bubbles), abrasion, adhesion, fretting and fatigue wear processes. Tribocorrosion is often linked to the synergy resulting from the coupling of mechanical and environmental effects. Many of these mechanoelectrochemical interactions are time-dependent and nonlinear. Tribocorrosion research aims to address the need to select or design new surfaces for future equipment, as well as to minimize the operating costs and extend the life of existing machinery and medical devices. For example, passive alloys (i.e., most engineering alloys) rely on a 1 to 10 nm-thick surface film for their protection against aggressive and corrosive environments. This film is formed by reaction with the environment, but wear can lead to the local rupture or complete removal of the film and permanent deformation of the substrate. This can in turn lead to areas of the substrate being exposed to the aggressive environment, and unless repassivation (repair or self-healing) mechanisms can reform the passive film to inhibit corrosion processes, then an accelerated anodic dissolution will occur within these sites.

On the positive side, tribocorrosion phenomena can be used as a manufacturing process, such as in the chemical mechanical polishing (CMP) of silicon wafers. An enhanced material removal for CMP processes has been achieved using electrochemical techniques [2]. The coupling of mechanical and environmental effects can also create surfaces of specific reaction layers on materials which could inhibit corrosion and/or wear. Examples of this are self-lubricating and/or self-healing surface layers [3].

Unpredictable interactions can result between wear and corrosion when the surfaces in contact have complex, multiple-phase microstructures that can lead to microgalvanic activity and selective phase corrosion (a localized attack), as well as three-body wear modes. Examples of such surfaces include composites or surfaces that undergo compositional changes induced by tribological interactions. For instance, the presence of carbides in a metallic surface, typically formed for improved wear resistance, establishes a microgalvanic corrosion cell as the carbide is likely to be cathodic with respect to the surrounding metallic matrix [4]. This can result in a preferential anodic dissolution of the metallic matrix close to or at the matrix/carbide interface, and thereby accelerate carbide removal from surfaces and reduce the antiwear properties of the surface.

15.1.2   Erosion

Erosion is one of several wear modes involved in tribocorrosion. Solid particle erosion is a process by which discrete small solid particles, with inertia, strike the surface of a material, causing damage or material loss to its surface. This is often accompanied by corrosion due to the environment. A major environmental factor with significant influence on erosion–corrosion rates is that of flow velocity, but this should be set in the context of the overall flow field as other parameters such as wall shear stress, wall surface roughness, turbulent flow intensity and mass transport coefficient (this determines the rate of movement of reactant species to reaction sites and thus can relate to corrosion wall wastage rates). For example, a single value of flow velocity, referred to as the "critical velocity," is often quoted to represent a transition from flow-induced corrosion to enhanced mechanical–corrosion interactive erosion–corrosion processes. It is also used to indicate the resistance of the passive and protective films to mechanical breakdown [5].

The velocity profiles and transverse momentum transfer close to the solid/liquid interface dictate wall shear stress levels and mass transport efficiencies, both of which are important drivers for erosion–corrosion. Therefore, critical velocity values are very geometry-specific and cannot be readily applied to predict component service life in generic flow systems.

15.2   Particle–Surface Interactions

When a corrosion rate is partially or wholly controlled by mass transfer of reactants to, or products from, the surface then local conditions under erosion may well influence the mass transfer kinetics (measured by the mass transfer coefficient k_m). Under such conditions the corrosion will be controlled by the mass-transfer (typically diffusion processes) and the driving concentration gradient (relative concentrations of active species near the surface compared to free stream concentrations) [6]. Both, the mass transfer and concentration gradient will be affected by solid particle impingement that influences the local fluid flow field and increases surface roughness. Erosion will also increase the surface area wetted by the corrosive electrolyte, and could establish microgalvanic cells on the surface with damaged areas being anodic to the passive and (cathodic) unaffected areas. Erosion will also increase the dislocation density in affected surface areas, which can lead to potential differences between eroded and less-eroded areas being established and the formation of anodic and cathodic sites that would generate a microgalvanic effect enhancing material loss from the surface. It is also likely that the corrosion kinetics will be altered on eroded areas.

15.3   Depassivation and Repassivation Kinetics

As suggested above, the depassivation of film surfaces is associated with the flow velocity. Figure 15.1 shows the trend for such a passive system as a function of flow velocity and

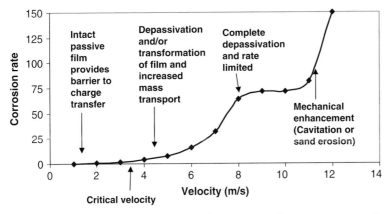

Figure 15.1 *Possible corrosion trends (with arbitrary units) for a passive metallic surface. Reprinted from Ref. [7]. Copyright (2007) with permission from Elsevier.*

its vulnerability to mechanical processes which can accelerate corrosion loss rates due to depassivation.

15.3.1   Depassivation

Depassivation in erosion–corrosion processes is assumed to be rapid and associated with the mechanical removal or stripping of the passive surface layers through particle, cavitation bubble or liquid droplet impact. Figure 15.2 illustrates the level of damage to passive films as a function of the solid particle impingement angle. Mechanical removal and/or rupture of the passive layer enables charge transfer to proceed at varying rates, such that parent metal dissolution is likely. "Recovery" or repassivation aspects of the passive film are therefore important but are very system-dependent.

Figure 15.3 shows the current response over time for individual solid particle impacts on the naturally passivating system of stainless steel. Some impacts will only result in partial passive layer removal or cracking which will influence repassivation kinetics and possibly the composition and thickness of the regrown layer.

Figure 15.4 is an attempt to map surface responses to the whole range of erosion and corrosion mechanisms from erosion dominated through erosion–corrosion, corrosion–erosion zones to corrosion-dominated mechanisms. To identify which zone a particular application might be operating under is difficult without some data on the erosion and corrosion performance of the surface in question. However, the sand particle energy is directly related to the likely mechanical erosion damage rates given by T-C, where T is the total loss from wear-corrosion and C is the corrosion loss in the absence of sand particles. This will define a position on the T-C axis of Figure 15.4 and the rate of depassivation, as this is likely to be linked to the kinetic energy of the sand particle E_k. Surface repassivation, on the other hand, is material-electrolyte-flow field-specific, and this will dictate the corrosion rate C. Three

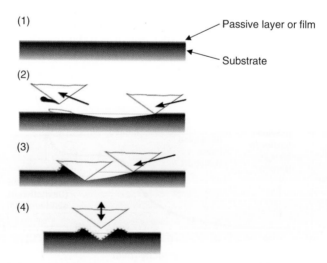

Figure 15.2 *Schematic showing the dependency of passive layer damage on the particle impact angle. The numbers (2) to (4) show the effect of increasing the impact angle. Reprinted from Ref. [7]. Copyright (2007) with permission from Elsevier.*

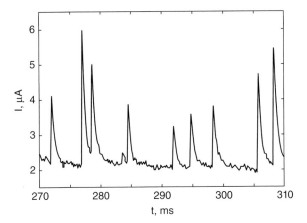

Figure 15.3 *Current transients in erosion–corrosion due to depassivation/repassivation. Reproduced from Ref. [8]. Reprinted by permission of the publisher (Taylor & Francis Ltd, http://www.tandf.co.uk/journals).*

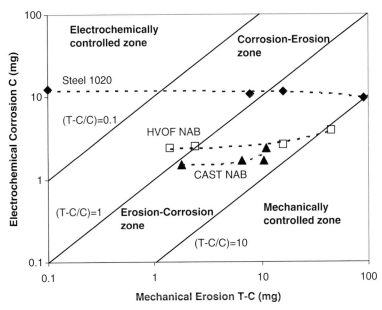

Figure 15.4 *Mapping electrochemical material loss against mechanical erosion rates for a nonpassivating surface carbon steel (AISI 1020) along with two potentially passivating surfaces of nickel aluminum bronze (NAB): one that has been thermally sprayed by high-velocity oxy-fuel deposition as a coating on carbon steel (□); and another which has been cast (▲). These results were obtained from jet impingement erosion–corrosion tests. Reprinted from Ref. [7]. Copyright (2007) with permission from Elsevier.*

systems are plotted on this mechanical versus corrosion map. Two are passivating systems based on cast and high-velocity oxy-fuel (HVOF) nickel–aluminium–bronze (NAB) surfaces in 3.5% NaCl solution, and have a positive slope (increase with increasing mechanical erosion effects). This increase is related to the repassivation kinetics being unable to repassivate the eroded surface under increasing erosion. The third system is carbon steel in saline solution, which is not passivating and has a slope near to zero or slightly negative. Here, the corrosion products formed on the surface are nonadherent and easily removed, such that the trend is dominated by the erosion rate while the corrosion rate remains relatively constant. These results were obtained using a free jet impingement erosion–corrosion rig.

15.3.2 Repassivation Rate

A set of reciprocating wear tests rubbing a 6 mm-diameter alumina ball against cast NAB in 3.5% NaCl solution at room temperature have been conducted by the author, with a stroke length of 21.5 mm between 1 and 12 Hz frequency and an applied load of 5 to 35 N. The electrochemical response is presented in Figure 15.5. The test comprised a 1 h static immersion period with the pin loaded onto the test surface without reciprocation, and this was followed by a 1 h wear-corrosion test. As seen, the level of increased current was dependent on a combination of load and speed. Figure 15.6 shows the currents from a set of experiments conducted at 35 N, but with varying frequencies. Clearly, the corrosion current increases with increasing frequency but this increase is nonlinear. A transition from a relatively low current between 1–5 Hz and higher currents above 10 Hz is seen. The interesting trend is seen at 8 Hz, where initially the current was low (40 μA) but rose steadily throughout the wear test period to 80 μA. This indicated a transition between

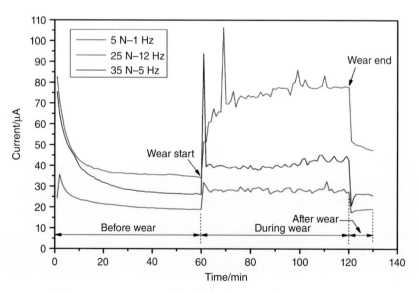

Figure 15.5 *Sliding wear-corrosion of nickel aluminum bronze in 3.5% NaCl solution. Stroke 21.5 mm, 6 mm-diameter alumina ball; reference electrode Ag/AgCl; counterelectrode graphite; Static 60 min; wear-corrosion 60 min and 10 min post wear-corrosion.*

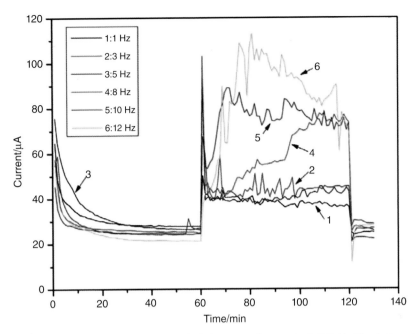

Figure 15.6 *Sliding wear-corrosion of nickel aluminum bronze in 3.5% NaCl solution at 35 N and room temperature.*

predominantly passivated to depassivated conditions in the wearing contact. This frequency coincided with the expected repassivation time for this material of between 150 and 200 ms. It is also unclear whether trends predicted for depassivated surfaces hold here. Equation (15.1) predicts wear current for nonpassivating surfaces and shows a dependence on load × the square root of frequency [9]. However, such trends are not seen here, nor between wear volume or friction as a function of frequency. Hence, temporal effects of repassivation are clearly important but wear mechanisms and friction levels are also varying to complicate analysis and illustrate the complex nature of tribocorrosion of sliding couples.

$$I_r = A_r i_r = K_w l f \sqrt{\frac{F_n}{H}} \int_0^{1/f} i_a dt \tag{15.1}$$

where I_r is the wear-accelerated current, A_r is the real repassivation area, i_r is the real repassivation current density, l is the sliding distance, f is the sliding frequency, F_n is the normal load, H is the hardness of the test material, K_w is a proportionality constant, and i_a is the anodic current density.

15.4 Models and Mapping

Erosion damage by solid particle impact or cavitation bubble collapse to an oxide or passive film will reveal the underlying nascent surface inducing a higher activity (higher corrosion current), for a limited duration, than for the intact oxide surface. Bozzini *et al.* [10] employed a simple approximate model using a "recovering target" concept. This erosion–corrosion model has the advantage in that it can be applied to both passivating and actively corroding conditions. The impacting particles are modeled with rigid monodispersed spheres of radius r_p. The particle impact process is assumed to be Poissonian with a frequency of impact, λ. It is assumed that each impact gives rise to an alteration of the corrosion rate through a localized change in corrosion current density for a period of time, relating to a recovery to the unaffected state. The effective corrosion current density, i_{corr} (nA cm^{-2}), at a given electrode potential (typically the corrosion potential) can be related to the mechanically affected corrosion component of the synergistic damage through a coefficient, f_a, (such that $0 \le f_a \le 1$) expressing the fraction of the corroding surface which is affected by the erosive action of impinging particles, by Equation (15.2):

$$i_{corr} = f_a i_a + (1 - f_a) i_u \qquad (15.2)$$

where the subscripts a and u indicate "affected" and "unaffected," respectively. The current densities i_a and i_u are characteristic for the corroding material in the absence and in the presence of the erosive action, and can be measured separately by means of suitable experiments. In general, the coefficient f_a can be defined in Equation (15.3):

$$f_a = \left(\frac{\text{no. of impacts}}{\text{control area}} \right) \times \left(\frac{\text{damaged area}}{\text{impact}} \right) \times \text{recovery time} = \lambda A_a \tau \qquad (15.3)$$

where A_a is the affected surface area and τ is the passive recovery time.

Equation (15.2) indicates an accelerating element to corrosion caused by film damage or removal, and is one element of the complex interactions between erosion and corrosion that needs to be understood. These interactions can be defined as follows. The total damage under wear-corrosion, T, can be represented as:

$$T = E + C + S \qquad (15.4)$$

where E is the pure erosion material loss, and C is the solids free flow corrosion rate [possibly derived from the corrosion current density i in Equation (15.2)]. The synergistic effect (interactive term), S, is referred to as ΔE_c or $(\Delta C_e + \Delta E_c)$, depending on the literature source, where ΔE_c is the enhanced erosion loss due to corrosion and ΔC_e is the enhanced corrosion due to erosion. The term ΔC_e can be partially expressed by Equation (15.2), but additional terms are required relating to the effect of the erodent deforming the surface and leading to increased corrosion activity. The S terms, and how they should be measured, are given by the ASTM G119-93 standard, which is a useful guide to evaluate synergy [11]. It is also important to note that the synergistic term S can be either positive or negative. Bozzini *et al.* [10] showed that annealed carbon steel has a more active corrosion potential before it is work-hardened, resulting in $i_u > i_a$, and by using Equation (15.2) shows a reduction (i.e., a negative synergy) in the overall corrosion rate with erosion present.

It has also been shown that synergistic effects, which result in damage due to separate corrosion and erosion processes, are normally greater than the sum of the individual damage processes and can accelerate material removal significantly (i.e., $T = 50 \times (C + E)$ for grey cast iron; see Wood and Hutton [12] and Figure 15.1).

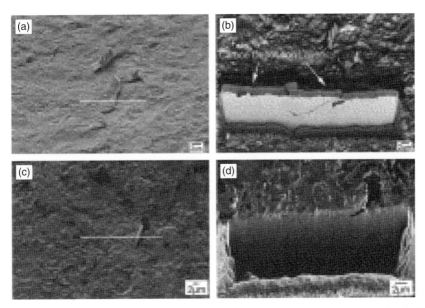

Figure 15.7 *(a, c) Secondary electron images of erosion–corrosion surfaces; (b, d) Corresponding location (white line) of the focused ion beam (FIB) cross-sections. Reprinted from Ref. [13]. Copyright (2007) with permission from Elsevier.*

However, the ability to predict or map erosion–corrosion relies on including the correct mechanisms and material responses (i.e., phase transformations).

The use of modern advanced surface analysis techniques such as focused ion beam (FIB) and tunneling electron microscopy (TEM) allows the detailed subsurface characterization of eroded samples at the nanoscale (see Figure 15.7) [13]. Slurry pot erosion and erosion–corrosion testing of cylindrical samples of 316 stainless steel have been conducted in abrasive silica slurry at 7 m s^{-1} velocity for 1 h. Conventional scanning electron microscopy showed that both erosion and erosion–corrosion surfaces exhibited similar surface features, including impact crater, particle embedment and crater lip formation. FIB cross-sectioning of similar areas from both surfaces revealed a nanograin surface layer as well as subsurface cracks, attributed to the flow of material from the edge of impact crater lips. Site-specific FIB–TEM lamella samples were created from cross-sections and showed how subtle differences could exist as a result of corrosion activity. Selected area diffraction patterns revealed how particle impact was responsible for the formation of martensite from the austenite microstructure. The presence of a corrosive NaCl solution resulted in the dissolution of the martensitic phase and a lamellar microstructure. The decrease in the volume fraction of martensite was believed to be responsible for reducing the work hardening rate and increasing elongation to failure, resulting in a longer lip formation and, ultimately, material removal. These studies have shown how subtle nanoscale microstructural changes can influence erosion–corrosion rates, and why such surface damage accumulation mechanisms need to be taken into account when developing theoretical models to predict material loss.

15.5   Electrochemical Monitoring of Erosion–Corrosion

As stated before, Sasaki and Burstein observed current transients for single particle impacts [8] on passive stainless steels, illustrating that the monitoring of such transients can provide direct information on erosion–corrosion processes. Therefore, techniques to detect the noise level (small perturbations) on electrochemical corrosion measurements are being developed, referred to as electrochemical noise analysis (ENA). Noise can be measured on both current (ECN) and potential (EPN) outputs, and subsequent analysis can yield corrosion resistance details, assuming that both measurements are in phase with each other. EPN measurements can be made with two electrode cells, while current noise typically requires three electrodes. Further details of this technique can be found in Ref. [14]. Such measurements can be made under flow corrosion and erosion–corrosion, and are now being analyzed to provide insights into synergistic processes and surface performance indicators. Figure 15.8 shows a corrosion flow cell designed for ENA, while Figure 15.9 shows typical EPN and ECN outputs for the stainless steel pipe section electrodes shown in Figure 15.8 and subjected to a flowing NaCl solution at Re = 2000. The features seen relate to metastable and stable pitting activity on the wetted surfaces of the stainless steel electrodes.

It is important to apply caution to any electrochemical measurement under erosion–corrosion conditions. Other issues that make electrochemical analysis and its comparison to synergy difficult include the possibility of local film currents between anodes/cathodes

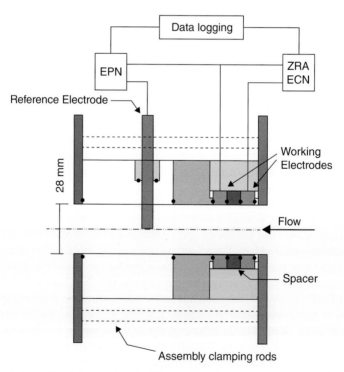

Figure 15.8 *Schematic of electrochemical flow cell and measurement set-up. Reprinted from Ref. [7]. Copyright (2007) with permission from Elsevier.*

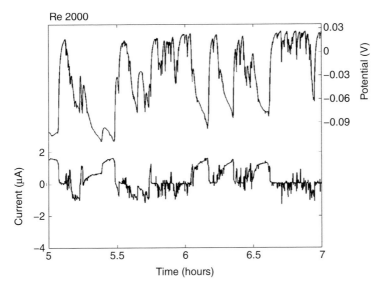

Figure 15.9 *Electrochemical current and potential noise for AISI 304L in 3.5 wt% NaCl solution at a transition Reynolds number of 2000. Reprinted from Ref. [7]. Copyright (2007) with permission from Elsevier.*

[15] which will not be seen by ECN measurements, and the effects of charging/recharging double-layer currents due to fluctuating local events.

15.6   Tribocorrosion within the Body: Metal-on-Metal Hip Joints

Currently, much attention is being focused on the metal-on-metal (MoM) joint implants that have, nonetheless, recently been banned by the UK National Health Service, notably with regards to their wear and corrosion (ion release) behaviors. The corrosion, wear and wear-corrosion behaviors of three materials (high-carbon CoCrMo, low-carbon CoCrMo, and UNS S31603) have been investigated by Yan *et al.* [16]. In the steady-state regime, 20–30% of the material degradation was attributed to corrosion-related damage; however, high-carbon CoCrMo showed excellent corrosion, wear and corrosion-wear resistance, and thus delivered the best overall performance in terms of a lower wear rate, a lower friction coefficient, and a higher resistance to corrosion.

Sinnett-Jones *et al.* [17] investigated the synergistic effects of corrosion and wear of a surgical-grade, cast F-75 cobalt-chromium-molybdenum (CoCrMo) alloy. Depassivation and repassivation processes identified micro-abrasion-corrosion methods which showed strong synergistic effects that ranged from negative to positive. The synergistic levels appeared to depend on the integrity of the passive films and the repassivation kinetics.

Although the proteins contained in joint fluids are thought to be involved in joint lubrication, their role in the tribocorrosion of metal implants is not well understood. However, initial studies have demonstrated conflicting trends, with corrosion either being enhanced or

292   *Developments in Electrochemistry*

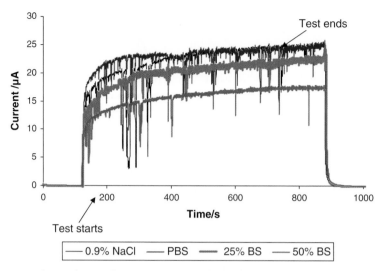

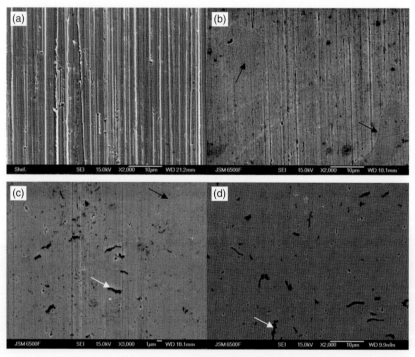

Figure 15.10   *Electrochemical current noise obtained in microabrasion–corrosion test on as-cast CoCrCo (1.0 g cm⁻³ SiC) in 0.9% NaCl, PBS solutions, 25% and 50% bovine serum (BS) at 37 °C. Reprinted from Ref. [17]. Copyright (2005) with permission from Elsevier.*

Figure 15.11   *SEM images of wear scars on as-cast CoCrMo under pure sliding conditions in (a) 0.9% NaCl, (b) PBS, (c) 25% bovine serum (BS), (d) 50% BS at 37 °C. The black arrows show a carbide standing proud of the surface, while white arrows show possible protein and wear debris on the worn surfaces. Polished wear scars were produced in proteinaceous solutions, indicating that proteins form a lubrication film that protects the surface. Reprinted from Ref. [18], with permission from Elsevier.*

reduced when protein is present and adsorbed onto the articulating surfaces. When Sun [18] studied the abrasion–corrosion of cast CoCrMo in either saline (0.9% NaCl), phosphate-buffered saline (PBS) or proteinaceous solutions of bovine serum, the *in-situ* current noise was clearly reduced in the presence of proteins (Figure 15.10). In a similar experiment, using the same test geometry, the abrasives were removed so that the alumina ball would abrade the CoCrMo surface under sliding wear mode. However, the presence of protein again caused reductions in both the mechanism and rate of wear (Figure 15.11).

15.7   Conclusions

As noted in this chapter, material performance under tribocorrosion conditions is highly system-dependent, and modest changes in environmental conditions can have a profound impact on wastage rates. The discussions in this chapter have confirmed that the current knowledge of the mechanisms involved, and of their controlling parameters, is inadequate to enable informed material or coating selection for such aggressive duties. Consequently, the selection of materials must be accompanied either by experience of good material performance under similar erosion–corrosion conditions, or by experimental data acquired from laboratory testing that has been proven to simulate field conditions to allow these materials and coatings to be screened. Today, electrochemical techniques are playing an increasingly important role in providing knowledge of the interface in tribocorrosion contacts. However, these techniques – as originally inspired by Martin Fleischmann – have yet to be fully deployed by tribologists and, consequently, this is very much "work in progress."

Acknowledgments

The authors would like to thank the National Centre for Advanced Tribology (nCATS) in Southampton, and in particular Doctors Julian Wharton and Terry Harvey for their help with illustrations and informative discussions.

References

(1) *Oxford English Dictionary Online* (2006) Oxford University Press.
(2) (a) Gao, F. and Liang, H. (2009) Material removal mechanisms in electrochemical-mechanical polishing of tantalum. *Electrochimica Acta*, **54**, 6808–6815; (b) Levy, A.V. (1995) *Solid particle erosion and erosion-corrosion of materials*. ASM International, Ohio, USA.
(3) Meng, H.C. and Ludema, K.C. (1995) Wear models and predictive equations: their form and content. *Wear*, **181–183**, 443–457.
(4) Thakare, M.R., Wharton, J.A., Wood, R.J.K. and Menger, C. (2008) Exposure effects of strong alkaline conditions on the microscale abrasion–corrosion of D-gun sprayed WC–10Co–4Cr coating. *Tribology International*, **41**(7), 629–639.
(5) Hu, X. and Neville, A. (2005) The electrochemical response of stainless steels in liquid-solid impingement. *Wear*, **258**(1–4), 641–648.
(6) Silverman, D.C. (2004) The rotating cylinder electrode for examining velocity-sensitive corrosion – a review. *Corrosion*, **60**(11), 1003–1023.

(7) Wood, R.J.K. (2007) Erosion-corrosion. *Comprehensive Structural Integrity*, **6**, 395–427.

(8) From Sasaki, K. and Burstein, G.T. (2000) Observation of a threshold impact energy required to cause passive film rupture during slurry erosion of stainless steel. *Philosophical Magazine Letters*, **80**(7), 489–493.

(9) Mischler, S., Debaud, S. and Landolt, D. (1998) Wear-accelerated corrosion of passive metals in tribocorrosion systems. *Journal of The Electrochemical Society*, **145**(3), 750–758.

(10) Bozzini, B., Ricotti, M.E., Boniardi, M. and Mele, C. (2003) Evaluation of erosion-corrosion in multiphase flow via CFD and experimental analysis. *Wear*, **255**, 237–245.

(11) ASTM Standard G 119-93 (1993) Standard Guide for Determining Synergism between Wear and Corrosion.

(12) Wood, R.J.K. and Hutton, S.P. (1990) The synergistic effect of erosion and corrosion: trends in published results. *Wear*, **140**, 387–394.

(13) Wood, R.J.K., Walker, J.C., Harvey, T.J. *et al.* (2013) Influence of the microstructure on the erosion and erosion-corrosion characteristics of 316 stainless steel. *Wear*, **306**, 254–262.

(14) Cottis, R. and Turgoose, S. (1999) *Electrochemical Impedance and Noise*. National Association of Corrosion Engineers, Houston, TX, USA.

(15) Oltra, R., Chapey, B. and Renuad, L. (1995) Abrasion-corrosion studies of passive stainless steels in acidic media: combination of acoustic emission and electrochemical techniques. *Wear*, **186–187**, 533–541.

(16) Yan, Y., Neville, A. and Dowson, D. (2006) Biotribocorrosion – an appraisal of the time dependence of wear and corrosion interactions: I. The role of corrosion. *Journal of Physics D: Applied Physics*, **39**(15), 3200–3205.

(17) Sinnett-Jones, P.E., Wharton, J.A. and Wood, R.J.K. (2005) Micro-abrasion-corrosion of a CoCrMo alloy in simulated artificial hip joint environments. *Wear*, **259**(2), 898–909.

(18) Sun, D., Wharton, J.A., Wood, R.J.K. *et al.* (2009) Microabrasion-corrosion of cast CoCrMo alloy in simulated body fluids. *Tribology International*, **42**, 99–110.

16

Hard Science at Soft Interfaces

Hubert H. Girault
Ecole Polytechnique Fédérale de Lausanne
Department of Chemistry and Chemical Engineering, Switzerland

During the early 1980s, Martin Fleischmann was pioneering electrochemical recording of ion fluxes through ion channels in artificial lipid layers, and this was later to become a very active field of research, particularly with respect to DNA sequencing. At the same time, he encouraged a more general interest in soft liquid–liquid interfaces. As a result, I was to begin a study as a PhD project and have never stopped being fascinated by the rich scientific aspects of the field. In the early days, the emphasis was placed mainly on the instrumentation to develop reliable four-electrode potentiostats with IR compensation able to record precise voltammetric data of charge transfer reactions (ion transfer, assisted ion transfer and heterogeneous electron transfer reactions) across the interface between two immiscible liquids. Now, the field has blossomed into understanding all their features and applying such interfaces in electrochemical technology.

16.1 Charge Transfer Reactions at Soft Interfaces

Electrochemistry at soft interfaces is a very interesting topic, as many different types of charge transfer reactions can take place in parallel and concomitantly. The different charge reactions include: (i) ion transfer reactions where the flux of ions crossing the interface gives rise to a current; (ii) assisted ion transfer reactions where the extraction of, for example, an aqueous ion by an organic soluble ionophore also gives rise to an ionic current; and (iii)

Developments in Electrochemistry: Science Inspired by Martin Fleischmann, First Edition.
Edited by Derek Pletcher, Zhong-Qun Tian and David E. Williams.
© 2014 John Wiley & Sons, Ltd. Published 2014 by John Wiley & Sons, Ltd.

heterogeneous electron transfer reactions between, for example, an aqueous electron donor and an organic soluble electron acceptor.

These reactions can be written as:

$$\text{Ion}^{\text{water}} \rightleftarrows \text{Ion}^{\text{organic}} \tag{16.1}$$

$$\text{Ion}^{\text{water}} + \text{Ligand}^{\text{organic}} \rightleftarrows \left[\text{Complexed ion}\right]^{\text{organic}} \tag{16.2}$$

$$\text{Donor}^{\text{water}} + \text{Acceptor}^{\text{organic}} \rightleftarrows \left[\text{Donor}^+\right]^{\text{water}} + \left[\text{Acceptor}^-\right]^{\text{organic}} \tag{16.3}$$

Usually, when thinking of electrochemical reactions, reactions are considered at a metal/electrolyte or semiconductor/electrolyte interface, but rarely about the interface between electrolyte solutions or, more recently, the electrolyte–ionic liquid interface or even the interface between two immiscible ionic liquids.

If it is accepted that an electrochemical reaction is one where the Gibbs energy of the reaction depends on the potential difference between two phases, as well as temperature and pressure (as do classical chemical reactions), then charge transfer reactions at soft interfaces are truly electrochemical reactions. Indeed, their characteristic is to be potential-dependent and controlled by the Galvani potential difference between the two phases in contact.

16.1.1 Ion Transfer Reactions

The distribution of ions in a biphasic system is given by the Nernst equation:

$$\Delta_o^w \phi = \Delta_o^w \phi_i^\ominus + \frac{RT}{z_i F} \ln\left(\frac{a_i^o}{a_i^w}\right) \tag{16.4}$$

with

$$\Delta_o^w \phi_i^\ominus = \frac{\Delta G_{\text{tr},i}^{\ominus,\text{w}\rightarrow\text{o}}}{z_i F} \tag{16.5}$$

where $\Delta_o^w \phi_i^\ominus$ is the standard transfer potential of the ion i defined as the Gibbs energy of transfer $\Delta G_{\text{tr},i}^{\ominus,\text{w}\rightarrow\text{o}}$ but expressed on a voltage scale [1].

As for a classical redox reaction on a solid electrode it is possible, by controlling the applied potential difference, to determine the ratio between reactants and products, here the partition of the ions between the two phases $\left(P_i = a_i^o / a_i^w\right)$. As for a redox reaction described by the Nernst equation in a better known form,

$$E = E_{\text{Ox/Red}}^\ominus + \frac{RT}{nF} \ln\left(\frac{a_{\text{Ox}}}{a_{\text{Red}}}\right) \tag{16.6}$$

the partition of ions in a biphasic system follows a two-state Fermi–Dirac distribution, the ions being either in one phase or in the other. By contrast to the partition of neutral molecules, the partition coefficient of ions is potential-dependent.

$$P_i = \frac{a_i^o}{a_i^w} = \exp\left[\frac{z_i F}{RT}\left(\Delta_o^w \phi - \Delta_o^w \phi_i^\ominus\right)\right] = P_i^\ominus \exp\left[\frac{z_i F}{RT}\Delta_o^w \phi\right] \tag{16.7}$$

If the thermodynamic aspects of ion partitioning are rather well understood, the kinetic aspects of the ion transfer reactions still pose challenging questions. Over the years, as methods to measure ion transfer reaction rates by electrochemical methods (e.g., chronoamperometry, AC impedance and, more recently, using micro- or nano-interfaces; see below) have been developed, the values for the standard rates measured have increased. Indeed, a significant challenge is the deconvolution of the mass transport limitation from the key step of ion transfer. Early models were based either on a phenomenological approach, treating the transfer as a transport process, or a Butler–Volmer type approach which considered an activated state based on a mixed interfacial solvation of the ion. The key issue here is the definition of the ion transfer reaction itself, as it is affected by the soft and dynamic nature of the interface. Unfortunately, there are no imaging techniques to observe liquid interfaces. Whereas, electrochemistry on solid electrodes has benefited greatly from the development of scanning probe and ultra-high-vacuum surface-specific techniques, somewhat of a blind-spot occurs when trying to observe a liquid–liquid interface. Apart from X-ray or even neutron-scattering experiments that provide some indirect information on the density profile across the interfaces, or some nonlinear spectroscopy techniques such as surface second harmonic generation or sum frequency generation that can provide some information on the dynamics of some surface modes, a great reliance must be placed on molecular dynamic simulations that provide "cartoon" views of the interfacial structure. All of these simulations tend to describe the interface as molecularly sharp with a certain corrugation. The time scales of these simulations are usually too short to provide global dynamic information, and the number of molecules that can nowadays be used is too small to model the presence of electrolytes and the role of ion–ion interactions.

As originally proposed by Benjamin [2], it is important to distinguish the transfer of an aqueous ion to the organic phase from that of an organic ion to the aqueous phase. In the former case, the ion is heavily hydrated and transfers to the organic phase with some water molecules forming a partial shell but, more importantly, forming an aqueous "tail" linking it to the water phase. Once this tail is broken, the ion progressively loses its water solvation molecules that then diffuse back to the aqueous phase. Inversely, the transfer of an organic ion occurs through a "harpoon" mechanism, where dipolar water molecules attracted by the charged species form a string to grab ions located in the vicinity of the interface and drag them into the aqueous phase.

The role of the solvent dynamics, and in particular of protrusions in ion transfer reactions, was modeled by Marcus who demonstrated the importance of the balance between too-convex protrusions that can efficiently grab ions and the mechanical resistance of the opposite phase to the formation of protrusions [3]. From a solvation viewpoint, it is clear that the Gibbs energy of solvation of the ions is not a step function located at the interface but rather a smooth variation taking place over few nanometers. From an electrical viewpoint, the potential dependence of the rate constant remains difficult to take into account. Indeed, the surface concentrations are potential-dependent according to a Boltzmann distribution, but the potential dependence of the transfer step is still a matter of debate. How much of the total applied potential difference, mainly spread between the two back-to-back diffuse layers, can act as a driving force is still an open question. From an experimental viewpoint, it is difficult to measure an ion transfer current that does not perturb the structure and/or the dynamics of the interface itself. Too-large currents can even cause emulsification of the interfacial region. For most purposes, it can certainly be said that ion transfer reactions are fast and "reversible" in the sense that they mostly appear as diffusion-controlled reactions.

16.1.2   Assisted Ion Transfer Reactions

The transfer of ions can be facilitated by the presence of ligands in the organic phase. While ion transfer reactions can be considered as pseudo-first-order with respect to the concentration of the ions transferred, assisted ion transfer reactions are more generally bimolecular reactions, although it is never clear if the complexation reaction is purely heterogeneous and occurs after a transfer step, as shown in Figure 16.1. The ligand in the organic phase can be a neutral ionophore molecule such as valinomycin for the assisted transfer of potassium, a charged species to drive ion pair extraction, or even a base to drive proton-pumping reactions for the formation of organic acids. As depicted schematically in Figure 16.1, the reaction mechanism depends on the partition coefficient of the ion, P_{I+}, of the ligand, P_L, of the complex itself, P_{IL+}, and the association constants in water, K_a^w, and in the organic phase K_a^o.

Extensive studies have been carried out to derive the voltammetric response of the different types of assisted ion transfer for different techniques, such as cyclic voltammetry and differential pulse voltammetry. This part of electrochemistry at soft interfaces has found many applications, mainly for the determination of complexation constants for liquid extraction [4], and these are of interest not only for nuclear reprocessing but also for the development of amperometric ion-selective electrodes [5].

These measurements are usually made by recording the voltammetric responses with the variation of the concentration of one of the species. In the case of the 1 : 1 extraction of an aqueous ion by interfacial complexation when the ion is in excess, the half-wave potential is given by

$$\Delta_o^w \phi_{\frac{1}{2}} = \Delta_o^w \phi_i^{\ominus} - \frac{RT}{F} \ln K_a^o + \frac{RT}{F} \ln \left(\frac{a_{iL}^o}{a_i^w a_L^o} \right) + \frac{RT}{F} \ln \left(\frac{D_{iL}^o}{D_L^o} \right) \tag{16.8}$$

where K_a^o is the association constant in the organic phase, defined by

$$K_a^o = \frac{a_{iL}^o}{a_i^o a_L^o} \tag{16.9}$$

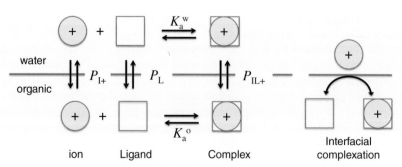

Figure 16.1 *Schematic of assisted ion-transfer reaction mechanism, where the empty squares symbolize the ligand and filled squares the complex.*

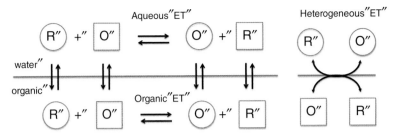

Figure 16.2 *Schematic electron transfer reaction mechanism, where the circles symbolize aqueous species and squares organic species.*

16.1.3   Electron Transfer Reactions

Electron transfer (ET) reactions at soft interfaces, similar to assisted ion transfer reactions, can occur either heterogeneously or homogeneously in the vicinity of the interface following the transfer of one of the reactants (see Figure 16.2).

The Nernst equation for a monoelectronic heterogeneous ET reaction is given by

$$\Delta_o^w \phi = \left[E^\ominus_{O_2/R_2} \right]^o_{SHE} - \left[E^\ominus_{O_1/R_1} \right]^w_{SHE} + \frac{RT}{F} \ln \left(\frac{a^w_{R_1} a^o_{O_2}}{a^w_{O_1} a^o_{R_2}} \right) \tag{16.10}$$

where $\left[E^\ominus_{O_1/R_1} \right]^w_{SHE}$ is the standard redox potential of the aqueous redox species expressed with respect to the Standard Hydrogen Electrode (SHE), and where $\left[E^\ominus_{O_2/R_2} \right]^o_{SHE}$ is the standard redox potential of the organic redox species but also expressed with respect to the aqueous SHE.

$$\left[E^\ominus_{O_2/R_2} \right]^w_{SHE} = \left(\tilde\mu^o_{O_2} - \tilde\mu^o_{R_2} \right) - \left(\tilde\mu^w_{H^+} - \tilde\mu_{H_2} \right) \tag{16.11}$$

where $\tilde\mu$ is the electrochemical potential. $\left[E^\ominus_{O_2/R_2} \right]^o_{SHE}$ is often the most difficult to determine; a method often used involves measuring the standard redox potential of the organic species, for example by voltammetry, with respect to the ferrocene scale, $\left[E^\ominus_{O_2/R_2} \right]^o_{Fc}$. In this way, it is only necessary to determine the standard redox potential of the ferrocinium/ferrocene couple in the organic phase with respect to the aqueous SHE by means of a thermodynamic cycle.

One very interesting class of ET reactions at soft interfaces are those that are photoinitiated. Following the pioneering studies of the Russian school, including those of Volkov [6] and Kuzmin [7], it has been shown that photosensitizers soluble in one phase are often adsorbed at the interface and can be quenched by electron donor or acceptors. This class of reaction offers interesting perspectives to design biomimetic approaches to artificial photosynthesis. Photoelectrochemistry at the interfaces between two immiscible electrolyte solutions (ITIES) is rather analogous to photoelectrochemistry at a semi-conductor electrode, where the potential drop within the semi-conductor should be considered.

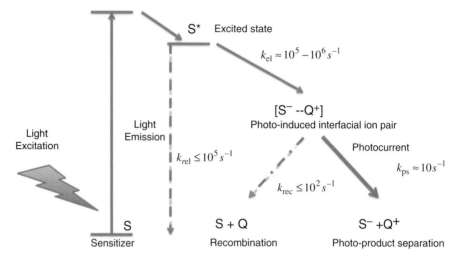

Figure 16.3   *Schematic photo-electron transfer reaction mechanism.*

Figure 16.3 shows a schematic representation of a photoelectron transfer reaction where a sensitizer (S*) in one phase is quenched by an electron donor (Q) in the adjacent phase. A charge-transfer complex $[S^- - Q^+]$ is formed at the interface. In a bulk solution, recombination often occurs due to the cage effect formed by the solvent molecules. At soft interfaces, the dissociation of the charge transfer complex into photoproducts can be favored by the presence of the static electric field, and this is still a very important point to quantify in the coming years.

In photo-ET reactions, the interface can be illuminated in a total internal reflection mode in such a way that only the sensitizers located in the evanescent wave, usually a few hundred nanometers thick, can be photoexcited. The interfacial quenching can be followed by monitoring the photocurrent for a given electric polarization of the interface. In a series of reports, intensity-modulated photocurrent spectroscopy (IMPS), originally developed for semi-conductor photoelectrochemistry, has been used to measure the kinetics of photo-ET at the ITIES [8]. By varying the frequency of the light intensity modulation from 10 kHz to 1 mHz, it is possible to cover the domain relevant to the different steps. By using quasi-elastic light scattering, the adsorption of different sensitizers at the interface has been measured, and by using light polarization-modulated reflectance it has been shown that their orientation can also be studied.

16.2   Electrocatalysis at Soft Interfaces

Until now, the methodology available to study charge transfer reactions at soft interfaces has been rather mature, and studies in the field have shifted to the study of catalyzed reactions such as the oxygen reduction reaction (ORR), hydrogen evolution reaction (HER), or even oxygen evolution reaction (OER). For this, two classes of catalysts have been used: (i) molecular catalysts; and (ii) nanoparticle solid catalysts. These two approaches draw their inspiration from classical molecular catalysis and from electrocatalysis, respectively.

16.2.1   Oxygen Reduction Reaction (ORR)

Oxygen reduction in biosystems is carried out by cytochrome oxidase using two iron heme groups and two copper centers. Overall, it is a four-electron process that reduces oxygen to water, as it is important in living systems to minimize the production of hydrogen peroxide and reactive oxygen species (ROS).

In the chemical industry, more than a megaton of hydrogen peroxide is produced yearly in a biphasic reaction scheme known as the anthraquinone auto-oxidation process, where reduced anthraquinone is used to reduce oxygen to H_2O_2 and where the anthraquinone is reduced again by hydrogen on a palladium catalyst.

The reduction of oxygen at soft interfaces was first reported by Schiffrin [9], and also by Kihara [10], using ferrocene and tetrachlorohydroquinone, respectively. It has been shown that decamethylferrocene was an efficient electron donor to reduce oxygen to hydrogen peroxide. When studying oxygen reduction in biphasic systems, it is important to ensure that the electron donor does not reduce H_2O_2, at least in the time scale of the experiment. Although tetrathiafulvalene meets this requirement, strangely enough this small organic molecule is able to reduce oxygen to water very selectively, albeit rather slowly [11]. One class of molecular catalysts that have been studied widely for oxygen reduction are the porphyrins and phthalocyanins. As it happens, these planar molecules usually adsorb at liquid–liquid interfaces and can act as a catalyst for oxygen reduction using lipophilic electrons donors and aqueous protons. Free-base porphyrins such as 5,10,15,20-tetraphenyl-21H,23H-porphine (H_2TPP) can bind oxygen and catalyze the two-electron reduction to H_2O_2. Cobalt porphyrins, such as 2,8,13,17-tetraethyl-3,7,12,18-tetramethyl-5-p-aminophenylporphyrin cobalt(II) (CoAP), have also been widely studied for this purpose [12]. Here, oxygen binds to the cobalt metal center to form (Co-O_2)AP, which protonates at the interface to form the adduct $[(Co-O_2H)AP]^+$. The latter is then reduced in a proton-concerted reaction by the lipophilic electron donor to yield H_2O_2 and the oxidized porphyrin that, in turn, is reduced in the bulk organic phase by the lipophilic electron donor, as shown in Figure 16.4.

One unique property of the ORR with porphyrins adsorbed at ITIES is that the reaction mechanisms depend strongly on aggregation and self-assembly. Whereas, isolated

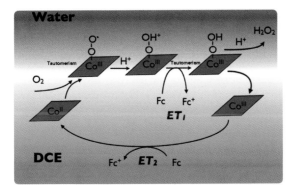

Figure 16.4   *Oxygen reduction. Reprinted with permission from Ref. [12]. Copyright (2010), American Chemical Society.*

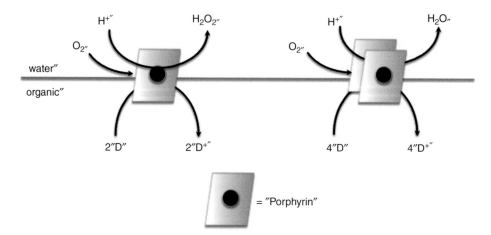

Figure 16.5   *Oxygen reduction by isolated or aggregated porphyrins.*

molecules favor the two-electron production of hydrogen peroxide, self-assembled or stacked porphyrins favor the four-electron reduction of oxygen to water (Figure 16.5). This was clearly demonstrated using both planar and neutral porphines or for self-assembled, oppositely charged cobalt tetramethylpyridinium porphyrin ($CoTMPyP_4^+$) and cobalt tetra-sulfonatophenyl porphyrin ($CoTPPS_4^-$). For the latter, the self-assembled cobalt porphyrins were as efficient as the synthetized co-facial (so-called Pacman) analogs such as $[Co_2(DPX)]$ (DiPorphyrins (DP) with a xanthenyl (X) bridge), to produce water with a selectivity of 80% [13].

Of course, it is possible to reduce oxygen at ITIES functionalized with platinum nanoparticles floating at the interface.

16.2.2   Hydrogen Evolution Reaction (HER)

It has been shown that metallocenes can reduce acids to form hydrogen, and this approach has been pursued to evolve hydrogen at soft interfaces. In a series of reports, it has been shown by the present author's group that cobaltocene and decamethylferrocene in 1,2-dichloroethane react with aqueous protons to form hydrogen, and that the process is much faster if the interface is polarized positively to favor formation of the metallocene hydride. The reaction is believed to proceed by protonation of the hydride to form a di-hydride that can release hydrogen, or by a homolytic pathway involving two hydrides. This former mechanism is therefore analogous to a Volmer–Heyrovsky mechanism in the classical HER, and the latter to a Volmer–Tafel reaction. Although, so far, ferrocene can be used for H_2 evolution, other metallocenes such as osmocene, decamethylosmocene and decamethylruthenocene can produce hydrogen under light excitation. Overall, the reaction can be written

$$2Mc^o + 2H^{+w} \rightarrow 2Mc^{+o} + H_2$$

where Mc stands for metallocene (see Figure 16.6).

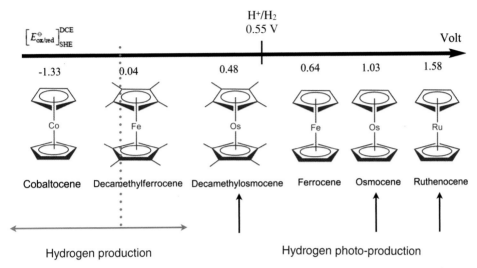

Figure 16.6 *Metallocene redox potential on the standard hydrogen electrode scale.*

The HER can be catalyzed using nanoparticulate catalysts. For example, platinum and palladium nanoparticles can be electrodeposited *in-situ* using the metallocene itself to reduce the corresponding aqueous salt, for example, $PtCl_4^{2-}$ or $PdCl_4^{2-}$. Other inorganic catalysts such as molybdenum disulfide (MoS_2), molybdenum boride (MoB) or molybdenum carbide (Mo_2C) can also be used [14]. These inorganic catalysts are insoluble in both phases but are deposited at the interface by sedimentation. In fact, Mo_2C was found to be a more powerful catalyst than platinum itself at the ITIES. Recently, a more efficient way to produce hydrogen at a liquid–liquid interface was demonstrated by using graphene or carbon nanotubes (CNTs) decorated by inorganic catalysts [15]. In this way, the interface is functionalized by electron conductors that can collect electrons all along the tube from the donors in the organic phase. Aqueous protons are reduced to hydrogen at the carbon-supported semi-conductor catalyst nanoparticles that act as nanoelectrodes, with the carbon–catalyst interface forming a Schottky barrier; this mechanism is shown in Figure 16.7.

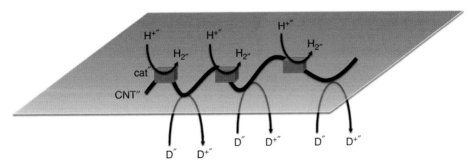

Figure 16.7 *Hydrogen production at soft interface functionalized by electron-conducting carbon nanotubes (CNT) decorated by nanoparticle catalysts (cat).*

16.3   Micro- and Nano- Soft Interfaces

Following Fleischmann's studies on microelectrodes, a series of micropipettes to support micro-liquid–liquid interfaces was developed. One of the major difficulties with the interface between two ITIES is the high resistance of the organic phase; however, by using micropipettes it became possible to study ion transfer reactions from aqueous to very poorly conductive organic phases [16]. As shown in Figure 16.8, the asymmetric diffusion fields – with linear diffusion in the solution located inside the pipette and spherical diffusion in the solution where the micropipette was immersed – yielded asymmetric voltammograms with a steady-state current for the ingress reaction and a peak-shaped current for the egress reaction.

Such voltammograms as shown in Figure 16.8 cannot be obtained by analytical solutions of the diffusion equations, and for this reason computer simulations of electrochemical processes have been developed. Much progress has been made in this respect; from the finite difference approaches coded in Fortran and running on a mainframe computer, to the finite element packages running on Sun workstations, and now to the Comsol Multiphysics programs running on a personal computer, the evolution of that field has been tremendous. Nonetheless, as the tool becomes increasingly performant the user must likewise be increasingly careful in the validation of the computational results. When possible, an analytical solution retains advantages.

For the case of ion transfer voltammetry at the tip of a micropipette, many groups – and in particular that of Shao at Peking University [17] – have developed the technology of pipette pulling to a point where it is now possible to reproducibly pull nanopipettes such that an opening at the tip may have a diameter of only few nanometers. The voltammograms obtained at nanopipettes are steady-state both for the ingress and egress reactions. In fact, if diffusion in the nanopipette is considered to be of a spherical nature occurring in a solid angle, and that diffusion outside the nanopipette is also spherical, it is possible to show that the steady-state current, I_{ss}, is given by:

$$I_{ss} = \pi n F D c r \theta \qquad (16.12)$$

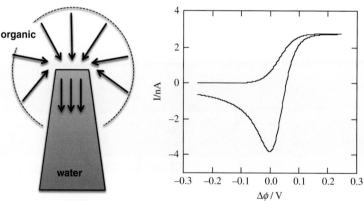

Figure 16.8   *Schematic ion ingress transfer at a micro-ITIES (left), and the current response for an ingress–egress voltammogram.*

where D is the diffusion coefficient, c is the concentration, r is the radius at the tip, and θ is the shank angle [17].

Another approach to realizing micro- or nano-liquid interfaces is to drill micro- or nano-holes (or even arrays) in polymers films or wafers and to support the interface in the holes. In this way, the diffusion fields are symmetric and steady-state voltammograms similar to those obtained on a solid microelectrode can be recorded. These supported micro-interfaces have found applications in ion sensing, and detectors for ion chromatography have also been realized.

16.4   Plasmonics at Soft Interfaces

Surface-enhanced Raman spectroscopy (SERS) was one of the major discoveries of Martin Fleischmann's career, and since the early experiments on pyridine on roughened silver electrodes, the plasmonic properties of SERS have by now been well established. Considering that the electron oscillations in the dipolar approximation of the Mie theory depend strongly on the refractive index of the surrounding solvent, metallic nanoparticles located at an ITIES may have a double resonance, for example, if the nanoparticle straddles the interface with each half oscillating at different frequencies. It has been shown that gold films formed at ITIES can form liquid mirrors, and that it is possible to observe a longitudinal propagation of the plasmon with an angular dependence similar to those observed for gold films evaporated on a solid support such as glass. Furthermore, it has been shown that it was possible to record SERS spectra at ITIES functionalized by silver or gold nanoparticles [18].

16.5   Conclusions and Future Developments

Soft interfaces present some very interesting properties. Compared to solid electrodes, they are defect-free in a sense but dynamic, not to say unstable. A rather extensive methodology has been developed over the years to study charge transfer reactions. The concepts developed – in particular the role of the potential difference to control ion partition – have been used to model ion-selective electrodes and to develop amperometric ion sensors. They have also been used to address phase-transfer catalysis from a more physical standpoint.

The grand challenge now is to develop applications. One goal of the author's research group is to develop a water-splitting process based on the use of biphasic reactions. In natural photosynthesis, water is oxidized by the photosystem II to produce oxygen, whereas at the other end of the Z-scheme NAD is reduced to NADH. Indeed, mother Nature does not rely on hydrogen as a fuel but on hydrides. The concept to be pursued preferentially consists of two polymer tubes running on the ground, mainly to avoid the use of expensive flat glass panels. Each tube contains an emulsion between an aqueous and an organic phase, and in both tubes the organic phase contains a mixture of electron acceptors and donors, such tetracyanodimethane (TCNQ) and tetrathiafulvalene (TTF). The first tube is dedicated to the photo-production of hydrogen and contains a sensitizer and nanoparticle solid combining an electron conductor (e.g., a CNT) and a nanoparticle catalyst, as shown in Figure 16.9. The net reaction in tube 1 is:

$$4TTF^o + 4H^{+w} + 4A^{-w} \xrightarrow{h\nu} 2H_2 + 4TTF^{+o} + 4A^{-o}$$

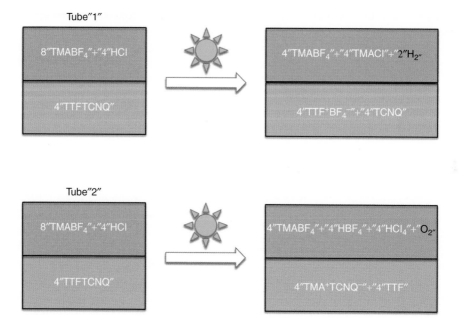

Figure 16.9 *Photo-reduction of protons in Tube 1 and photo-oxidation of water in Tube 2, both starting from an aqueous solution containing a salt composed of weakly hydrophilic ions (tetramethylammonium tetrafluoroborate) and hydrochloric acid.*

where A^- is a mildly hydrophilic anion that can be extracted in the organic phase by TTF^+ such as BF_4^-. The second tube is dedicated to the water photo-oxidation and contains a sensitizer and another nanoparticle catalyst. The net reaction is:

$$4TCNQ^0 + 2H_2O + 4C^{+w}A^{-w} \xrightarrow{h\nu} O_2 + 4TCNQ^{-0} + 4C^{+0} + 4H^{+w} + 4A^{-w}$$

where C^+ is a mildly hydrophilic cation that can be extracted by $TCNQ^-$ such as tetramethylammonium and A^- is a weakly hydrophilic anion such as BF_4^-. These two photo-reactions are summarized in Figure 16.9.

At the end of each tube, the solutions are degassed to recover hydrogen and oxygen, and the nanoparticles are recovered. The two immiscible phases are then separated on a decanting filter. The two aqueous phases are mixed according to the reactions shown in Figure 16.10, while the two organic phases are mixed to reset the sacrificial species ready for another cycle. The resulting salt C^+A^- is then extracted back into water to be mixed with the two aqueous phases. The resetting reaction in the organic reads:

$$4TTF^{+0}A^{-0} + 2C^{+0}TCNQ^{-0} \longrightarrow 4TTF + 4TCNQ + 2C^{+w}A^{-w}$$

The net reaction is the sum of all these steps, and can written as:

$$2H_2O \xrightarrow{h\nu} 2H_2 + O_2$$

The major advantage of this approach from an economical viewpoint is to be membrane-free and not to rely on flat glass panels, as do most other photoelectrochemical cells

Aqueous phase mixing after degassing

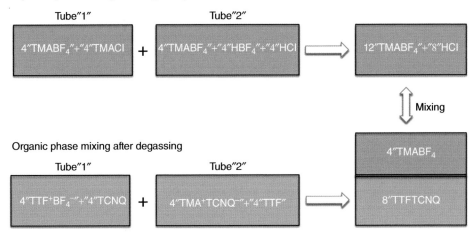

Figure 16.10   Mixing and resetting reactions.

(PECs). The major challenge is the yield of the photosensitized reaction in the presence of reversible electrons, rather than the classically used triethylamine as sacrificial electron donor or persulfate as sacrificial electron acceptors.

All in all, electrochemistry at soft interfaces is more than just a curiosity as often heard from "conventional electrochemists." Martin Fleischmann, together with Sir Graham Hills, David Schiffrin and Roger Parsons, were those who encouraged the author to explore this new field, and this review is dedicated to their vision and their friendship.

References

(1) Girault, H.H. (2010) Electrochemistry at liquid-liquid interfaces. *Electroanalytical Chemistry*, **23**, 1–104.
(2) Benjamin, I. (1993) Mechanism and dynamics of ion transfer across a liquid-liquid interface. *Science*, **261**, 1558–1560.
(3) Marcus, R.A. (2000) On the theory of ion transfer rates across the interface of two immiscible liquids. *Journal of Chemical Physics*, **113** (4), 1618–1629.
(4) Stockmann, T.J. and Ding, Z.F. (2011) Uranyl ion extraction with conventional PUREX/TRUEX ligands assessed by electroanalytical chemistry at micro liquid/liquid interfaces. *Analytical Chemistry*, **83** (19), 7542–7549.
(5) Samec, Z., Samcova, E. and Girault, H.H. (2004) Ion amperometry at the interface between two immiscible electrolyte solutions in view of realizing the amperometric ion-selective electrode. *Talanta*, **63** (1), 21–32.
(6) Kandelaki, M., Volkov, A., Levin, A. and Boguslavsky, L. (1983) Oxygen evolution in the presence of chlorophyll adsorbed at the octane water interface. *Bioelectrochemistry and Bioenergetics*, **11** (2-3), 167–172.
(7) Zaitsev, N., Gorelik, O., Kotov, N. *et al.* (1988) A photoelectrochemical effect at the polarizable interface between liquid electrolyte-solutions in protoporphyrin quinone systems. *Soviet Electrochemistry*, **24** (10), 1243–1247.
(8) Fermin, D., Duong, H., Ding, Z. *et al.* (1999) Photoinduced electron transfer at liquid/liquid interfaces - Part II. A study of the electron transfer and recombination dynamics by intensity

modulated photocurrent spectroscopy (IMPS). *Physical Chemistry Chemical Physics*, **1**, 1461–1467.

(9) Cunnane, V., Geblewicz, G. and Schiffrin, D. (1995) Electron and ion transfer potentials of ferrocene and derivatives at a liquid-liquid interface. *Electrochimica Acta*, **40** (18), 3005–3014.

(10) Ohde, H., Maeda, K., Yoshida, Y. and Kihara, S. (2000) Redox reactions between molecular oxygen and tetrachlorohydroquinone at the water | 1,2-dichloroethane interface. *Journal of Electroanalytical Chemistry*, **483**, 108–116.

(11) Olaya, A.J., Ge, P., Gonthier, J.F. *et al.* (2011) Four-electron oxygen reduction by tetrathiafulvalene. *Journal of the American Chemical Society*, **133** (31), 12115–12123.

(12) Su, B., Hatay, I., Trojanek, A. *et al.* (2010) Molecular electrocatalysis for oxygen reduction by cobalt porphyrins adsorbed at liquid/liquid interfaces. *Journal of the American Chemical Society*, **132** (8), 2655–2662.

(13) Olaya, A.J., Schaming, D., Brevet, P.F. *et al.* (2012) Self-assembled molecular rafts at liquid|liquid interfaces for four-electron oxygen reduction. *Journal of the American Chemical Society*, **134** (1), 498–506.

(14) Scanlon, M.D., Bian, X., Vrubel, H. *et al.* (2013) Low-cost industrially available molybdenum boride and carbide as "platinum-like" catalysts for the hydrogen evolution reaction in biphasic liquid systems. *Physical Chemistry Chemical Physics*, **15** (8), 2847–2857.

(15) Ge, P., Scanlon, M.D., Peljo, P. *et al.* (2012) Hydrogen evolution across nano-Schottky junctions at carbon supported MoS_2 catalysts in biphasic liquid systems. *Chemical Communications*, **48** (52), 6484–6486.

(16) Taylor, G. and Girault, H.H.J. (1986) Ion transfer reactions across a liquid-liquid interface supported on a micropipette tip. *Journal of Electroanalytical Chemistry*, **208**, 179–183.

(17) Li, Q., Xie, S., Liang, Z. *et al.* (2009) Fast ion-transfer processes at nanoscopic liquid/liquid interfaces. *Angewandte Chemie*, **48** (43), 8010–8013.

(18) Cecchini, M.P., Turek, V.A., Paget, J. *et al.* (2013) Self-assembled nanoparticle arrays for multiphase trace analyte detection. *Nature Materials*, **12** (2), 165–171.

17

Electrochemistry in Unusual Fluids

Philip N. Bartlett
University of Southampton, Chemistry, UK

The vast majority of electrochemical studies have been conducted in aqueous solution, followed by studies in a limited number of high-dielectric nonaqueous solutions (e.g., acetonitrile, DMSO, propylene carbonate, dimethyl formamide) and, more recently, in ionic liquids. Nevertheless, electrochemists have always been interested in the possibilities offered by more unusual media, including the opportunity to have a wider potential window and to study electrochemical reactions at higher potentials, to extend the scope of electroanalysis to new analytes and media, or the deposition of a wider range of materials. To some extent, electrochemistry at extreme conditions of temperature [1] or pressure [2] offers some of these same challenges and possibilities.

The development and application of microelectrodes by Martin Fleischmann during the 1980s opened up significant opportunities for the extension of electrochemistry away from studies in "conventional" media characterized by high ionic conductivity at temperatures and pressures close to ambient. The low iR drop at microelectrodes made possible studies in solutions with no deliberately added electrolyte [3] and at high potentials in inert solvents [4]. It also opened up the possibility to conduct electrochemistry of gas-phase species [5] by partition into thin adventitious or deliberately deposited electrolyte films between a closely spaced microelectrode and counterelectrode [6, 7]. Microelectrodes also offer the significant advantages of steady-state voltammetric behavior at slow scan rates, which can be of significant benefit in studies under conditions where it is difficult to implement forced convection electrodes, such as the rotation disk or channel geometries.

There are four states of matter: solid, liquid, gas, and plasma. Electrochemistry in the solid and liquid states has been, and still is, widely studied. In contrast, electrochemical

Developments in Electrochemistry: Science Inspired by Martin Fleischmann, First Edition.
Edited by Derek Pletcher, Zhong-Qun Tian and David E. Williams.
© 2014 John Wiley & Sons, Ltd. Published 2014 by John Wiley & Sons, Ltd.

studies and applications in the gas phase and plasma are much less common as these are unusual fluids for electrochemistry and challenging to work with. Yet despite this – as will be shown in this chapter – they are of great interest to electrochemists and could be technologically useful.

17.1   Electrochemistry in Plasmas

A plasma is a state of matter made up of a (roughly) neutral mixture of positive and negative particles. As such, it is inherently electrically conducting in a similar way to the conventional electrolyte solutions used in electrochemistry. Plasmas can be produced in a variety of ways, for example by electrical discharge or by radiofrequency excitation. In these cases, the system is far from thermal equilibrium and the charge carriers are molecular or atomic ions and free electrons with the electron temperature (in Kelvin) tens, hundreds or thousands of times greater than the gas temperature. Plasmas of this type are widely used in materials processing for cleaning and for deposition [8], and several groups have studied the effects of electrolytic deposition from a plasma. For example, Vennekamp and Janek [9] studied the galvanostatic and potentiostatic deposition of AgCl on Ag electrodes in an inductively coupled chlorine plasma, while Kawabuchi and Magari [10] and Ogumi *et al.* [11] used low-temperature plasmas to electrodeposit metal oxides, Uchimoto *et al.* [12] deposited metal halides, and Richmonds *et al.* have recently described the use of an atmospheric pressure plasma as a gaseous, metal-free electrode for electron transfer to aqueous solution [13].

The link between arc plasmas and electrochemistry was explored by Vijh [14], who studied 32 metals and compared the electrochemical description of the plasma–metal interface with that of treating the interface as a boundary between two plasmas, one for the metal and the other for the arc. Vijh concluded that the interface could be described as a metal/electrolyte interface with a characteristic interfacial potential distribution which depended on the choice of metal.

Over the past 12 years, Caruana and coworkers have made a series of reports investigating electrochemistry in flame plasmas. Flames represent a convenient and relatively controlled way to generate a plasma. In the flame plasma there is a thermal equilibrium between the electrons and the ions and neutral species. In Caruana's studies the flames are produced using a Méker burner fed with a premixed flow of methane, oxygen and nitrogen gas. The methane flame is a weak plasma which is overall neutral. The negative charged species are mainly electrons, while the positive species are predominantly molecular ions (CHO^+, H_3O^+) with concentrations as high as 10^{13} cm^{-3} (equivalent to ~16 nM). This type of flame has a calculated adiabatic temperature of around 2870 K [15, 16]. In later studies, Caruana and colleagues changed to using hydrogen instead of methane; with hydrogen, the calculated adiabatic temperature is around 2300 K [17, 18].

Metal ions can be introduced into the flame by injecting an aerosol of the appropriate aqueous metal salt. The metal ions can be present in the flame as M^+, $M^+ \cdot H_2O$ or $M^+ \cdot 2H_2O$. When considering the electrochemistry here it is important to realize that, in the flame plasma, the mobility of the electrons in much higher than that of the ions (by a factor of ~1000). For example, the mobilities of Cs^+ and H_3O^+ in the flame plasma are 5.0 cm^2 V^{-1} s^{-1} and 8.0 cm^2 V^{-1} s^{-1}, respectively, whereas the mobility for electrons is 4000 cm^2 V^{-1} s^{-1} [16].

In their early studies, Caruana *et al.* measured the potential difference in a single flame between two dissimilar metal electrodes (chosen from Ti, Mo, Nb, Hf, Ta, and W, because of their high melting points and low ionization potentials) [15]. Consequently, stable, reproducible potentials were observed and it was possible also to demonstrate a Nernstian-like relationship between the potential and the composition of a Pt/Rh alloy electrode. Preliminary voltammetric studies were also carried out at a Pt working electrode, using a three-electrode configuration for flames containing iron and copper ions. The results of these initial experiments confirmed that the flame acts rather like an electrolyte solution and can be used as a medium for electrochemistry.

By designing the burner so that there are two separate gas mixtures – each of which can be seeded with a different metal salt or different concentrations of metal salt feeding two flames that are in contact – it is possible to achieve a greater control over the experiment. Caruana's group have used this arrangement to study the electrochemical diffusion potential in the flame [16]. This is the equivalent of the classical electrochemical measurement of a liquid junction with transference. The potential arises because of the difference in the mobilities of the different ions on either side of the junction. Given the significant differences in mobility of the cations and electrons in the flame, it can be expected that significant effects will be seen under the correct conditions. Experimentally, Caruana *et al.* showed that the potential difference varied with the concentration of Cs^+ ions in the two flames, and demonstrated a good agreement with the model developed by Henderson [19] for the classical liquid junction. They also showed that the diffusion potential was independent of the flow rate, as expected, and varied with the concentration gradient in the flame. Similar results were obtained by Goodings *et al.* for a hydrogen flame, using a different flame geometry and doping the flame with Na^+ and methane [20].

Using the same approach with two flames, Caruana *et al.* have studied the overall cell potential in the case where the ionic species introduced into the two flames are different [18, 21]. In this case, Li, K and Cs were used [18] as well as Cu [21]. The total cell potential is composed of contributions from the mixed potentials at the two electrodes and the diffusion potential at the junction of the two dissimilar flames. The mixed potential at each electrode surface is determined by the two surface reactions:

$$e^-_{(g)} \rightarrow e^-_{(el)} \tag{17.1}$$

$$M^+_{(g)} + e^-_{(el)} \rightarrow M_{(g)} \tag{17.2}$$

where $e^-_{(g)}$ is the electron in the plasma and $e^-_{(el)}$ the electron in the metal electrode. This leads to the following expression for the overall cell potential [18]

$$E_{cell} = \frac{\alpha \left(\Delta\phi^{o\prime}_{M/M^+} - \Delta\phi^{o\prime}_{N/N^+} \right)}{(\alpha_{e^-} + \alpha)} + \frac{RT}{F(\alpha_{e^-} + \alpha)} \ln \left(\frac{n^\alpha_N n^\alpha_{M^+} n^{\alpha_{e^-}}_{e^-_{Lw}}}{n^\alpha_M n^\alpha_{N^+} n^{\alpha_{e^-}}_{e^-_{Rw}}} \right)$$

$$+ \frac{RT}{F(\alpha_{e^-} + \alpha)} \ln \left(\frac{i_{0_{N^+}} i_{0_{e^-_R}}}{i_{0_{M^+}} i_{0_{e^-_L}}} \right) + \Delta\phi_{diff} \tag{17.3}$$

where $\Delta\phi^{o\prime}$ are the standard potentials, α are the transfer coefficients and i_0 the exchange currents for the different electrode reactions, n is the concentrations of the species, and the subscripts refer to the different species involved, M/M^+, N/N^+ and e^-. The subscripts R,

L, Rw and Lw refer to the right and left flames and right and left walls, respectively. $\Delta\phi_{diff}$ is the diffusion potential.

According to this model the overall cell potential depends on four terms. The first term contains the standard potentials and describes the dependence on the identity of the chemical species in the two flames. The second term describes the Nernstian dependence on the concentrations of the different species at the electrode surface. The third term describes the potential dependence on the electrode kinetics, and the final term is the diffusion potential.

This electrochemical model was shown to be consistent with the experimental results and, as expected, the diffusion potential was found to make a significant contribution to the overall cell potential because of the significant difference in mobility for the electrons and cations. In contrast, changing the cation from Li^+ to Cs^+ had little effect on the diffusion potential. One of the experimental advances in this study was the use of boron-doped diamond electrodes, which have the advantage of slowly eroding in the flame and therefore resisting surface contamination better than Pt electrodes.

As in all electrochemistry, for voltammetric studies it is very helpful to have a reference electrode that provides a stable, reproducible potential. For flame electrochemistry, Fowowe *et al.* [22] investigated the use of an electrode made from a powdered mixture of metal and a corresponding metal oxide packed in a recrystallized ceramic tube. Of the various metals studied, the Ti/TiO_2 system was found to give the best performance. The availability of a suitable reference electrode allows reproducible voltammetric studies. Using a conventional three-electrode system, Elahi *et al.* [17] have investigated the electrochemistry of heteropolyanions in flames; the experimental arrangement is shown in Figure 17.1. The reference and counter electrode are placed in one flame and the working electrode in the other; the redox species to be studied are then injected, in the form of an aqueous aerosol, into the flame impinging on the working electrode.

Figure 17.2 shows the resulting voltammograms for $(NH_4)_6H_2W_{12}O_{40}$, $(NH_4)_6Mo_7O_{24}$ and NH_4VO_3 nebulized into the flame to generate the molecular oxides such as WO_3 and MO_3. The voltammograms are rather curious, as they show several peaks (rather than waves) at different potentials depending on the species, with the same peaks, in the same direction, observed on the cathodic and anodic scans. This is clearly very different from the type of voltammetry usually seen in solution. Similar results were obtained when the reference electrode was changed to yttria-stabilized zirconia (YSZ), which has the

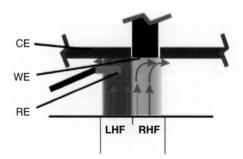

Figure 17.1   *Schematic cross-section of the three electrodes in the flame, showing the position of the working electrode in a hole in the platinum electrode shield. Reproduced with permission from Ref. [17]. Copyright © 2012 WILEY-VCH Verlag GmbH & Co. KGaA, Weinheim.*

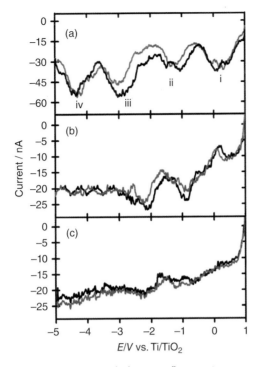

Figure 17.2 Cyclic voltammetry recorded in a flame. Aqueous solutions containing 1×10^{-3} mol dm^{-3} (a) $(NH_4)_6H_2W_{12}O_{40}$, (b) $(NH_4)_6Mo_7O_{24}$ and (c) NH_4VO_3 were nebulized in the right-hand flame. Forward (black) and backward (red) scans at 1 V s^{-1}, starting from 1 V. Reproduced with permission from Ref. [17]. Copyright © 2012 WILEY-VCH Verlag GmbH & Co. KGaA, Weinheim.

advantage of greater stability [23]. The peaks in the voltammetry were shown to depend on the concentration of redox species introduced into the flame, the area of the working electrode and also the scan rate, with some peaks increasing and others decreasing with increasing scan rate [23]. In general, similar results were observed at Pt, Au, and graphite working electrodes.

To explain the unusual shapes of the voltammetric peaks, the authors suggested that the results reflected the overlap in the electronic energy states between the electrode and the species in the gas phase (see Figure 17.3). The important point to note here is that, unlike the conventional solution case, there is no solvent to provide a thermal bath around the reacting ion. Therefore, Caruana *et al.* suggested that electron transfer only occurs when there is resonance between the Fermi level in the metal, unoccupied molecular orbital; as the potential sweeps more negative from A to B to C in Figure 17.3, the current initially increases to a maximum but then falls again in the absence of solvent to carry away the excess energy. On the reverse scan, the peak is again seen when the Fermi level comes into resonance with the peak reduction potential E_R. These are clearly intriguing experiments, and it will be interesting to see how these investigations into amperometric electrochemistry develop.

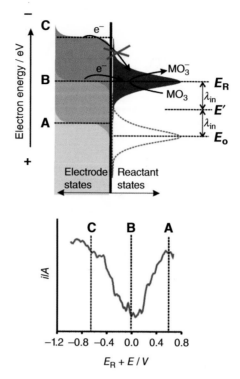

Figure 17.3 *Energy states on an absolute scale, for the gas-phase reactant species (MO$_3$) and the electrons in the solid at different Fermi energies A, B, and C, showing that only when the Fermi energy is at B, electrons transfer to the reactant. At potentials lower (A) or higher (C) than B, electron transfer is forbidden. The lower figure shows the current versus potential response for the reduction of MO$_3$ (the cathodic current is a downwards response). Reproduced with permission from Ref. [17]. Copyright © 2012 WILEY-VCH Verlag GmbH & Co. KGaA, Weinheim.*

Although, at present, there is not a full understanding of electrochemistry in flames, it is clear that there are interesting potential applications. Thus, Caruana's group have shown that electroreduction in a flame containing Cu$^+$ can be used to deposit Cu onto diamond electrodes [24]. In this case, the wide potential window in the flame (>10 V) suggests the possibility of depositing a wide range of technologically interesting materials. Caruana *et al.* have also demonstrated the analytical possibilities of electrochemistry in flames by using this approach to characterize bioaerosols and other airborne particles [25–27].

17.2   Electrochemistry in Supercritical Fluids

In supercritical fluids (SCFs), the gas molecules of the fluid act as the solvent. Strictly speaking, an SCF is defined as any fluid above its critical temperature (T_c) and pressure (p_c) (Figure 17.4). However, once the fluid is significantly far from its critical point, so that its density is significantly below its critical density, it will not be a useful solvent. Density

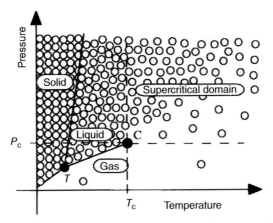

Figure 17.4 *Schematic representation of a phase diagram illustrating the density evolution from the liquid to the gas without crossing the T–C line which corresponds to the liquid–gas equilibrium line, C being the critical point and T being the triple point. P_c and T_c are the critical pressure and temperature, respectively. Reproduced from Ref. [29] with permission from The Royal Society of Chemistry.*

is a critical parameter of the SCF for electrochemical applications, because it determines the solubility and degree of dissociation of ionic species [28]. SCFs close to the critical density are unique solvents that combine the properties of liquids and gases and are used in many applications [29–31]. As a result of this interest, the properties of SCFs have been studied in some detail [32–34].

The densities and transport properties of SCFs close to their critical temperature and pressure lie between those of gases and liquids. In order to be used as a solvent for electrochemistry, the SCF needs to have a liquid-like density, to ensure high solvation, and a gas-like diffusion coefficient and viscosity for high mass transport. However, the diffusion coefficient and viscosity will change as the density of the SCF changes with temperature and pressure since, unlike conventional solvents, the physical properties of SCFs can be tuned by changing the temperature and/or pressure [34]. Thus, it is necessary to strike a balance between the three properties when choosing the conditions (T and p) to carry out electrochemistry in a SCF. It is also important (as will be seen below) to remember that the phase behavior will be altered when significant amounts of electrolyte are dissolved in the fluid, and T_c and p_c for the solvent cannot be relied on alone to ensure that the mixture is in a single-phase, supercritical state. It is essential to properly characterize the phase behavior for the actual system. In order to use SCFs for electrochemistry it is necessary to understand the phase behavior of the mixture, and to appreciate how this links to the solubility and conductivity.

The earliest studies of electrochemistry (as opposed to corrosion) in SCFs were almost certainly carried out by Williams and Naiditch who, in 1970, described a two-electrode experiment in which they formed dendritic deposits of silver on a platinum electrode from "dense gaseous solutions" of $AgNO_3$ in ammonia [35]. As these experiments were carried out in sealed glass vessels only slightly (7 K) above the critical temperature for NH_3, it is not certain that the fluid was actually in the supercritical (sc) – as opposed to close to supercritical – state. Over a decade later in 1981, Silvestri *et al.* reported the results of a preliminary study of SCFs as solvents for electrosynthesis when they studied CO_2, bromotrifluoromethane, hydrogen chloride and ammonia [36]. In this case, it was found that "... a solution of tetrabutylammonium iodide in CO_2 was a poor conductor in the liquid, as well as in the supercritical state. Bromotrifluoromethane, in which the electrolyte was practically insoluble, also proved to be a very poor conductor". However, Silvestri and coworkers were more successful with the more polar, higher dielectric solvents, scHCl and $scNH_3$, and reported long-term electrolyses for Ag and Fe electrodes in $scNH_3$. Moreover, although the scHCl was difficult to work with (because it is highly corrosive), elemental iodine was produced via the electrolysis of KI in scHCl.

These studies were followed, between 1984 and 1997, by a more sustained series of investigations made by Bard and coworkers, using $scNH_3$ [37–39], water [40–43], acetonitrile [44,45], and SO_2 [46]. During the course of these studies, techniques were developed for working with SCFs (and particularly corrosive SCFs such as water) and the electrochemistry of both inorganic and organic redox systems was investigated. The studies included Cu deposition [43], halide oxidation [41], oxygen reduction [41], and hydroquinone oxidation [41], all from scH_2O, and the electrochemistry of solvated electrons and organic redox couples in $scNH_3$ [37–39].

In all of these early studies, polar fluids with high dielectric constants were used in order to dissolve and dissociate the electrolyte. These polar fluids generally have high critical temperatures and pressures (see Table 17.1), and can be highly corrosive; indeed, corrosion in scH_2O is a significant problem in the nuclear industry. Less-polar SCFs such as $scCO_2$ are much more attractive to work with because they have lower critical temperatures and pressures and are much less corrosive; however, they represent a significant challenge to the electrochemist because they have very low dielectric constants. As a result, it is extremely difficult – if not impossible – to dissolve and dissociate sufficient electrolyte to achieve useful solution conductivities. For example, in one of the few studies in $scCO_2$ Abbott and Harper [47], when using a hydrophobic electrolyte (tetradecylammonium tetraphenylborate), found some conductivity ($\sim 10^{-6}$ S cm^{-1}) and were able to see some, poorly resolved, voltammetry for the nickel complex, $TDDA_2Ni(mnt)_2$.

One way to overcome this problem is to mix $scCO_2$ with a more polar cosolvent, such as methanol or acetonitrile. Although these mixtures will still form a single-phase SCF

Table 17.1   *Critical temperatures, critical pressures, density at the critical temperature and pressure and dielectric constant of some supercritical fluids [28, 37, 48–50].*

	T_c (K)	p_c (bar)	ρ_c (g cm^{-3})	ε
H_2O	647.3	218.3 atm	0.32	6
NH_3	405.6	112.5	0.24	3–4
CO_2	304.2	72.9	0.47	1.4
CHF_3	299.29	48.32	0.5265	
CH_2F_2	351.26	57.82	0.424	4.9
CH_2FCF_3	374.21	40.49	0.51190	3.5
H_2O (l)	—	—	1.0	78.54
CH_3CN (l)	—	—	0.7857	37.5
CH_2Cl_2	—	—	1.327	9.08

at the appropriate T and p, the presence of the cosolvents will significantly increase the solubility of the ionic species and, in turn, the conductivity that can be achieved. Of course, in these mixed systems the solvation of ions is likely to be predominantly by the more polar component. Grinberg and Mazin [51] have reviewed the electrochemistry in $scCO_2$ and $scCO_2$ with various cosolvents covering the period up to 1998. In the present author's studies [52], comparisons were made of the behavior of mixtures of CO_2 with $[^nBu_4N][BF_4]$ and either methanol (CH_3OH) or acetonitrile (CH_3CN) as a cosolvent. These experiments showed that, at similar temperatures and pressures, the solubility of $[^nBu_4N][BF_4]$ was about fivefold higher in $CH_3CN + CO_2$ than in $CH_3OH + CO_2$.

When using SCFs – and particularly when using mixed-solvent systems – it is important to characterize the phase behavior of the system since the addition of significant concentrations (>1 mM) of electrolyte are likely to alter the phase behavior. This can be achieved conveniently by using a view cell in which the different phases can be characterized as a function of temperature, pressure, and composition [53].

As an example, Figure 17.5 shows the results of a study of the phase behavior of a ternary $CO_2–[^nBu_4N][BF_4]–CH_3CN$ system. For the mixture shown in Figure 17.5 there is a single supercritical phase region in the top left-hand corner of the phase diagram that gives way to two- and three-phase regions as the pressure is decreased. In principle, addition of the redox species will also alter the phase behavior, but in practice the effects are generally small, not least because the concentrations of the redox species are much smaller. Thus, the addition of about 3.5×10^{-4} mole fraction of a redox species produces only slight shifts (<2 MPa at fixed temperature) in the supercritical phase boundary for the system shown in Figure 17.5 [52].

An alternative approach to the use of a cosolvent is to start with a fluid with a higher dielectric constant. Hydrofluorcarbons (HFCs) have been used as supercritical solvents for electrochemistry as they are polar and give higher dielectric fluids whilst retaining

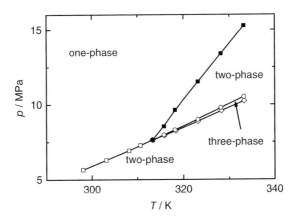

Figure 17.5 *p–T phase diagram of the mixture of CO_2 (1) + CH_3CN (2) + $[^nBu_4N][BF_4]$ (3) with $x_1 = 0.871$, $x_2 = 0.129$, and $x_3 = 8.3 \times 10^{-4}$. (■) and (□), the phase boundary between one-, and two-phase region; (○) and (◇) mark the region where three phases are in coexistence. The pressure difference between the two three-phase boundaries is less than 0.3 MPa at 333 K. Reproduced by permission of the PCCP Owner Societies.*

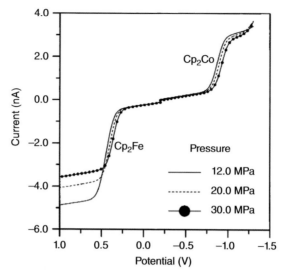

Figure 17.6 *Comparison of ferrocene and cobaltocenium waves as a function of pressure in CCl_2F_2. Conditions: 90 µM ferrocene, 90 µM cobaltocenium, 10.0 mM [$^nBu_4N$][BF_4], 388 K; 25 µm-diameter Pt disk electrode; scan rate 20 mV s^{-1}. For clarity, only the forward sweep of each voltammogram is displayed, with the ferrocene wave recorded first and the cobaltocenium wave immediately after, each wave originating from −0.20 V. Reprinted with permission from Ref. [54]. Copyright 1994, American Chemical Society.*

reasonable critical temperatures and pressures (see Table 17.1). Early studies using scHFCs were carried out by Olsen and Tallman [54] using microelectrodes to minimize the effects of the iR drop. In this case, the electrochemistry of ferrocene and cobaltocenium in scCHClF$_2$ containing millimolar tetrabutylammonium tetrafluoroborate electrolyte was studied, and good voltammetry was obtained for ferrocene and cobaltocenium in scCCl$_2$F$_2$ (Figure 17.6). Similar results were obtained by Goldfarb and Corti [55–57] for the electrochemistry of decamethylferrocene in scCHF$_3$ containing tetrabutylammonium hexafluorophosphate electrolyte.

Abbott and colleagues also recognized the potential of scHFCs for electrochemistry, and conducted a series of studies in this area between 1998 and 2007 [58–66]. For example, when using tetrabutylammonium perchlorate electrolyte in liquid CHF$_3$CH$_3$F and CH$_2$F$_2$ at 298 K and 10 bar, they demonstrated electrochemical windows of 9.4 and 5.8 V, respectively [63].

Even when using the more polar SCFs, the issue of electrolyte conductivity is very important for electrochemical studies. Conductivity depends on the solubility of the electrolyte, the dissociation of the ions, and the mobility of the ions in the SCF. In most of the reported studies, commercially available electrolytes were employed (e.g., tetrabutylammonium tetrafluoroborate) that are conventionally used for nonaqueous electrochemistry. For example, when Olsen and Tallman [67] measured the equivalent conductivity for concentrations of tetrabutylammonium tetrafluoroborate between 6 and 13 mM in scCHClF$_2$ at 388 K, and for pressures from 10 to 24 MPa, they found a nonlinear increase in equivalent conductivity with the square-root of electrolyte concentration. Such behavior is indicative

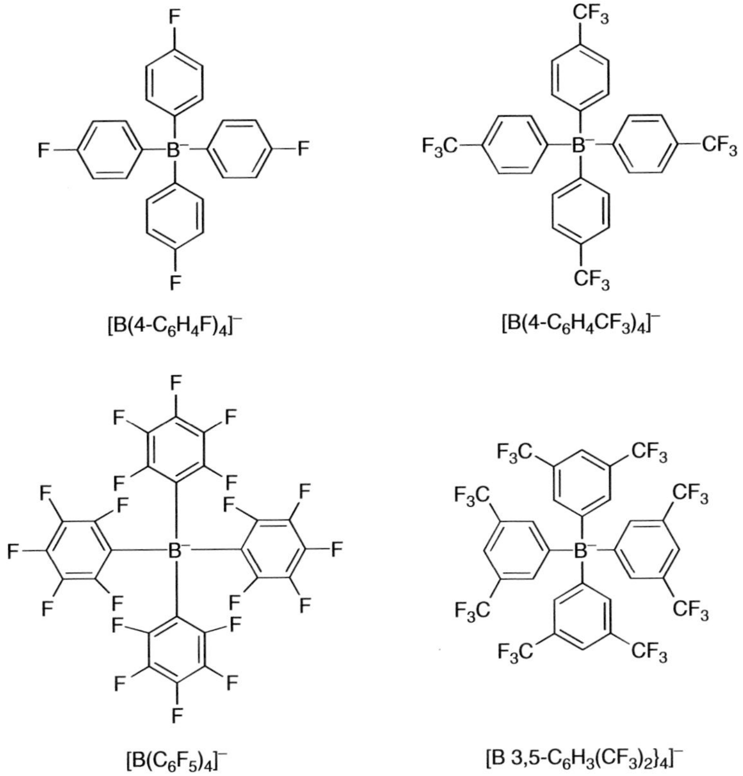

Figure 17.7 *The structures of some different fluorinated aryl borates.*

of the formation of triple ions, as predicted by the Fuoss–Kraus equation [68]. Both, Abbott and Eardley [62] and Goldfarb and Corti [56] also found evidence of the contribution of triple ions for tetrabutylammonium tetrafluoroborate in $scCH_2F_2$ and tetrabutylammonium perchlorate in $scCHF_3$, respectively.

At Southampton and Nottingham, investigations have been made into using the tetraalkylammonium salts of fluorinated aryl borates as electrolytes. The fluorinated aryl borates are essentially bulkier versions of $[BF_4]^-$ and were originally developed as anions for organometallic synthesis [69, 70]; the structures of the ions are shown in Figure 17.7. Subsequently, the conductivity of the different salts was measured in $scCO_2/CH_3CN$ (Figure 17.8) [52, 71]. In terms of the effect of the anions on molar conductivity, the general trend was $[BF_4]^- \sim [B(4\text{-}C_6H_4F)_4]^- < [B(4\text{-}C_6H_4CF_3)_4]^- \sim [B(C_6F_5)_4]^- < [B\{3,5\text{-}C_6H_3(CF_3)_2\}_4]^-$, with the highest molar conductivity (22–26 S cm^2 mol^{-1}) achieved using $[B\{3,5\text{-}C_6H_3(CF_3)_2\}_4]^-$; $[^nBu_4N][B\{3,5\text{-}C_6H_3(CF_3)_2\}_4]$ is about a 10-fold improvement over $[^nBu_4N][BF_4]$.

Studies of conductivity as a function of concentration for the $[B\{3,5\text{-}C_6H_3(CF_3)_2\}_4]^-$ salts in $scCO_2/CH_3CN$ also showed evidence of the contribution from triple ions.

These same electrolytes are also effective in $scCH_2F_2$ (Figure 17.9; Table 17.2), and are more conducting than $[^nBu_4N][BF_4]$ under the same conditions [72]. However, the increase is less significant than it is for $scCO_2/CH_3CN$.

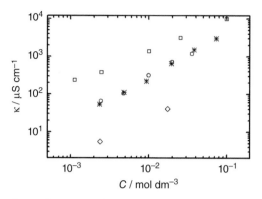

Figure 17.8 *Electrical conductivity of some supporting electrolytes in scCO$_2$/CH$_3$CN (x_{CO2} : x_{CH3CN} = 0.89 : 0.11) at 328.15 K and 200 bar. ◇, [^{n}Bu$_4$N][BF$_4$]; ∗, [^{n}Bu$_4$N][B{3,5-C$_6$H$_3$(CF$_3$)$_2$}$_4$]; ○, [NBu$_3$R$_f$][B{3,5-C$_6$H$_3$(CF$_3$)$_2$}$_4$]; □, [^{n}Bu$_4$N][BF$_4$] in liquid CH$_3$CN at ambient condition (R$_f$ is CF$_3$(CF$_2$)$_7$(CH$_2$)$_3$). Reproduced from Ref. [71].*

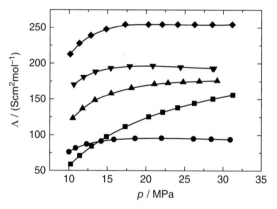

Figure 17.9 *Molar conductivity (Λ) of the supporting electrolytes in CH$_2$F$_2$ at 363.15 K and pressures up to 32 MPa. The molar concentration of the electrolytes is ~9.0 mol m^{-3}. ■, [^{n}Bu$_4$N][BF$_4$]; •, [NR$_f$Bu^n$_3$][B{3,5-C$_6$H$_3$(CF$_3$)$_2$}$_4$],; ▲, [^{n}Bu$_4$N][B(C$_6$F$_5$)$_4$]; ▼, [^{n}Bu$_4$N][B{3,5-C$_6$H$_3$(CF$_3$)$_2$}$_4$]; ♦, Na[B{3,5-C$_6$H$_3$(CF$_3$)$_2$}$_4$], (R$_f$ is CF$_3$(CF$_2$)$_7$(CH$_2$)$_3$). Reproduced from Ref. [72] by permission of The Royal Society of Chemistry.*

Table 17.2 *Molar conductivity of different supporting electrolytes (~9.0 mol m^{-3}) in CH$_2$F$_2$ at 363 K and 20 MPa.*

Electrolyte	Λ (S cm^2 mol^{-1})
[^{n}Bu$_4$N][BF$_4$]	124
[NR$_f$Bu^n$_3$][B{3,5-C$_6$H$_3$(CF$_3$)$_2$}$_4$]	95
[^{n}Bu$_4$N][B(C$_6$F$_5$)$_4$]	170
[^{n}Bu$_4$N][B{3,5-C$_6$H$_3$(CF$_3$)$_2$}$_4$]	197
Na[B{3,5-C$_6$H$_3$(CF$_3$)$_2$}$_4$]	255

Reproduced from Ref. [72] by permission of The Royal Society of Chemistry.

In addition to studies of electrolyte conductivity, further studies have been conducted of other fundamental issues such as diffusion, effects of solvation, and the structure of the double layer. Goldfarb and Corti [57] carried out a detailed study of the effect of T and p on the diffusion coefficient for decamethylferrocene in $scCHF_3$, taking account of the effects of ion pairing and correlating the results on the basis of the Stokes–Einstein model. Abbott *et al.* [65, 66] used a quartz crystal microbalance to measure the viscosity of $scCH_2F_2$ and found that this could change significantly (8- to 32-fold increases) at pressures close to the critical value when an electrolyte was added. Abbott and coworkers also investigated the effects of ion-pairing on the redox potential of ferrocene carboxylic acid in $scCH_2F_2$ [59]; in fact, Abbott *et al.* are to date the only group to investigate the structure of the double layer in an SCF [60, 64].

17.2.1 Applications of SCF Electrochemistry

The wide potential window, high mass transport rate, low viscosity and lack of surface tension make SCFs an attractive medium for the electrodeposition of nanoscale (< 20 nm) structures and devices. However, few data are available relating to electrodeposition from SCFs. Both, Williams and Naiditch [35] and Silvestri *et al.* [36] have deposited Ag from supercritical or near-supercritical NH_3, while MacDonald *et al.* [43] have deposited Cu from scH_2O. There are also two reports describing the electrodeposition of conducting polymers from a SCF. This is a rather different situation, because in this case a neutral species (the heterocyclic monomer) is oxidized to produce a charged polymer which precipitates on the electrode with the counterion from solution. Atobe *et al.* [73] electropolymerized both pyrrole and thiophene from $scCHF_3$ containing tetrabutylammonium hexafluorophosphate, while and Yan *et al.* [74] deposited poly(pyrrole) films from $scCO_2/CH_3CN$ containing tetrabutylammonium hexafluorophosphate.

Part of the present author's development of SCF electrochemistry for the deposition of nanoscale structures and devices have included studies of the electrodeposition of copper [71, 75]. The cyclic voltammetry of $[Cu(CH_3CN)_4][BF_4]$ in $scCO_2/MeCN$ with $[^nBu_4N][BF_4]$ electrolyte is shown in Figure 17.10. In this case, $[Cu(CH_3CN)_4][BF_4]$ is a good choice for two reasons: (i) as the CH_3CN ligand is also the cosolvent, no additional species are added; and (ii) the use of a Cu(I) complex avoids the presence of Cu(II) and the problem of comproportionation between Cu(0) and Cu(II). The relative stability of the Cu(I) and Cu(II) redox states depends heavily on the ligand environment; hence, although it is possible to electroplate copper from the $[Cu(hfac)_2]$ (hfac = hexafluoroacetylacetonate) from $scCO_2/CH_3CN$ the comproportionation reaction

$$[Cu(hfac)_2] + Cu(0) + 8CH_3CN \rightarrow 2[Cu(CH_3CN)_4]^+ + 2hfac \qquad (17.4)$$

complicates the process.

The voltammetry in Figure 17.10 shows the characteristic features expected for electrodeposition with a mass transport-limited reduction of Cu(I) to Cu(0) at negative potentials, accompanied by a stripping peak for Cu on the return scan. The stripping peak is sharp and undistorted, showing that iR drop is insignificant in these experiments. The inset in Figure 17.10 shows a plot of the mass transport-limited currents at a microdisk electrode for

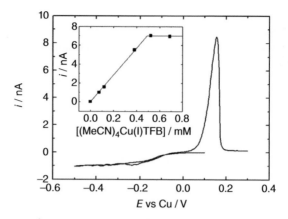

Figure 17.10 *Copper voltammetry was performed in scCO₂ with 12.1 wt% CH₃CN and [ⁿBu₄N][BF₄] (20 mM) at 310 K and 172.4 bar. Electrodes were: 25 μm-diameter platinum disk working electrode, 0.5 mm-diameter platinum disk pseudo reference, and 0.5 mm-diameter platinum wire counterelectrode. The sweep rate was 20 mV s⁻¹. Reproduced from Ref. [71].*

different concentrations of $[Cu(CH_3N)_4][BF_4]$. This is a very sensitive way to determine the diffusion coefficient and solubility of the redox active species, since the mass transport-limiting current at the microdisk is given by:

$$I_L = 4nFaDc \tag{17.5}$$

where n is the number of electrons transferred and a is the radius of the microdisk electrode. Assuming that the diffusion coefficient, D, does not change with concentration, the limiting current should be proportional to the concentration of species dissolved in the solution. From the inset in Figure 17.10 it can be seen that the current increases linearly with concentration at low concentrations, but then sharply reaches a plateau at higher concentrations, indicating that the solution has become saturated. From the intersection of the two lines a solubility of $[Cu(CH_3CN)_4][BF_4]$ of 0.49 mM is obtained. From the slope of the initial portion of the curve where the limiting current varies with the concentration, by using Equation (17.5) a diffusion coefficient of 3.5×10^{-5} cm² s⁻¹ can be obtained, which is approximately 1.6-fold larger than the value measured for the same complex in liquid acetonitrile at the same temperature (2.2×10^{-5} cm² s⁻¹).

By using this system it is possible to electroplate smooth, reflective copper films onto macroelectrodes (1 cm²). Figure 17.11 shows scanning electron microscopy (SEM) images of films deposited at different potentials. The morphology of the deposits varied with the applied potential so that, at high overpotentials – when the deposition is mass transport-limited – the films were rough and dendritic, whereas at low overpotentials the films were smooth, shiny, and adherent. Chemical characterization showed that the films were of high purity, with resistivity measurements of 4×10^{-6} Ω·cm for the best films. This was close to values reported for copper electrodeposition from aqueous solution ($1.75–2 \times 10^{-6}$ Ω·cm, depending on the plating bath additives [76]).

This same electrolyte solution can be used to deposit copper into nanoscale (<10 nm) pores [71]. Figure 17.12 shows transmission electron microscopy (TEM) images and

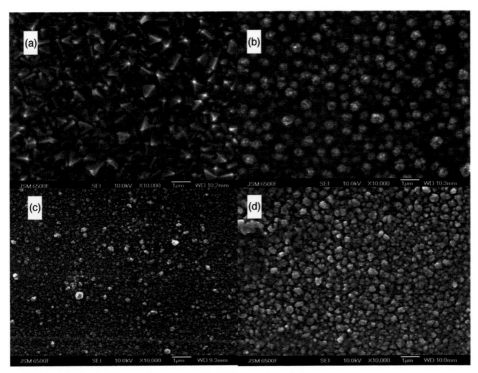

Figure 17.11 *SEM images of copper films deposited from a saturated solution of [Cu(CH₃CN)₄][BF₄] in 87.4 wt% CO₂, 13 wt% acetonitrile, 0.6 wt% [ⁿBu₄N][BF₄] at 309–311 K and 138 bar at (a) −0.6 V, (b) −0.9 V, (c) −1.5 V, and (d) −1.7 V versus Pt. Reproduced from Ref. [71].*

characterization data for copper deposited into a mesoporous silica template with a regular hexagonal array of approximately 3 nm-diameter cylindrical pores approximately 6 nm apart. The TEM image in Figure 17.12a shows the regular array of cylindrical pores in cross-section on the left-hand side of the image. The energy-dispersive X-ray (EDX) analysis (Figure 17.12b) shows strong copper (as well as the expected silicon and oxygen) signals, while the selected area electron diffraction (SAED) analysis (Figure 17.12c) shows diffuse rings attributed to silica, together with rings consistent with the expected diffraction peaks for copper. A similar approach can be applied for silver deposition from scCO₂/CH₃CN using [Ag(CH₃CN)₄][BF₄] [77].

The use of CH₃CN as a cosolvent in scCO₂/CH₃CN has the disadvantage that it limits the potential window. It is therefore desirable to use other SCFs which could extend the potential window and allow the deposition of a wider range of functional materials, including magnetic and semiconducting materials. A first step in this direction was to explore the deposition of elemental germanium, which is a semiconductor with an indirect band gap of 0.66 eV and an important technological material. As single crystalline SiGe layers are used in integrated electronics [78], the electrodeposition of nanostructured germanium is of great interest. Consequently, several possible Ge(II) and Ge(IV) reagents have been examined

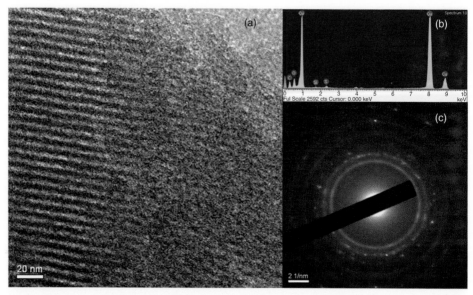

Figure 17.12   *(a) TEM image of copper nanorods electrodeposited into silica mesopores from a solution of scCO$_2$ with 12.1 wt% CH$_3$CN, 20 mM [^{n}Bu$_4$N][B{3,5-C$_6$H$_3$(CF$_3$)$_2$}$_4$] and 2 mM [Cu(CH$_3$CN)$_4$][B{3,5-C$_6$H$_3$(CF$_3$)$_2$}$_4$] at 311 K and 172.4 bar. The working electrode was a 0.8 × 0.5 cm ITO on glass slide modified with an approximately 250 nm-thick film of mesoporous silica; the counterelectrode was a large, coiled, copper wire, and the reference electrode was a 0.5 mm-diameter copper disc; (b) EDX spectrum recorded from the TEM sample in panel (a); (c) Selected area electron diffraction (SAED) pattern recorded on one of the copper rods. Reproduced from Ref. [71].*

for use in ssCO$_2$/CH$_3$CN and scCH$_2$F$_2$ [79]. Among these, the most promising results so far have been obtained using GeCl$_4$. At 357 K and 15 MPa, binary mixtures of CH$_2$F$_2$ with GeCl$_4$ and [NBu^n$_4$]Cl form a single-phase, supercritical system. These conditions can be used for the bulk deposition of Ge, whereby the as-deposited material is amorphous but can be crystallized under high-intensity laser illumination [79].

Beyond the area of nanostructured materials and electrodeposition, opportunities also exist for SCF electrochemistry in electrosynthesis, although at present these have been barely explored. Chanfreau *et al.* [80] have studied CO$_2$-DMF expanded liquid and scCO$_2$ as solvents for electrocarboxylation, motivated by a desire to develop a "Green Chemistry" route; however, the conversion was poor in the absence of DMF and problems were encountered with the low conductivity of the solution with the electrolytes used. Sun *et al.* [81] have studied the electro-oxidation of benzyl alcohol to benzaldehyde in scCO$_2$/[Bmim][PF$_6$]/CH$_3$CN solutions, and obtained a high product selectivity but a low rate of conversion. Interest has also been shown in the cathodic reduction of CO$_2$ [82–84], in which context electrochemistry in scCO$_2$ is an interesting option because of the low solubility of CO$_2$ in protic and aprotic solvents and the high concentration of CO$_2$ in the supercritical phase. Abbott and Eardley [61] studied CO$_2$ reduction in a mixed SCF of 1,1,1,2-tetrafluoroethane/CO$_2$ at Pt and Pb electrodes in both liquid and supercritical states.

Better results were obtained at Pt than at Pb electrodes, with a higher faradaic efficiency for the formation of oxalate in the supercritical state. One very interesting recent contribution in this area was made by Méndez *et al.* [85], who used a liquid–liquid interface between $scCO_2$ and water to carry out the photoreduction of CO_2 to CO using ruthenium tris(2,2'-bipyridyl) as the dye, and nickel cyclam as the catalyst.

17.3 Conclusions

As has been seen, electrochemistry in plasmas and SCFs is possible but presents significant challenges. Nevertheless, both areas offer interesting challenges and opportunities with prospects for novel technological applications. Notably, both areas have benefitted – and will continue to benefit – from innovations originated by Martin Fleischmann.

Acknowledgments

These studies were part of the Supercritical Fluid Electrodeposition project (www. scfed.net), a multidisciplinary collaboration of British universities investigating the fundamental and applied aspects of supercritical fluids funding by a Programme Grant from the EPSRC (EP/I013394/1).

References

(1) Wildgoose, G.G., Giovanelli, D., Lawrence, N.S. and Compton, R.G. (2004) High-temperature electrochemistry: a review. *Electroanalysis*, **16**, 421–433.
(2) Giovanelli, D., Lawrence, N.S. and Compton, R.G. (2004) Electrochemistry at high pressures: a review. *Electroanalysis*, **16**, 789–810.
(3) Bond, A.M., Fleischmann, M. and Robinson, J. (1984) Electrochemistry in organic solvents without supporting electrolyte using Pt microelectrodes. *Journal of Electroanalytical Chemistry*, **168**, 291–312.
(4) Cassidy, J., Khoo, S.B., Pons, S. and Fleischmann, M. (1985) Electrochemistry at very high potentials: the use of ultramicroelectrodes in the anodic oxidation of short chain alkanes. *The Journal of Physical Chemistry*, **89**, 3933–3935.
(5) Ghoroghchian, J., Sarfarazi, F., Dibble, T. *et al.* (1986) Electrochemistry in the gas phase. Use of ultramicroelectrodes for the analysis of electroactive species in gas mixtures. *Analytical Chemistry*, **58**, 2278–2282.
(6) Fang, Y. and Leddy, J. (1995) Voltammetry in gas phase environments. *Journal of Electroanalytical Chemistry*, **384**, 5–17.
(7) Reed, R.A., Geng, L. and Murray, R.W. (1986) Solid-state voltammetry of electroactive solutes in polyethylene oxide polymer-films on microelectrodes. *Journal of Electroanalytical Chemistry*, **208**, 185–193.
(8) Hess, D.W. (1999) Plasma assisted oxidation, anodization, and nitridation of silicon. *IBM Journal of Research and Development*, **42**, 127–145.
(9) Vennekamp, M. and Janek, J. (2005) Control of the surface morphology of solid electrolyte films during field-driven growth in a reactive plasma. *Physical Chemistry Chemical Physics*, **7**, 666–677.
(10) Kawabuchi, K. and Magari, S. (1980) Growth-morphology of UO_2, ZrO_2, HfO_2 and ThO_2 crystallites grown from oxide plasmas. *Journal of Crystal Growth*, **49**, 81–85.
(11) Ogumi, Z., Uchimoto, Y. and Takehara, Z.-I. (1995) Electrochemistry using a plasma. *Advanced Materials*, **7**, 323–325.

(12) Uchimoto, Y., Okada, T., Ogumi, Z. and Takehara, Z. (1994) Vapour phase electrolytic deposition: a novel method for preparation of orientated thin films. *Journal of the Chemical Society, Chemical Communications*, 585–586.

(13) Richmonds, C., Witzke, M., Bartling, B. *et al.* (2011) Electron-transfer reactions at the plasma-liquid interface. *Journal of the American Chemical Society*, **133**, 17582–17585.

(14) Vijh, A.K. (1986) Electrode-potentials and interface plasmons in metal gaseous electrolyte (i.e. plasma) interphase region. *Materials Chemistry and Physics*, **14**, 47–56.

(15) Caruana, D.J. and McCormack, S.P. (2000) Electrochemistry in flames: a preliminary communication. *Electrochemistry Communications*, **2**, 816–821.

(16) Caruana, D.J. and McCormack, S.P. (2001) Electrochemical diffusion potential in the gas phase. *Electrochemistry Communications*, **3**, 675–681.

(17) Elahi, A., Fowowe, T. and Caruana, D.J. (2012) Dynamic electrochemistry in flame plasma electrolyte. *Angewandte Chemie, International Edition*, **51**, 6350–6355.

(18) Hadzifejzovic, E., Galiani, J.A.S. and Caruana, D.J. (2006) Plasma electrochemistry: Potential measured at boron doped diamond and platinum in gaseous electrolyte. *Physical Chemistry Chemical Physics*, **8**, 2797–2809.

(19) Morf, W.E. (1977) Calculation of liquid-junction potentials and membrane-potentials on basis of Planck theory. *Analytical Chemistry*, **49**, 810–813.

(20) Goodings, J.M., Guo, J. and Laframboise, J.G. (2002) Electrochemical diffusion potential in a flame plasma: theory and experiment. *Electrochemistry Communications*, **4**, 363–369.

(21) Caruana, D.J. and McCormack, S.P. (2002) Electrochemical redox potential in flame plasma. *Electrochemistry Communications*, **4**, 780–786.

(22) Fowowe, T., Hadzifejzovic, E., Hu, J. *et al.* (2012) Plasma electrochemistry: development of a reference electrode material for high temperature plasma. *Advanced Materials*, **24**, 6305–6309.

(23) Elahi, A. and Caruana, D.J. (2013) Plasma electrochemistry: voltammetry in a flame plasma electrolyte. *Physical Chemistry Chemical Physics*, **15**, 1108–1114.

(24) Hadzifejzovic, E., Stankovic, J., Firth, S. *et al.* (2007) Plasma electrochemistry: electroreduction in a flame. *Physical Chemistry Chemical Physics*, **9**, 5335–5339.

(25) Caruana, D.J. and Yao, J. (2003) Gas phase electrochemical detection of single latex particles. *Analyst*, **128**, 1286–1290.

(26) Sarantaridis, D. and Caruana, D.J. (2010) Potentiometric detection of model bioaerosol particles. *Analytical Chemistry*, **82**, 7660–7667.

(27) Sarantaridis, D., Hennig, C. and Caruana, D.J. (2012) Bioaerosol detection using potentiometric tomography in flames. *Chemical Science*, **3**, 2210–2216.

(28) Darr, J.A. and Poliakoff, M. (1999) New directions in inorganic and metal-organic coordination chemistry in supercritical fluids. *Chemical Reviews*, **99**, 495–541.

(29) Cansell, F., Chevalier, B., Demourgues, A. *et al.* (1999) Supercritical fluid processing: a new route for materials synthesis. *Journal of Materials Chemistry*, **9**, 67–75.

(30) Cooper, A.I. (2000) Polymer synthesis and processing using supercritical carbon dioxide. *Journal of Materials Chemistry*, **10**, 207–234.

(31) Yang, J., Hasell, T., Smith, D.C. and Howdle, S.M. (2009) Deposition in supercritical fluids: from silver to semiconductors. *Journal of Materials Chemistry*, **19**, 8560–8570.

(32) Souvignet, I. and Olesik, S.V. (1998) Molecular diffusion coefficients in ethanol/water/carbon dioxide mixtures. *Analytical Chemistry*, **70**, 2783–2788.

(33) McHugh, M.A. and Krukonis, V.J. (1994) *Supercritical Fluid Extraction: Principles & Practice*, Butterworth-Heinemann, Boston.

(34) Clifford, T. (1999) *Fundamentals of Supercritical Fluids*, Oxford University Press, Oxford.

(35) Williams, R.A. and Naiditch, S. (1970) Electrodeposition of silver from dense gaseous solutions of silver nitrate in ammonia. *Physics and Chemistry of Liquids*, **2**, 67–75.

(36) Silvestri, G., Gambino, S., Filardo, G. *et al.* (1981) Electrochemical processes in supercritical phases. *Angewandte Chemie, International Edition*, **20**, 101–102.

(37) Crooks, R.M. and Bard, A.J. (1987) Electrochemistry in near-critical and supercritical fluids. 4. Nitrogen-heterocycles, nitrobenzene, and solvated electrons in ammonia at temperatures to 150-degrees-C. *The Journal of Physical Chemistry*, **91**, 1274–1284.

(38) Crooks, R.M. and Bard, A.J. (1988) Electrochemistry in near-critical and supercritical fluids. 5. The dimerization of quinoline and acridine radical-anions and dianions in ammonia from −70 degrees-C to 150 degrees-C. *Journal of Electroanalytical Chemistry*, **240**, 253–279.

(39) Crooks, R.M., Fan, F.R.F. and Bard, A.J. (1984) Electrochemistry in near-critical and supercritical fluids. 1. Ammonia. *Journal of the American Chemical Society*, **106**, 6851–6852.

(40) Flarsheim, W.M., Bard, A.J. and Johnston, K.P. (1989) Pronounced pressure effects on reversible electrode-reactions in supercritical water. *The Journal of Physical Chemistry*, **93**, 4234–4242.

(41) Flarsheim, W.M., Tsou, Y.M., Trachtenberg, I. *et al.* (1986) Electrochemistry in near-critical and supercritical fluids. 3. Studies of Br^-, I^-, and hydroquinone in aqueous-solutions. *The Journal of Physical Chemistry*, **90**, 3857–3862.

(42) Liu, C.Y., Snyder, S.R. and Bard, A.J. (1997) Electrochemistry in near-critical and supercritical fluids. 9. Improved apparatus for water systems (23–385 degrees-C). The oxidation of hydroquinone and iodide. *The Journal of Physical Chemistry B*, **101**, 1180–1185.

(43) McDonald, A.C., Fan, F.R.F. and Bard, A.J. (1986) Electrochemistry in near-critical and supercritical fluids. 2. Water – experimental-techniques and the copper(II) system. *The Journal of Physical Chemistry*, **90**, 196–202.

(44) Cabrera, C.R. and Bard, A.J. (1989) Electrochemistry in near-critical and supercritical fluids. 8. Methyl viologen, decamethylferrocene, $Os(bpy)_3{}^{2+}$ and ferrocene in acetonitrile and the effect of pressure on diffusion-coefficients under supercritical conditions. *Journal of Electroanalytical Chemistry*, **273**, 147–160.

(45) Crooks, R.M. and Bard, A.J. (1988) Electrochemistry in near-critical and supercritical fluids. 6. The electrochemistry of ferrocene and phenazine in acetonitrile between 25 degrees-C and 300 degrees-C. *Journal of Electroanalytical Chemistry*, **243**, 117–131.

(46) Cabrera, C.R., Garcia, E. and Bard, A.J. (1989) Electrochemistry in near-critical and supercritical fluids. 7. SO_2. *Journal of Electroanalytical Chemistry*, **260**, 457–460.

(47) Abbott, A.P. and Harper, J.C. (1996) Electrochemical investigations in supercritical carbon dioxide. *Journal of the Chemical Society, Faraday Transactions*, **20**, 3895–3898.

(48) James, A.M. and Lord, M.P. (1992) *Macmillan's Chemical and Physical Data*, The MacMillan Press, London.

(49) Abbott, A.P., Eardley, C.A. and Tooth, R. (1999) Relative permittivity measurements of 1,1,1,2-tetrafluoroethane (HFC 134a), pentafluoroethane (HFC 125), and difluoromethane (HFC 32). *Journal of Chemical & Engineering Data*, **44**, 112–115.

(50) Deguchi, S. and Tsujii, K. (2007) Supercritical water: a fascinating medium for soft matter. *Soft Matter*, **3**, 797–803.

(51) Grinberg, V.A. and Mazin, V.M. (1998) Electrochemical processes in liquid and supercritical carbon dioxide. *Russian Journal of Electrochemistry*, **34**, 223–229.

(52) Bartlett, P.N., Cook, D.C., George, M.W. *et al.* (2010) Phase behaviour and conductivity study on multi-component mixtures for electrodeposition in supercritical fluids. *Physical Chemistry Chemical Physics*, **12**, 492–501.

(53) Licence, P., Dellar, M.P., Wilson, R.G.M. *et al.* (2004) Large-aperture variable-volume view cell for the determination of phase-equilibria in high pressure systems and supercritical fluids. *Review of Scientific Instruments*, **75**, 3233–3236.

(54) Olsen, S.A. and Tallman, D.E. (1994) Voltammetry of ferrocene in subcritical and supercritical chlorodifluoromethane. *Analytical Chemistry*, **66**, 503–509.

(55) Goldfarb, D.L. and Corti, H.R. (2000) Electrochemistry in supercritical trifluoromethane. *Electrochemistry Communications*, **2**, 663–670.

(56) Goldfarb, D.L. and Corti, H.R. (2004) Electrical conductivity of decamethylferrocenium hexafluorophosphate and tetrabutylammonium hexafluorophosphate in supercritical trifluoromethane. *The Journal of Physical Chemistry B*, **108**, 3358–3367.

(57) Goldfarb, D.L. and Corti, H.R. (2004) Diffusion of decamethylferrocene and decamethylferrocenium hexafluorophosphate in supercritical trifluoromethane. *The Journal of Physical Chemistry B*, **108**, 3368–3375.

(58) Abbott, A.P., Corr, S., Durling, N.E. and Hope, E.G. (2004) Equilibrium reactions in supercritical difluoromethane. *The Journal of Physical Chemistry B*, **108**, 4922–4926.

(59) Abbott, A.P. and Durling, N.E. (2001) Effect of ionic equilibria on redox potentials in super-critical difuoromethane. *Physical Chemistry Chemical Physics*, **3**, 579–582.

(60) Abbott, A.P. and Eardley, C.A. (1999) Double layer structure in a supercritical fluid. *The Journal of Physical Chemistry B*, **103**, 6157–6159.

(61) Abbott, A.P. and Eardley, C.A. (2000) Electrochemical reduction of CO_2 in a mixed supercritical fluid. *The Journal of Physical Chemistry B*, **104**, 775–779.

(62) Abbott, A.P. and Eardley, C.A. (2000) Conductivity of $(C_4H_9)_4N\,BF_4$ in liquid and supercritical hydrofluorocarbons. *The Journal of Physical Chemistry B*, **104**, 9351–9355.

(63) Abbott, A.P., Eardley, C.A., Harper, J.C. and Hope, E.G. (1998) Electrochemical investigations in liquid and supercritical 1,1,1,2-tetrafluoroethane (HFC 134a) and difluoromethane (HFC 32). *Journal of Electroanalytical Chemistry*, **47**, 1–4.

(64) Abbott, A.P. and Harper, J.C. (1999) Double layer capacitance and conductivity studies of long chain quaternary ammonium electrolytes in supercritical carbon dioxide. *Physical Chemistry Chemical Physics*, **1**, 839–841.

(65) Abbott, A.P., Hope, E.G. and Palmer, D.J. (2005) Effect of electrolyte concentration on the viscosity and voltammetry of supercritical solutions. *Analytical Chemistry*, **77**, 6702–6708.

(66) Abbott, A.P., Hope, E.G. and Palmer, D.J. (2007) Effect of solutes on the viscosity of supercritical solutions. *The Journal of Physical Chemistry B*, **111**, 8114–8118.

(67) Olsen, S.A. and Tallman, D.E. (1996) Conductivity and voltammetry in liquid and supercritical halogenated solvents. *Analytical Chemistry*, **68**, 2054–2061.

(68) Fuoss, R.M. and Karus, C.A. (1933) Properties of electrolytic solutions. IV. The conductance minimum and the formation of triple ions due to the action of coulomb forces. *Journal of the American Chemical Society*, **55**, 2387–2399.

(69) Strauss, S.H. (1993) The search for larger and more weakly coordinating anions. *Chemical Reviews*, **93**, 927–942.

(70) Krossing, I. and Raabe, I. (2004) Noncoordinating anions - Fact or fiction? A survey of likely candidates. *Angewandte Chemie, International Edition*, **43**, 2066–2090.

(71) Ke, J., Su, W.T., Howdle, S.M. *et al.* (2009) Electrodeposition of metals from supercritical fluids. *Proceedings of the National Academy of Sciences of the United States of America*, **106**, 14768–14772.

(72) Bartlett, P.N., Cook, D.C., George, M.W. *et al.* (2011) Phase behaviour and conductivity study of electrolytes in supercritical hydrofluorocarbons. *Physical Chemistry Chemical Physics*, **13**, 190–198.

(73) Atobe, M., Ohsuka, H. and Fuchigami, T. (2004) Electrochemical synthesis of polypyrrole and polythiophene in supercritical trifluoromethane. *Chemistry Letters*, **33**, 618–619.

(74) Yan, H., Sato, T., Komago, D. *et al.* (2005) Electrochemical synthesis of a polypyrrole thin film with supercritical carbon dioxide as a solvent. *Langmuir*, **21**, 12303–12308.

(75) Cook, D., Bartlett, P.N., Zhang, W.J. *et al.* (2010) The electrodeposition of copper from super-critical CO_2/acetonitrile mixtures and from supercritical trifluoromethane. *Physical Chemistry Chemical Physics*, **12**, 11744–11752.

(76) Vas'ko, V.A., Tabakovic, I., Riemer, S.C. and Kief, M.T. (2004) Effect of organic additives on structure, resistivity, and room-temperature recrystallization of electrodeposited copper. *Microelectronic Engineering*, **75**, 71–77.

(77) Bartlett, P.N., Perdjon-Abel, M., Cook, D. *et al.* (2014) The electrodeposition of silver from supercritical carbon dioxide/acetonitrile. *ChemElectroChem*, **1**, 187–194.

(78) Lee, M.L., Fitzgerald, E.A., Bulsara, M.T. *et al.* (2005) Strained Si, SiGe, and Ge channels for high-mobility metal-oxide-semiconductor field-effect transistors. *Journal of Applied Physics*, **97**, 011101.

(79) Ke, J., Bartlett, P.N., Cook, D. *et al.* (2012) Electrodeposition of germanium from supercritical fluids. *Physical Chemistry Chemical Physics*, **14**, 1517–1528.

(80) Chanfreau, S., Cognet, P., Camy, S. and Condoret, J.-S. (2008) Electrocarboxylation in super-critical CO_2 and CO_2-expanded liquids. *Journal of Supercritical Fluids*, **46**, 156–162.

(81) Sun, N., Hou, Y., Wu, W. *et al.* (2013) Electro-oxidation of benzyl alcohol to benzaldehyde in supercritical CO_2 with ionic liquid. *Electrochemistry Communications*, **28**, 34–36.

(82) Finn, C., Schnittger, S., Yellowlees, L.J. and Love, J.B. (2012) Molecular approaches to the electrochemical reduction of carbon dioxide. *Chemical Communications*, **48**, 1392–1399.

(83) Kuhl, K.P., Cave, E.R., Abram, D.N. and Jaramillo, T.F. (2012) New insights into the electrochemical reduction of carbon dioxide on metallic copper surfaces. *Energy & Environmental Science*, **5**, 7050–7059.

(84) Whipple, D.T. and Kenis, P.J.A. (2010) Prospects of CO_2 utilization via direct heterogeneous electrochemical reduction. *The Journal of Physical Chemistry Letters*, **1**, 3451–3458.

(85) Méndez, M.A., Voyame, P. and Girault, H.H. (2011) Interfacial photoreduction of supercritical CO_2 by an aqueous catalyst. *Angewandte Chemie, International Edition*, **50**, 7391–7394.

18

Aspects of Light-Driven Water Splitting

Laurence Peter
University of Bath, Department of Chemistry, UK

If mankind is to satisfy its increasing thirst for energy without a massive increase in greenhouse gas emissions, the twenty-first century must become the "solar century." According to the *BP Statistical Review of World Energy* [1], global primary energy consumption in 2012 was equivalent to around 12.5 gigatonnes (Gtoe) of oil. This rate of primary energy consumption corresponds to a thermal power output of 16 TW. Some 87% of all energy consumption was associated with burning carbon-based fuels (oil, gas and coal), 7% was generated as hydroelectricity, 4% was derived from nuclear power stations, and only 2% came from renewables (e.g., wind, geothermal, solar, biomass and biofuels). With the rapid expansion of Asian economies and the world's population predicted to increase by over 30% to reach nearly 10 billion by 2050, it is clear that energy consumption will at least double from its current value, posing enormous problems in terms of climate change and sustainability. Several authoritative scenarios predict that 10% of the world's electricity will be generated by photovoltaic modules by 2050, but this leaves the problem of providing green fuels for transport, which currently accounts for around one-quarter of total energy consumption. Since it is difficult to compete with the high energy density of liquid fossil fuels by using batteries, there is a pressing need to develop a carbon neutral fuel economy. The challenge is therefore to develop *solar fuels* – that is, the chemical storage of solar energy.

Although Nature has been locking up solar energy in chemical compounds for at least three billion years since the evolution of photosynthetic organisms, it is only during the

Developments in Electrochemistry: Science Inspired by Martin Fleischmann, First Edition.
Edited by Derek Pletcher, Zhong-Qun Tian and David E. Williams.
© 2014 John Wiley & Sons, Ltd. Published 2014 by John Wiley & Sons, Ltd.

past 40 years or so that serious attempts have been made to develop systems for artificial photosynthesis, with efforts essentially having been focused on two approaches.

The first approach involves designing and arranging molecules to mimic the various components and functions (light absorption, electron and proton transfer, etc.) of the photosynthetic apparatus, often with substantial simplification. Here, the problem is often long-term stability. In simple terms, the photosynthetic process is light-driven water splitting, and the only way that the photosynthetic center can survive the extreme oxidizing conditions associated with molecular oxygen production is by continual self-repair (Photosystem II is replaced several times a day in higher plants). The introduction of self-repair into artificial photosynthetic systems is a daunting task.

The second approach, which is the main focus of this chapter, involves using stable inorganic materials (mainly semiconducting oxides) for light-driven water splitting. Here, the similarities with photosynthesis are reduced to the three steps: light harvesting; charge separation; and electron transfer. These processes are the basic components of "semiconductor photoelectrochemistry," a topic that the author was first encouraged to investigate by Martin Fleischmann. This chapter outlines the basic physics and chemistry involved in light-driven water splitting at semiconductor electrodes, and also provides a brief overview of some of the problems that arise when trying to apply "traditional" models to light-driven water-splitting reactions. For an up to date survey of the area with chapters by leading researchers in the field, the reader is referred to a recent book [2].

18.1   A Very Brief History of Semiconductor Electrochemistry

Three key names associated with the development of semiconductor electrochemistry are Gerischer [3], Memming [4], and Pleskov [5], all of whom contributed to placing the subject on a firm theoretical and experimental basis. Their investigations into single-crystal semiconductors such as the Group II–VI compounds ZnO, CdS and the Group III–V materials GaAs and GaP laid the foundation for an upsurge of interest in regenerative photoelectrochemical solar cells, triggered by the 1973 oil crisis. These cells were based on mimicking conventional solid-state solar cells by using semiconductor electrodes in contact with redox electrolytes. Although impressive efficiencies were obtained over the next 10 years [6], problems with complexity and stability eventually led to a decline in interest. Today, the only remaining regenerative photoelectrochemical solar cell device that is the subject of intense research is the dye-sensitized solar cell (DSC), as developed by Grätzel's group at EPFL in Lausanne [7, 8]. Since regenerative photoelectrochemical devices generate electrical power, they can be considered as alternatives to conventional photovoltaic cells. By contrast, photoelectrolysis cells (PECs) involve nonregenerative reactions at the electrodes, in the most common case the hydrogen evolution reaction (HER) and the oxygen evolution reaction (OER). Either, or both, of these reactions can be driven by light. The first report to draw attention to the possibility of using semiconductor oxides for photoelectrochemical water splitting appeared as a very short note in *Nature* in 1972, by Fujishima and Honda [9]. Their photoelectrolysis cell consisted of an n-type single-crystal rutile (TiO_2) anode illuminated with ultraviolet (UV) light and a platinum black cathode. Although not widely appreciated at the time, the separated anode and cathode compartments contained solutions with different pH values, providing an additional voltage bias that assisted photoelectrolysis.

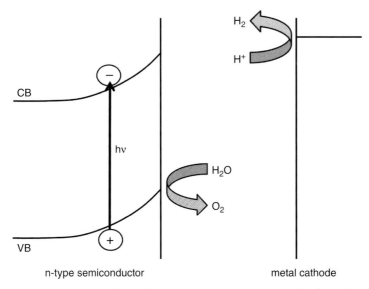

Figure 18.1 *Basic processes taking place in the Fujishima and Honda photoelectrolysis cell.*

The basic processes occurring in the Fujishima and Honda PECs are illustrated in Figure 18.1. When the semiconductor is immersed in the electrolyte and a suitable bias voltage is applied, a space charge region or depletion region forms at the semiconductor electrolyte junction, leading to the band bending, as shown in Figure 18.1. Electron–hole pairs created by the absorption of photons with energy greater than the band gap of rutile (3.0 eV) are separated in the space charge region that is formed at the semiconductor/electrolyte junction: the holes move to the interface and electrons to the back contact. The holes, which are electron vacancies in the valence band of the TiO_2, can accept electrons from water, leading ultimately to oxidation to molecular oxygen. Electrons pass from the conduction band of the TiO_2 to the cathode, where hydrogen evolution occurs. The criteria for this overall light-driven water-splitting process to occur efficiently are explored in the next section.

Although water splitting using the small UV component of sunlight is clearly not an efficient process, Fujishima and Honda's report marked the starting point for concerted efforts during the 1970s and 1980s to find suitable semiconductor materials that would satisfy the thermodynamic, kinetic and stability criteria for light-driven (or light-assisted) water splitting. In spite of the development of an 11% efficient pre-commercial device by Texas Instruments for the photoelectrolysis of HBr and HI using p- and n-type silicon microspheres [10], progress towards a commercially viable water-splitting device has been slow. Quite apart from the scientific criteria mentioned above, the cost targets are very demanding: Parkinson and Turner [11] have recently reviewed the techno-economic aspects of solar hydrogen production, and concluded that a solar-to-hydrogen efficiency of at least 15% is required in order to produce hydrogen at a commercially competitive price. Nevertheless, the past few years have seen a resurgence in activity on solar fuels in general, with most of the emphasis on the production of hydrogen by water splitting, either in photoelectrolysis cells or, to a lesser extent, using dispersed semiconductor powders or colloids. Today, solar fuels are an important research priority; for example, in the United States the Joint Center for Artificial Photosynthesis (JCAP) is the world's largest research program devoted to solar fuel generation technology; other centers in the United States include the Center for Bio-Inspired Solar Fuel Production at Arizona State. A new Solar Fuels program has been started at the Helmholtz Center in Berlin, and there are several initiatives elsewhere in Europe, including the Nordic initiative for solar fuel development and The European Science Foundation's EuroSolarFuels program.

18.2   Thermodynamic and Kinetic Criteria for Light-Driven Water Splitting

At room temperature and pressure, the cell voltage must exceed 1.23 V in order for the electrolysis of water to occur (this voltage corresponds to the standard potential the O_2/H_2O couple). In practice, electrolysis requires an additional voltage to overcome the activation energy barriers associated with the HER and OER. This additional voltage corresponds to the sum of the current density-dependent overpotentials (kinetic and mass transport) at the cathode and anode, plus the ohmic voltage drop in the electrolysis system. A typical polymer electrolyte membrane (PEM) electrolyzer requires a voltage of 2.0 V to operate at a current density of 1 A cm^{-2}. In a photoelectrolysis cell of the type shown in Figure 18.1, the thermodynamic criteria are that the free energy of holes in the TiO_2 and the free energy of electrons in the Pt cathode must be adequate to oxidize and reduce water, respectively.

In semiconductor physics, the equilibrium free energies of electrons and holes are expressed in terms of the corresponding Fermi energies or (equivalently) their electrochemical potentials. In order to enable comparison with the electron energy levels in the electrolyte, it is useful to define the *redox Fermi levels* on the vacuum scale using the approximate relationship

$$E_{F,redox} = -4.5eV - qU_{redox} \qquad (18.1)$$

where U_{redox} is the standard reduction potential of the redox couple concerned and q is the elementary charge. In our case we are interested in the H^+/H_2 and O_2/H_2O potentials, 0 and +1.23 V, respectively. The Fermi level in the semiconductor is related to the Fermi Dirac function, f_{FD}, which describes the probability of occupation of energy levels.

$$f_{FD} = \frac{1}{1 + \exp\left(\dfrac{E_F - E}{k_B T}\right)} \qquad (18.2)$$

The Fermi–Dirac function can be approximated by the Boltzmann function if $(E_F - E)$ is much greater than $k_B T$. The Fermi energy E_F determines n and p, the concentrations of electrons in the conduction and valence bands, respectively.

$$n = N_C \exp\left(\frac{E_F - E_C}{k_B T}\right) \qquad p = N_V \exp\left(\frac{E_V - E_F}{k_B T}\right) \qquad (18.3)$$

where N_C and N_V are the densities of states in the conduction and valence bands and E_C and E_V are the conduction and valence band energies. In the case of metals, the Fermi energy corresponds to the uppermost occupied energy level in the partly filled conduction band (probability of occupation $\frac{1}{2}$).

In photoelectrolysis cells, the concentrations of electrons and holes are changed from their dark equilibrium values due to creation of electron–hole pairs by absorption of photons. Since thermal equilibration with lattice vibrations is much faster than the electron–hole recombination of electrons, electrons and holes can be considered to be in thermal equilibrium with the lattice, even though they are metastable states. Therefore, Fermi–Dirac statistics can still be used by defining quasi-Fermi levels ($_nE_F$ and $_pE_F$) for electrons and holes in terms of their steady-state concentrations under illumination, n^* and p^*.

$$n^* = N_C \exp\left(\frac{_nE_F - E_C}{k_BT}\right) \qquad p^* = N_V \exp\left(\frac{E_V - _pE_F}{k_BT}\right) \qquad (18.4)$$

It is these quasi-Fermi levels that are important in defining the thermodynamic criteria for light-driven water splitting. In the case of an n-type semiconductor such as rutile, the concentration of electrons in the dark is high (typically $>10^{16}$ cm^{-3}), whereas the concentration of holes is vanishingly small since the law of mass action applies.

$$np = N_C N_V \exp\left(\frac{-E_g}{k_BT}\right) \qquad (18.5)$$

where E_g is the band gap energy. Illumination creates equal numbers of holes and electrons, but because $n > p$ for a n-type semiconductor in the dark, $n^* \approx n$ whereas $p^* \gg p$ under illumination. As a consequence, the quasi-Fermi level for holes is displaced downwards toward the valence band energy E_V. Notice that this downward displacement from E_F to $_pE_F$ corresponds to an increase in the free energy of holes that arises from the entropic (i.e., concentration) term in their chemical potential. The separation of Fermi levels under illumination corresponds to the generation of an internal *photovoltage*: it is this photovoltage that can be used to drive water splitting.

Now, we are in a position to define the thermodynamic criteria for light-driven water splitting under short-circuit conditions in a two-electrode photoelectrolysis cell. $_nE$ needs to lie above the H$^+$/H$_2$ redox Fermi level, and $_pE_F$ needs to lie below the O$_2$/H$_2$O redox Fermi level. As the electron Fermi level for n-type semiconductors is close to the conduction band, this means in practice that the conduction band energy should lie well above the H$^+$/H$_2$ redox potential. At the same time, the valence band needs to be sufficiently far below the O$_2$/H$_2$O Fermi level to ensure that water oxidation by holes is feasible. In addition to satisfying these thermodynamic criteria, the semiconductor needs to be stable under illumination. Many semiconductors are either oxidized or reduced under illumination as a consequence of the reaction of holes or electrons with the crystal lattice. For example, n-type ZnO is oxidized by photogenerated holes to form Zn^{2+} and oxygen.

The positions on the vacuum energy scale of the conduction and valence bands of oxide semiconductors immersed in aqueous electrolytes depend on pH [12]. Generally, the bands move upwards in energy by 59 meV per (increasing) pH unit. This pH dependence, which arises from changes in the surface dipole potential associated with acid–base equilibria at the oxide surfaces, is the same as that of the *reversible* hydrogen and oxygen redox Fermi levels, so that changing the pH does not affect the band positions relative to the reversible hydrogen and oxygen levels. Figure 18.2 illustrates the positions of the conduction and

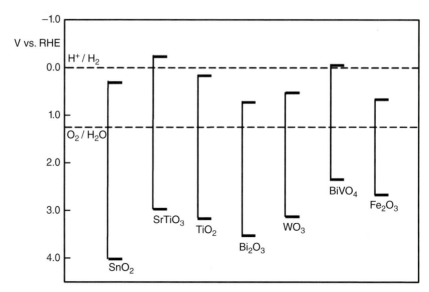

Figure 18.2   *Positions of conduction and valence band of some oxide semiconductors on the reversible hydrogen (RHE) scale.*

valence bands of some oxides relative to the reversible hydrogen redox Fermi level. It can be seen that only wide band gap oxides such as $SrTiO_3$ actually satisfy the criteria. In the majority of cases, it is necessary to apply an additional voltage bias to lift the Fermi level of the cathode high enough for proton reduction to occur. One way to achieve this voltage bias is to use a tandem cell in which a solar cell placed behind the photoanode absorbs light at photon energies below the band gap of the oxide, and thus generates a bias voltage that is applied to the electrolysis cell [13]. This is an example of "photo-assisted electrolysis"; photoanode materials used in this arrangement include Fe_2O_3 [14] and WO_3 [15].

18.3   Kinetics of Minority Carrier Reactions at Semiconductor Electrodes

Although, so far, the discussion has centered on the thermodynamic criteria for water splitting, it is also necessary to consider the kinetics of electron transfer. Any practical photoelectrochemical water-splitting device must operate at a current density that is determined by the efficiency of the device and the fraction of the incident photon flux that is absorbed by the semiconductor(s). If, for example, a minimum band gap of 2.0 eV is needed to guarantee a sufficiently high photovoltage to drive water splitting, then the maximum possible photocurrent density would be approximately 15 mA cm^{-2} at 1 sun (AM 1.5). For a practical device, therefore, the quasi-Fermi levels for electrons and holes need to lie above and below the hydrogen and oxygen Fermi levels by an amount that corresponds to the overpotentials for the HER and OER respectively at this current density. In practice, the larger overpotential is associated with the four-hole oxidation of water: the HER at catalytic electrodes requires only low overpotentials at the current densities corresponding to 1 sun. This discussion highlights two points: (i) that the requirement to overcome the overpotential for light-driven OER means that $_pE_F$ at the surface of the semiconductor needs to be several hundred meV below the oxygen Fermi level; and (ii) that catalysis of the OER is likely to be a central issue as the four-electron–four-proton process is generally slow.

Electrochemists tend to take for granted the fact that the rate of electron transfer at an electrode varies with applied potential, as expressed by the Butler Volmer or Tafel equations, and this potential dependence is, after all, the basis for most electrochemical techniques. In terms of the energy level picture developed above, changing the potential of a metal electrode by some amount, ΔU, corresponds to shifting the Fermi level of the metal relative to the electrolyte redox level by a corresponding amount, $-q\Delta U$. The electrochemical potential of electrons in the metal is given by

$$\bar{\mu}_{n,m} = \mu_{n,m} - q\Phi_m \tag{18.6}$$

where μ_n is the chemical potential and Φ_m is the inner or Galvani potential of the metal. The electrochemical potential of electrons in a redox electrolyte involving a one-electron transfer is

$$\bar{\mu}_{redox} = \left(\mu_R - \mu_O\right) - q\Phi_{sol} \tag{18.7}$$

The equilibrium inner potential difference between the metal and the solution is therefore

$$\Delta\Phi_{m,sol} = \mu_{n,m} - (\mu_O - \mu_R) \tag{18.8}$$

Remembering that the Fermi level is equivalent to the electrochemical potential of electrons, it can be seen that changing the potential difference across the metal solution interface changes the relative position of the metal and redox Fermi levels. In effect, the potential change is located across the Helmholtz layer, and the activation energy for cathodic electron transfer is changed by some amount $\alpha q\Delta\Phi_{m,sol}$, where α is the transfer coefficient. Changing the potential of a metal electrode will therefore alter the *potential energy* of electrons. Electrochemists are familiar with the concept of overpotential, which is used to describe the deviation of $\Delta\Phi_{m,sol}$ from its equilibrium value.

In the case of illuminated semiconductor electrodes, the situation is different, however. A *space charge layer* is formed at the semiconductor/electrolyte junction when the electrode potential is made more positive (for an n-type semiconductor) or more negative (for a p-type semiconductor) than the flat band potential E_{fb}. The space charge region arises from removal of the majority carriers, leaving the immobile ionized donor (positively charged) or acceptor (negatively charged) species. If a space charge region is present, minority carriers (holes for an n-type semiconductor and electrons for a p-type semiconductor) move to the solid/liquid interface where they can react. The electrochemical potential of these minority carriers is equivalent to their quasi-Fermi level [16]. The extent of the space charge region is generally much larger than the width of the Helmholtz layer, and as a consequence the majority of any change in electrode potential appears across the space charge region, with only a very small change in the potential drop across the Helmholtz layer. It follows that the activation energy – and hence the rate constant of electron transfer involving minority carriers – is virtually independent of potential. Changes in rate – and hence current density – are associated primarily with changes in the concentration of carriers at the surface. For this reason, photocurrents at semiconductor electrodes depend on light intensity, but in the ideal case they become independent of voltage when the width of the space charge region exceeds the penetration depth of the light (photocurrent saturation region). In terms of the quasi-Fermi levels, it is the entropic logarithmic term in the electrochemical potential that matters. For photogenerated holes, for example,

$$\bar{\mu}_p = \mu_p^0 + k_B T \ln \frac{p}{N_V} + q\Phi_{sc} \tag{18.9}$$

[cf. Equation (18.4)]. In practical terms, this means that the key parameter as far as photo-electrochemical kinetics is concerned is the minority carrier quasi-Fermi level. In principle, the overpotential in this case can be defined as the difference between the quasi-Fermi level of the reacting minority carriers and the redox Fermi level. It is worth noting that the electrochemical potential of the redox electrolyte also depends on the ratio of concentrations of the oxidized and reduced species, as expressed by the Nernst equation in the form

$$\bar{\mu}_{redox} = \bar{\mu}_{redox}^0 + \frac{k_B T}{q} \ln \frac{n_R}{n_O} \tag{18.10}$$

The analogy with Equation (18.9) will be apparent.

18.4   The Importance of Electron–Hole Recombination

Photoelectrolysis involves carrying out one or both of the water-splitting half-reactions with minority carriers (in principle, an n-type photoanode and a p-type photocathode can be used). In the bulk of a semiconductor, where the concentration of majority carriers (e.g., electrons for a n-type semiconductor) is high, minority carriers are generally short-lived because they are lost by rapid recombination with majority carriers. Nonradiative recombination generally takes place through defects that give rise to an energy level in the band gap, and hole lifetimes for n-type oxide materials such as TiO_2 are on the order of nanoseconds. By contrast, minority carrier lifetimes in ultrapure silicon may be as high as milliseconds, which raises the question of why it is possible at all to drive the water-splitting reactions at oxide photoanodes with such short-lived charge carriers. Bulk recombination is a pseudo-first-order process because majority carriers are in large excess. This allows a *minority carrier lifetime*, τ_{min}, to be defined that is inversely proportional to the majority carrier concentration, which in turn is determined by the doping density. In order to help minority carriers survive for long enough to drive water splitting, it is necessary to lower the concentration of majority carriers. In principle, this could be achieved by reducing the doping, but this increases the resistance of the semiconductor, leading to ohmic losses. The key to success is the space charge region, where the local concentration of majority carriers depends on the electrostatic potential, so that their concentration at the surface is many orders of magnitude lower than in the bulk. If the potential drop across the space charge layer is $\Delta\phi_{SC}$, the concentration of majority carriers at the semiconductor surface is given (for an n-type semiconductor) by

$$n_{x=0} = n_{bulk} e^{-\frac{\Delta\phi_{SC}}{k_B T}} \tag{18.11}$$

It follows that increasing $\Delta\phi_{SC}$ by 600 mV, for example, lowers $n_{x=0}$ by around ten orders of magnitude, and this gives minority carriers that reach the surface the chance to live for long enough to take part in the rather slow reactions involved in water splitting. In the simplest model, where the diffusion length of minority carriers $L_{min} = \sqrt{D_{min}\tau_{min}}$ is much smaller than W, the width of the space charge region, only electron–hole pairs created in the space charge layer are separated, leading to water splitting. The wavelength-dependent external quantum efficiency (EQE) of the water-splitting process is therefore approximated by the fraction of the incident light that is absorbed in the space charge region

$$EQE(\lambda) = 1 - e^{-\alpha W(\lambda)} \qquad (18.12)$$

where $\alpha(\lambda)$ is the absorption coefficient.

This sounds good; provided that W is larger than $1/\alpha$, the EQE should be high. However there is another problem. The surface of a semiconductor in contact with an electrolyte solution is usually non-ideal.[1] Defects and chemical species formed by interaction with the electrolyte give rise to electron energy levels that lie in the band gap. These surface states can promote recombination of electrons and holes – a process referred to as *surface recombination* [17–20]. Even if such states are absent before the semiconductor is illuminated, they may be formed under illumination by the reaction of minority carriers with the semiconductor surface or with the electrolyte. The desired reaction in photoelectrolysis – electron transfer from or to water – therefore has to compete with surface recombination. The longer that minority carriers remain at the surface waiting to take part in water splitting, the greater the danger that they will be lost by recombination. This means that slow electron transfer will lead to low water-splitting efficiencies.

Surface states not only promote electron–hole recombination, they also give rise to non-ideal behavior of the semiconductor electrolyte junction because they can store electronic charge and therefore change the potential drop across the Helmholtz layer. In extreme cases, the electrode may behave more like a metal, with the majority of any change in applied potential appearing across the Helmholtz layer rather than across the space charge region; this effect is referred to as "Fermi level pinning."

18.5 Fermi Level Splitting in the Semiconductor–Electrolyte Junction

Since photogenerated minority carriers recombine very rapidly in the neutral bulk of the semiconductor, it is only in the space charge region that a substantial change in minority carrier concentration occurs under illumination. As the concentration of minority carriers is highest near the surface, this is where the splitting of the Fermi levels is also largest. In the case of an n-type photoanode that is being used for water splitting, the quasi-Fermi level for holes will move downward under illumination until it is sufficiently low to drive the oxidation of water. In practice, this means that that it must lie below the oxygen Fermi level by an amount corresponding to the overpotential associated with current flow, as shown in Figure 18.3. It is worth noting at this point that it is the gradient of the quasi-Fermi level

---

[1] A rare exception is the hydrogen terminated (111) surface of silicon in acidic fluoride solutions, where the surface state density is extremely small.

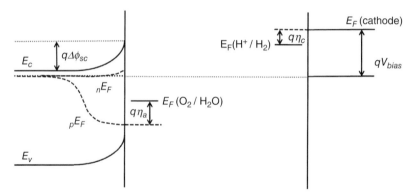

Figure 18.3   *Photoassisted electrolysis using an n-type photoanode. A bias voltage V_{bias} is used to raise the Fermi level of the cathode above the reversible hydrogen potential by an amount corresponding to the cathodic overpotential η_c. The bias also induced a band bending $q\Delta\phi_{SC}$ in the space charge region of the photoanode. The broken lines show the quasi-Fermi levels for holes and electrons (see text for details).*

(or electrochemical potential) and not (as is often assumed) the gradient of electrostatic potential which is the "driving force" for hole transport [21], as the current density at any point is given by

$$j_p(x) = p(x)\,\mu_p \frac{\partial \overline{\mu}_p(x)}{\partial x} \tag{18.13}$$

where

$$\frac{\partial \overline{\mu}_p(x)}{\partial x} = \frac{k_B T}{p(x)}\frac{\partial p(x)}{\partial x} + q\frac{\partial \phi(x)}{\partial x} \tag{18.14}$$

The first term in Equation (18.14) corresponds to diffusion, and the second term to migration. The steady-state profile of holes across the space charge region is the result of a balance between the effect of the electric field, which pushes holes towards the surface, and the effect resulting in a steep gradient of concentration which tends to drive holes back in the other direction.[2]

It is often assumed that the quasi-Fermi level of electrons in the space charge region does not deviate substantially from the dark Fermi level, but this is only an approximation because, as shown above, the equilibrium concentration of electrons can fall to very low values near the interface. As the local electron density can be increased significantly by illumination, the local quasi-Fermi level of electrons will increase towards the surface (not shown in Figure 18.3). If surface recombination occurs, there will be a flux of electrons into the surface states, casing the quasi-Fermi level to pass through a maximum. The change in

---

[2] Note, however, that the separation of the electrochemical potential into the "chemical" and "electrical" terms is entirely notional: the two components cannot be separated experimentally, in spite of the fact that physicists like to talk about "drift-diffusion" models.

slope reflects the fact that, near the surface, electrons move towards the surface whereas deep in the space charge region they move towards the back contact.

The quasi-thermodynamic treatment outlined here leads to the prediction that a minimum light intensity is required to drive the minority carrier Fermi level to the point where electron transfer becomes thermodynamically feasible. However, as pointed out by Gregg and Nozik [22], such a threshold would be almost impossible to detect since the initial redox Fermi level for water splitting will be far from the standard value because the concentration of product (oxygen in the case of a photoanode) will be very small.

18.6   A Simple Model for Light-Driven Water-Splitting Reaction

Calculations of the electron and hole concentration profiles across the space charge region can be carried out by numerical methods developed for semiconductor junctions. If the reaction of minority carriers with the electrolyte or solvent is slow, then very high concentrations can build up at the surface, and this is certainly the case in water-splitting reactions which have high activation energies. The build-up of minority carriers predicted by numerical modeling [23] can be detected in some cases by microwave reflectance methods: an excellent example is light-driven hydrogen evolution on p-Si [24]. In this case, the build-up of electrons close to the interface is so extreme that it leads to a redistribution of the potential drop across the space charge and Helmholtz regions that can be detected as a photoinduced capacitance and a delayed onset of the photocurrent–voltage curve [24].

Numerical modeling (or indeed, modeling of any type) has not been used very much for light-driven water-splitting reactions; the vast majority of reports have been only semi-quantitative and focused on the materials or performance aspects. A very simple analytical model developed in the author's group considers the kinetics of charge transfer and recombination in terms of nominal surface concentrations (cm^{-2}) of electrons and holes. This model, which is illustrated in Figure 18.4, has been applied successfully for low levels of illumination to predict the transient and periodic photocurrent response to pulsed or modulated illumination [25, 26]. Here, an examination is made of how robust the model is when dealing with water splitting at typical solar intensities.

To simplify the analysis, recombination in the space charge region is ignored, and surface electron–hole recombination is formulated in terms of the surface concentration of majority carriers present in the dark. Consider an n-type semiconductor electrode with a space charge region. If the diffusion length of holes, L_p, is much smaller than W, the flux of holes, J_p, into the surface under steady-state conditions is given by [cf. Equation (18.12)]:

$$J_p(\lambda) = I_0(1 - e^{-\alpha W(\lambda)}) \tag{18.15}$$

where I_0 is the incident photon flux. Holes reaching the surface may either accept electrons from the electrolyte (oxidation), or they may combine with electrons so that

$$\frac{dp_{surf}}{dt} = 0 = J_p - k_{tr}p_{surf} - k_{rec}p_{surf} \tag{18.16}$$

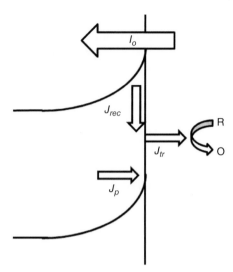

Figure 18.4 *Simple kinetic model showing the photogenerated hole flux, J_p, the hole transfer flux, $J_{tr} = k_{tr}n_{surf}$ and the electron flux $J_{rec} = k_{rec}n_{surf}$ due to surface recombination. See the text for steady-state and transient solutions based on this model.*

where k_{tr} and k_{rec} are first-order rate constants for charge transfer and surface recombination, respectively. Surface recombination is assumed to be pseudo first-order if the surface electron concentration is much higher than the surface hole concentration (i.e., $_nE_F$ is much closer to the conduction band than $_pE_F$ is to the valence band). It follows from Equation (18.16) that the steady-state surface hole concentration and the corresponding steady-state photocurrent, $j(\infty)$, are given by

$$p_{surf} = \frac{J_0}{k_{tr} + k_{rec}} \qquad j(\infty) = qk_{tr}p_{srf} = qJ_p\frac{k_{tr}}{k_{tr} + k_{rec}} \qquad (18.17)$$

Here, k_{rec} depends on the electron concentration at the surface, which is determined by $\Delta\phi_{SC}$, the potential drop, across the space charge region, σ, the capture cross-section for recombination and v_{th} the thermal velocity of electrons.

$$k_{rec} = \sigma v_{th}n_{bulk}e^{-\frac{\Delta\phi_{SC}}{k_BT}} \qquad (18.18)$$

In order to see how useful this simple model is, some reasonable values can be placed into the above expressions. Typical values of k_{tr} for water oxidation by holes at oxide electrodes such as Fe_2O_3 are remarkably small – of the order of 1–10 s^{-1} [26]. If the potential drop across the space charge region is small, say 0.24 V, the equilibrium (dark) concentration of electrons at the surface will be approximately four orders of magnitude lower than the bulk value, which is determined by the doping density (ca. 10^{18} cm^{-3} for Fe_2O_3). Taking an electron thermal velocity of 10^4 cm s^{-1} and a recombination cross-section of 10^{-17} cm^2 gives [cf. Equation (18.18)] $k_{rec} = 10^{-17} \times 10^4 \times 10^{18} \times 10^{-4} = 10$ s^{-1}. This is of the same order of magnitude as k_{tr}. In general, the fraction, f, of the hole flux J_p that is used to produce oxygen is given by the ratio

$$f = \frac{k_{tr}}{k_{tr} + k_{rec}} \qquad (18.19)$$

Suppose the hole flux J_p is on the order of 10^{16} cm^{-2} s^{-1} (corresponding to a relatively modest current density of 1.6 mA cm^{-2}). If k_{tr} and k_{rec} are both of the order of 10 s^{-1}, the surface concentration of holes will be 5×10^{14} cm^{-2} [cf. Equation (18.17)]. This illustrates the severe limitations of this type of model: this surface hole concentration corresponds to something like one hole for each surface atom. In fact of course, the holes are not located *at* the surface but *near* the surface. If it is supposed that most of the holes are located within about 10 nm of the surface, for example, this would correspond roughly to a volume concentration of 5×10^{20} cm^{-3}. This would mean that the semiconductor has become *degenerate* at the surface: in other words, the hole Fermi level lies inside the valence band and Fermi–Dirac statistics apply; in essence, the surface has become *metallic*. Clearly the model is looking very questionable. Furthermore, the model began by assuming that surface recombination is pseudo-first-order – that is, the electrons are in excess. However, this is not the case; the dark concentration of electrons at the surface was calculated above to be only 10^{14} cm^{-3}, which is many orders of magnitude *lower* than the hole concentration. This means that there is what is known as a "type inversion" at the surface so that $p^* \gg n^*$ and the model breaks down completely. Following this line of argument, it was found that the simple model is only useful for the very slow hole reactions that are typical of light-driven water splitting if: (i) the incident light flux is very much lower than for typical solar intensities; and (ii) the potential is close to the flatband potential so that the potential drop in the space charge region is small. Nevertheless, the model does show that slow photoelectrochemical processes are likely to lead to extremely non-ideal behavior because a minority carrier build-up will distort the potential distribution across the semiconductor electrolyte junction. This effect is referred to as "light-induced Fermi level pinning."

18.7 Evidence for Slow Electron Transfer During Light-Driven Water Splitting

One of the most well-studied oxide systems is hematite (α-Fe$_2$O$_3$) [27]. Both photoelectrochemical [26, 28, 29] and spectroscopic [30] measurements have indicated that the oxidation of water by photogenerated holes in Fe$_2$O$_3$ is a slow process. This can be seen, for example, from the slow relaxation of the photocurrent when illumination is switched on and off, as illustrated in Figure 18.5. The explanation of the observed decay and overshoot is as follows. When the light is switched on, holes move towards the interface, generating a displacement current, qJ_p. As the surface hole concentration begins to build up while holes wait to receive electrons from water, electrons begin to move to the surface, where they recombine with the holes. The addition of a constant hole current and the increasing electron current leads to a decay in the photocurrent until a steady state is reached. When the light is switched off, the hole flux disappears, but electrons continue to flow to the

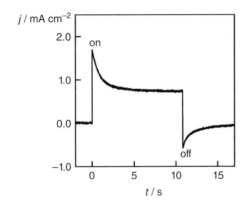

Figure 18.5   *Transient photocurrent response of thin film α-Fe$_2$O$_3$ electrode at 0 V versus Ag|AgCl, showing the decay and overshoot characteristic of surface electron–hole recombination. Electrolyte 1.0 M NaOH.*

surface to react with the remaining holes until they disappear, either by electron transfer from water or by recombination.

The general solution of Equation (18.16) leads to the following normalised expressions for the on and off photocurrent transients

$$\frac{j(t) - j(\infty)}{j(0) - j(\infty)} = e^{-t/\tau} \tag{18.20}$$

$$j_\infty/j_0 = \frac{k_{tr}}{(k_{tr} + k_{rec})} \tag{18.21}$$

$$\frac{j(t > t_0)}{j(0) - j(t_0)} = e^{-t(k_{tr}+k_{rec})} \tag{18.22}$$

Here, $j(0)$ is the "instantaneous" photocurrent observed when the light is switched on. It corresponds to the displacement current qJ_p. $j(\infty)$ is the steady-state photocurrent, that is, the plateau following the initial decay. $j(t/t_0)$ is the off transient observed when the light is switched off at a $t = t_0$. Inspection of these equations reveals that the two unknown parameters k_{tr} and k_{rec} can be extracted from the analysis of the photocurrent transients. In practice, it is often better to use small-amplitude perturbations of the illumination of the type used in intensity-modulated photocurrent spectroscopy (IMPS), because large amplitude changes in light intensity can lead to other effects such as changes in band bending. Alternatively, small-amplitude square-wave modulation can be superimposed on a large background illumination level.

Values of k_{tr} and k_{rec} obtained using low-intensity small-amplitude intensity modulation (IMPS) are illustrated in Figure 18.6 for a hematite electrode that had been treated with a cobalt(II) solution to enhance performance [26]. It can be seen that k_{tr} is indeed of the order of 10 s^{-1}, and it is almost independent of potential; k_{rec}, by contrast, changes by over three orders of magnitude as the potential is made more positive. The crossover point of the two plots is found to correspond to the half-way point on the rising photocurrent–voltage curve, as predicted by Equation (18.17). Interestingly, however, the potential dependence

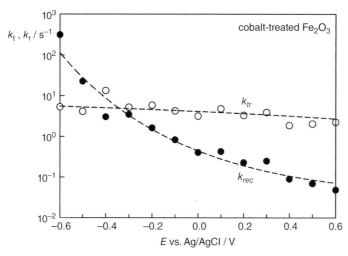

Figure 18.6 *Potential dependence of the rate constant for interfacial electron transfer (k_{tr}) and surface recombination (k_{rec}) measured by IMPS for a cobalt-treated Fe_2O_3 electrode. Note that the plots cross at –0.35 V versus Ag | AgCl. At this point, half of the holes reaching the surface are lost by recombination. At more positive potentials (> –0.2 V versus Ag | AgCl), recombination becomes unimportant. Electrolyte 1.0 M NaOH.; illumination 455 nm, 11 mW cm^{-2}. The treatment involved placing a drop of 10 mM cobalt nitrate solution on the electrode surface for 60 s, followed by rinsing with pure water for 30 s.*

of k_{rec} is far from the "ideal" value of 59 mV per decade, which indicates that the simple pseudo-first-order dependence employed in the model is not valid. This may indicate that the assumption of a flat electron quasi-Fermi level is incorrect. The dependence may also reflect the fact that the model considers only a simple one-hole oxidation reaction, whereas in fact a four-hole process is involved [25].

18.8 Conclusions

This brief excursion into light-driven water-splitting reactions should suffice to show that this is an area where much remains to be done. While current research is largely focused on materials and device architectures, there is a pressing need for better modeling of the mechanisms and kinetics of the reactions, particularly of the light-driven OER. At the same time, density functional theory modeling of the oxide–solution interface is beginning to shed new light on the possible rate-determining steps in oxygen evolution. Finally, further studies are required to identify new surface catalysts that can promote a faster electron transfer without acting as recombination centers.

Acknowledgments

The author is grateful to Martin Fleischmann and Graham Hills for tempting him to return to Southampton from Heinz Gerischer's group in Berlin in order to commence work on photoelectrochemistry, an area in which he has enjoyed working for the past 35 years.

References

(1) British Petroleum (2013) *BP Statistical Review of World Energy June 2013*, BP plc.
(2) Lewerenz, H.-J. and Peter, L.M. (eds) (2013) *Photoelectrochemical Water Splitting: Materials, Processes and Architectures*, Royal Society of Chemistry, Cambridge, UK.
(3) Kolb, D., Heller, A. and Rajeshwar, K. (2010) The life and work of heinz Gerischer. *ECS Interface*, **Fall**, 37–40.
(4) Memming, R. (2001) *Semiconductor Electrochemistry*, Wiley-VCH, Weinheim.
(5) Pleskov, Y.V. and Gurevich, Y.Y. (1986) *Semiconductor Electrochemistry*, Consultants Bureau, New York.
(6) Pleskov, Y.V. (1990) *Solar Energy Conversion. A Photoelectrochemical Approach*, Springer, Berlin.
(7) O'Regan, B. and Gratzel, M. (1991) A low-cost, high-efficiency solar-cell based on dye-sensitized colloidal TiO_2 films. *Nature*, **353**, 737–740.
(8) Kalyansundranam, K. (2010) *Dye-Sensitized Solar Cells*, EPFL Press, Lausanne.
(9) Fujishima, A. and Honda, K. (1972) Electrochemical photolysis of water at a semiconductor electrode. *Nature*, **238**, 37–38.
(10) Khaselev, O. and Turner, J.A. (1998) A monolithic photovoltaic-photoelectrochemical device for hydrogen production via water splitting. *Science*, **280**, 425–427.
(11) Parkinson, B. and Turner, J. (2013) The Potential Contribution of Photoelectrochemistry in the Global Energy Future, in *Photoelectrochemical Water Splitting: Materials, Processes and Architectures* (eds H-J. Lewerenz and L.M. Peter), Royal Society of Chemistry, Cambridge, UK.
(12) Morrison, S.R. (1980) *Electrochemistry of Semiconductor and Metal Electrodes*, Plenum Press, New York.
(13) Brillet, J., Cornuz, M., Le Formal, F. *et al.* (2010) Examining architectures of photoanode-photovoltaic tandem cells for solar water splitting. *Journal of Materials Research*, **25**, 17–24.
(14) Sivula, K., Le Formal, F. and Grätzel, M. (2011) Solar water splitting: progress using hematite (alpha-Fe_2O_3) photoelectrodes. *ChemSusChem*, **4**, 432–449.
(15) Solarska, R., Jurczakowski, R. and Augustynski, J. (2012) A highly stable, efficient visible-light driven water photoelectrolysis system using a nanocrystalline WO_3 photoanode and a methane sulfonic acid electrolyte. *Nanoscale*, **4**, 1553–1556.
(16) Reineke, R. and Memming, R. (1992) Comparability of redox reactions at n-type and p-type semiconductor electrodes 1. The quasi-Fermi level concept. *The Journal of Physical Chemistry*, **96**, 1310–1317.
(17) Peter, L.M., Li, J. and Peat, R. (1984) Surface recombination at semiconductor electrodes. Part 1. Transient and steady state photocurrents. *Journal of Electroanalytical Chemistry and Interfacial Electrochemistry*, **165**, 29–40.
(18) Li, J., Peat, R. and Peter, L.M. (1984) Surface recombination at semiconductor electrodes. Part II. photoinduced near-surface recombination centres in p-gallium phosphide. *Journal of Electroanalytical Chemistry*, **165**, 41–59.
(19) Li, J. and Peter, L.M. (1985) Surface recombination at semiconductor electrodes. Part III. Steady-state and intensity modulated photocurrent response. *Journal of Electroanalytical Chemistry*, **193**, 27–47.
(20) Li, J. and Peter, L.M. (1985) Surface recombination at semiconductor electrodes. Part IV. Steady-state and intensity modulated photocurrents at n-Gallium arsenide electrodes. *Journal of Electroanalytical Chemistry*, **193**, 27–47.
(21) Würfel, P., Wurfel, U. (2009) *Physics of Solar Cells. From Basic Principles to Advanced Concepts*, Wiley-VCH, Weinheim.
(22) Gregg, B.A. and Nozik, A.J. (1993) Existence of a light-intensity threshold for photoconversion processes. *The Journal of Physical Chemistry*, **97**, 13441–13443.
(23) Cass, M.J., Duffy, N.W., Peter, L.M. *et al.* (2003) Microwave reflectance studies of photo-electrochemical kinetics at semiconductor electrodes. 1. Steady-state, transient, and periodic responses. *The Journal of Physical Chemistry B*, **107**, 5857–5863.

(24) Cass, M.J., Duffy, N.W., Peter, L.M. *et al.* (2003) Microwave reflectance studies of photoelectrochemical kinetics at semiconductor electrodes. 2. Hydrogen evolution at p-Si in ammonium fluoride solution. *The Journal of Physical Chemistry B*, **107**, 5864–5870.

(25) Peter, L.M., Ponomarev, E.A. and Fermin, D.J. (1997) Intensity-modulated photocurrent spectroscopy: reconciliation of phenomenological analysis with multistep electron transfer mechanisms. *Journal of Electroanalytical Chemistry*, **427**, 79–96.

(26) Peter, L.M., Wijayantha, K.G.U. and Tahir, A.A. (2012) Kinetics of light-driven oxygen evolution at alpha-Fe_2O_3 electrodes. *Faraday Discussions*, **155**, 309–322.

(27) Brillet, J., Grätzel, M. and Sivula, K. (2010) Decoupling feature size and functionality in solution-processes porous hematite electrodes for water splitting. *Nano Letters*, **10**, 4155–4160.

(28) Wijayantha, K.G.U., Saremi-Yarahmadi, S. and Peter, L.M. (2011) Kinetics of oxygen evolution at alpha-Fe_2O_3 photoanodes: a study by photoelectrochemical impedance spectroscopy. *Physical Chemistry Chemical Physics*, **13**, 5264–5270.

(29) Cummings, C.Y., Marken, F., Peter, L.M. *et al.* (2012) Kinetics and mechanism of light-driven oxygen evolution at thin film alpha-Fe_2O_3 electrodes. *Chemical Communications*, **48**, 2027–2029.

(30) Pendlebury, S.R., Cowan, A.J., Barroso, M. *et al.* (2102) Correlating long-lived photogenerated hole populations with photocurrent densities in hematite water oxidation photoanodes. *Energy & Environmental Science*, **5**, 6304–6312.

19

Electrochemical Impedance Spectroscopy

Samin Sharifi-Asl and Digby D. Macdonald
University of California at Berkeley, Departments of Materials Science and Engineering and
Nuclear Engineering, USA

Martin Fleischmann was an electrochemical impresario, who conducted his exploration of the subject with all of the skill and knowledge of a master symphony maestro. His many, impressive accomplishments are cataloged in this book and in the scientific literature, and serve as a beacon for those who follow in his footsteps. Although Martin did not use impedance methods extensively in his own work, he did use electrochemical impedance spectroscopy (EIS) to explore electrochemical processes at electrodes of different geometries [1,2]. However, it was his masterful treatment of electrochemical reaction mechanisms that forges the link between Martin's work and that of others who seek to define reaction mechanisms using impedance techniques.

In this chapter, the point defect model (PDM), describing the formation and breakdown of passive films, is reviewed and developed. It is shown how important model parameters can be extracted from experimental impedance data and used to calculate the steady-state barrier layer thickness and passive current density as a function of voltage. In particular, the model is used to define the mechanism of the formation of Cu_2S on Cu in sulfide-containing brine. The present studies were conducted to provide a scientific basis for estimating the lifetimes of copper canisters in crystalline rock repositories in Sweden for the disposal of high level nuclear waste (HLNW).

Developments in Electrochemistry: Science Inspired by Martin Fleischmann, First Edition.
Edited by Derek Pletcher, Zhong-Qun Tian and David E. Williams.
© 2014 John Wiley & Sons, Ltd. Published 2014 by John Wiley & Sons, Ltd.

19.1   Theory

The PDM was developed by Macdonald and coworkers as a mechanistically based model that could be tested analytically against experiment [3–5]. The PDM is now highly developed and, to the present authors' knowledge, there are no known conflicts with experiment, where confluence between theory and experiment has been first demonstrated. Indeed, the model has predicted new phenomena that have subsequently been observed, including the photo-inhibition of passivity breakdown (PIPB) [6–9], and has provided a theoretical basis for designing new alloys from first principles [10, 11]. The PDM has been used previously to interpret electrochemical impedance data by optimizing the model on the experimentally determined real and imaginary components of the interphasial (metal/passive film/solution) impedance, with considerable success [12–16]. An earlier version of the model has been used extensively to analyze the data obtained in this program for carbon steel in simulated concrete pore water, and these analyses will be discussed at length in a later report. With the exception of one recent publication [17], all of the previous work from the present authors' laboratory used the commercial DataFit [18] software for optimization, which employs the Levenberg–Marquardt [19] method of minimization, in order to optimize the model onto the experimental data and to estimate values for various model parameters. The optimization procedures described below were performed using the same model as previously; however, the optimization was performed with a newer method of optimization, namely differential evolution (DE), using custom software [20], which resolves many of the issues associated with parameter optimization of functions of this type. The quality of solution is vastly improved (several orders of magnitude reduction in the chi-squared error over gradient-based methods). An overview of Evolutionary Algorithm methods is presented in Ref. [21]. Although gradient-based methods are computationally much faster than evolutionary methods (such as DE), without operator experience and the requirement for non-intuitive knowledge about a highly dimensional system, they are not operationally more efficient. The person-hours saved more than makes up for any shortcomings in terms of computational speed.

19.2   The Point Defect Model

As noted above, the PDM was developed to provide an atomic-scale description of the formation and breakdown of passive films. The physico-chemical basis of the PDM is shown in Figure 19.1.

Briefly, the model postulates that defect generation and annihilation reactions occur at the metal/barrier layer (m/bl) and the barrier layer/outer layer (bl/ol) interfaces, which are separated by a few nanometers, and that these reactions essentially establish the point defect concentrations within the barrier layer. The charged point defects are envisioned to move across the film in a direction that is consistent with their charge and the sign of the electric field (oxygen vacancies and metal interstitials from the m/bl interface, where they are generated, to the bl/s interface, where they are annihilated, and cation vacancies in the reverse direction). The electric field strength ($\varepsilon = 1\text{–}5 \times 10^6$ V cm^{-1}) is such that the defects move by migration, not by diffusion. Finally, it is assumed that the

Metal / Barrier Oxide Layer	Outer Layer/Solution
(1) $m + V_M^{\chi'} \xrightarrow{k_1} M_M + v_m + \chi e'$	(4) $M_M \xrightarrow{k_4} M^{\delta+} + V_M^{\chi'} + (\delta - \chi)e^-$
(2) $m \xrightarrow{k_2} M_i^{\chi+} + v_m + \chi e^-$	(5) $M_i^{\chi+} \xrightarrow{k_5} M^{\delta+} + (\delta - \chi)e^-$
(3) $m \xrightarrow{k_3} M_M + \frac{\chi}{2}V_{\ddot{O}} + \chi e^-$	(6) $V_{\ddot{O}} + H_2O \xrightarrow{k_6} O_O + 2H^+$
	(7) $MO_{\chi/2} + \chi H^+ \xrightarrow{k_7} M^{\delta+} + \frac{\chi}{2}H_2O + (\delta - \chi)e^-$
$\mid$	$\mid$
(**x = L**)	(**x = 0**)

Figure 19.1 *Interfacial defect generation/annihilation reactions that are postulated to occur in the growth of anodic barrier oxide films according to the Point Defect Model. m ≡ metal atom; $V_M^{\chi'}$ ≡ cation vacancy on the metal sublattice of the barrier layer; $M_i^{\chi+}$ ≡ interstitial cation; M_M ≡ metal cation on the metal sublattice of the barrier layer; $V_{\ddot{O}}$ ≡ oxygen vacancy on the oxygen sublattice of the barrier layer; O_O ≡ oxygen anion on the oxygen sublattice of the barrier layer; $M^{\delta+}$ ≡ metal cation in solution.*

potential drop across the bl/s interface ($\varnothing_{f/s}$) is a linear function of the applied voltage and pH

$$\varnothing_{f/s} = \alpha V + \beta pH + \varnothing_{f/s}^0 \tag{19.1}$$

such that the potential drop across the m/bl interface is given as

$$\varnothing_{m/f} = (1 - \alpha)V - \beta pH - \varepsilon L - \varnothing_{f/s}^0 \tag{19.2}$$

where L is the thickness of the barrier layer and $\varnothing_{f/s}^0$ is a constant. The form of Equation (19.1) is dictated by an electrical double-layer theory for oxide/solution interfaces. Other assumptions and postulates are contained in the original publications [6,7].

The electron current density, I, which is sensed in an external circuit, is given by:

$$I = F\left\{\chi k_1 C_v^L + \chi k_2 + \chi k_3 + (\delta - \chi)k_4 + (\delta - \chi)k_5 C_i^0 + (\delta - \chi)k_7\right\} \tag{19.3}$$

where C_v^L is the concentration of cation vacancies at the m/bl interface and C_i^0 is the concentration of cation interstitials at the bl/ol (bl/s) interface. Note that Equation (19.3) does not depend upon the concentration of oxygen vacancies, or upon the rate constant for Reaction (**6**) in Figure 19.1. Thus, no relaxations in the impedance response involve oxygen vacancies, but this is essentially an artifact of considering Reaction (**3**) in Figure 19.1, to be irreversible. If this reaction was considered to be reversible, then a relaxation involving oxygen vacancies would be present. Furthermore, the concentration of H^+ is considered to be constant, corresponding to a well-buffered solution, and is included in the definition of k_7, as indicated in Equation (19.6). Parenthetically, it should be noted that the inclusion of reversible reactions would allow the PDM to also account for the reduction of passive films, albeit at a considerable cost in mathematical complexity.

Using the method of partial charges, the rate constants for the reactions are found to be of the form [7]:

$$k_i = k_i^0 exp\left[a_i\left(V - R_{ol}I\right) - b_iL\right] \quad i = 1, 2, 3 \tag{19.4}$$

$$k_i = k_i^0 \exp\left[a_i\left(V - R_{ol}I\right)\right] \quad i = 4, 5 \tag{19.5}$$

and

$$k_7 = k_7^0 \exp\left[a_7\left(V - R_{ol}I\right)\right]\left(\frac{C_{H^+}}{C_{H^+}^0}\right)^n \tag{19.6}$$

where n is the kinetic order of barrier layer dissolution with respect to H^+. In deriving these expressions theoretically, it is assumed that a resistive outer layer, R_{ol}, exists on the surface of the barrier layer and that the passive current flows through the outer layer to a remote cathode, which is the normal experimental configuration. Because of this, the potential that exists at the bl/ol interface must be corrected from that applied at the reference electrode located at the outer layer/solution interface by the potential drop across the outer layer, where R_{ol} (Ω cm^{-2}) is the specific resistance of the outer layer. The coefficients in the rate constant expressions are summarized in Table 19.1.

It can be assumed that the applied potential changes sinusoidally around some mean value ($\overline{V}$) in accordance with Equation (19.7):

$$V = \overline{V} + \delta V = \overline{V} + \Delta V e^{j\omega t} \tag{19.7}$$

where ω is an angular frequency and ΔV is the amplitude (the bar over a letter refers to the corresponding value under steady-state conditions). Accordingly, in the linear approximation the independent variables have the following response $f = \overline{f} + \Delta f e^{j\omega t}$, where f represents the current density, I, and values on which I depends, namely, L, C_i^0, C_v^L, and the various rate constants.

The task, then, is to calculate the faradic admittance, Y_F, which is defined as:

$$Y_F = \frac{1}{Z_F} = \frac{\delta I}{\delta V} = \frac{\Delta I}{\Delta V} \tag{19.8}$$

where Z_F is the faradic impedance. Note that I is a function of the potential at the bl/ol interface (U), but the potential that is modulated is that at the outer layer/solution (ol/s) interface (V), or close to it, depending upon the exact placement of the tip of the Luggin probe. The two potentials are related by

$$U = V - R_{ol}I \tag{19.9}$$

It is evident that,

$$\frac{1}{Y_F} = \frac{1}{Y_F^0} + R_{ol} \quad \text{or} \quad Y_F = \frac{Y_F^0}{1 + R_{ol}Y_F^0} \tag{19.10}$$

where Y_F^0 is the admittance calculated in the absence of the outer layer, assuming that the potential at the bl/ol interface is $\overline{U}$ under steady-state conditions. It can be seen that $Y_F \rightarrow Y_F^0$ as $R_{ol} \rightarrow 0$ and $Y_F \rightarrow 1/R_{ol}$ for $Y_F^0 \rightarrow \infty$; that is, the interphasial impedance becomes controlled by the outer layer in the limit of an infinitely large outer layer specific resistance or an infinitely small barrier layer admittance.

Table 19.1 Coefficients for the rate constants for the reactions that generate and annihilate point defects at the m/bl interface [Reactions (**1**)–(**3**)] in Figure 19.1 and at the bl/s interface [Reactions (**4**)–(**6**)] in Figure 19.1, and for dissolution of the film [3–5]. $k_i = k_i^0 e^{a_i V} e^{b_i L} e^{c_i pH}$.

Reaction	$a_i\,(V^{-1})$	$b_i\,(cm^{-1})$	C_i	Units of k_i^0
(1) $m + V_M^{\chi'} \xrightarrow{k_1} M_M + v_m + \chi e'$	$\alpha_1 (1-\alpha)\chi\gamma$	$-\alpha_1 \chi K$	$-\alpha_1 \beta \chi \gamma$	$\dfrac{1}{s}$
(2) $m \xrightarrow{k_2} M_i^{\chi+} + v_m + \chi e'$	$\alpha_2 (1-\alpha)\chi\gamma$	$-\alpha_2 \chi K$	$-\alpha_2 \beta \chi \gamma$	$\dfrac{mol}{cm^2 s}$
(3) $m \xrightarrow{k_3} M_M + \dfrac{\chi}{2} V_{O^{..}} + \chi e'$	$\alpha_3 (1-\alpha)\chi\gamma$	$-\alpha_3 \chi K$	$-\alpha_3 \beta \chi \gamma$	$\dfrac{mol}{cm^2 s}$
(4) $M_M \xrightarrow{k_4} M^{\delta+} + (\delta - \chi) e'$	$\alpha_4 \alpha \delta \gamma$		$\alpha_4 \beta \delta \gamma$	$\dfrac{mol}{cm^2 s}$
(5) $M_i^{\chi+} \xrightarrow{k_5} M^{\delta+} + (\delta - \chi) e'$	$\alpha_5 \alpha \delta \gamma$		$\alpha_5 \beta \delta \gamma$	$\dfrac{cm}{s}$
(6) $V_{O^{..}} + H_2O \xrightarrow{k_6} O_O + 2H^+$	$2\alpha_6 \gamma$		$\alpha_6 \beta \delta \gamma$	$\dfrac{cm}{s}$
(7) $MO_{\chi/2} + \chi H^+ \xrightarrow{k_7} M^{\delta+} + \dfrac{\chi}{2} H_2O + (\delta - \chi) e'$	$\alpha_7 \alpha (\delta - \chi) \gamma$		$\alpha_7 (\delta - \chi) \beta \gamma$	$\dfrac{mol}{cm^2 s}$

The values of $\overline{U}$ and other steady-state values can be easily calculated. Assuming some arbitrary value of $\overline{U}$, it is possible to immediately calculate $\overline{k}_i$, $i = 4, 5, 7$ from Equations (19.5) and (19.6). From the rate equation for the change in thickness of the barrier layer, which is written as

$$\frac{dL}{dt} = \Omega k_3 - \Omega k_7 \tag{19.11}$$

we have $\overline{k}_3 = \overline{k}_7$, i.e.

$$L_{SS} = \left(\frac{a_7 - a_3}{b_3}\right) U + \left(\frac{C_7 - C_3}{b_3}\right) pH + \frac{1}{b_3} \ln\left[\left(\frac{k_7^0}{k_3^0}\right)\left(\frac{C_H}{C_H^0}\right)^n\right] \tag{19.12}$$

After that, the values $\overline{k}_i$ ($i = 1, 2$) can be calculated by using Equation (19.4).

The values of the steady-state concentrations $\overline{C}_i^0$ and $\overline{C}_v^L$ (concentrations of metal interstitials at the bl/ol interface and oxygen vacancies at bl/ol interface) can be found by equating the rates of formation and annihilation at the two interfaces to yield:

$$\overline{C}_v^L = \frac{\overline{k}_4}{\overline{k}_1} \tag{19.13}$$

$$\overline{C}_i^0 = \frac{\overline{k}_2}{\overline{k}_5} \tag{19.14}$$

and

$$\overline{C}_O^0 = \frac{\overline{k}_3}{\overline{k}_6} \tag{19.15}$$

Equations (19.13) to (19.15) follow from the condition that steady-state fluxes of cation vacancies, cation interstitials, and oxygen vacancies, are the same at the two interfaces.

Finally, it is possible to calculate the values of

$$\overline{I} = F\left\{\chi \overline{k}_1 \overline{C}_v^L + \chi \overline{k}_2 + \chi \overline{k}_3 + (\delta - \chi)\overline{k}_4 + (\delta - \chi)\overline{k}_5 \overline{C}_i^0 + (\delta - \chi)\overline{k}_7\right\} \tag{19.16}$$

and

$$\overline{V} = \overline{U} + R_{ol}\overline{I} \tag{19.17}$$

that is, the polarization behavior is calculated as the dependence $\overline{I}(\overline{V})$. As the actual value of $\overline{U}$, a value will be chosen at which $\overline{V}$ equals the prescribed value, because no outer layer is assumed to exist in the experiment or in this analysis. Practically, the task is reduced to the solution of a single equation $\overline{V} = \overline{U} + R_{ol}\overline{I}(\overline{U})$ relative to the unknown value $\overline{U}$ (the voltage at the bl/ol interface).

It can be seen that, if there is a code for calculating the admittance of the system in the absence of the outer layer, Y_F^0, then the admittance can be calculated in the presence of the outer layer, Y_F, by using Equation (19.10), assuming that Y_F^0 is calculated at the steady state applied potential that equals $\overline{U}$ (but not $\overline{V}$).

19.2.1 Calculation of Y_F^0

As follows from Equation (19.3), we have, in the linear form:

$$Y_F^0 = \frac{\delta I}{\delta U} = \frac{\Delta I}{\Delta U} = I_U + I_L \frac{\Delta L}{\Delta U} + I_v^L \frac{\Delta C_v^L}{\Delta U} + I_i^0 \frac{\Delta C_i^0}{\Delta U} \tag{19.18}$$

where

$$I_U = F \left\{ \chi a_1 \bar{k}_1 \bar{C}_v^L + \chi \bar{k}_2 a_2 + \chi \bar{k}_3 a_3 + (\delta - \chi) \bar{k}_4 a_4 + (\delta - \chi) \bar{k}_5 a_5 \bar{C}_i^0 + (\delta - \chi) \bar{k}_7 a_7 \right\} \tag{19.19}$$

$$I_L = F \left\{ \chi b_1 \bar{k}_1 \bar{C}_e^L + \chi \bar{k}_2 b_2 + \chi \bar{k}_3 b_3 \right\} \tag{19.20}$$

$$I_v^L = F \chi \bar{k}_1 \tag{19.21}$$

$$I_i^0 = F(\delta - \chi) \bar{k}_5 \tag{19.22}$$

Here, it is assumed that $U = \overline{U} + \delta U = \overline{U} + \Delta U e^{j\omega t}$ and the four terms on the right side are identified as arising from relaxations with respect to the applied potential V, the thickness of the barrier layer with respect to the voltage at the bl/ol interface, U, cation vacancies, $\bar{C}_v^L$, and cation interstitials, $\bar{C}_i^0$, respectively. Note the absence of a term for the relaxation of oxygen vacancies because, again, the concentration of oxygen vacancies does not appear in the current [Equation (19.3)], as a consequence of assuming Reaction (**3**) in Figure 19.1 to be irreversible.

It is now possible to calculate $\frac{\Delta L}{\Delta U}$, $\frac{\Delta C_v^L}{\Delta U}$ and $\frac{\Delta C_i^0}{\Delta U}$. It is convenient to start with $\frac{\Delta L}{\Delta U}$. The rate of change of the thickness of the barrier layer is described by Equation (19.11). Accordingly, by taking the total differential, we have

$$\frac{d\delta L}{dt} = j\omega \Delta L e^{j\omega t} = \Omega \delta k_3 - \Omega \delta k_7 = \Omega(\bar{k}_3 a_3 \delta U - \bar{k}_3 b_3 \delta L) - \Omega \bar{k}_7 a_7 \delta V \tag{19.23}$$

or

$$L_U \equiv \frac{\Delta L}{\Delta U} = \frac{\Omega(\bar{k}_3 a_3 - \bar{k}_7 a_7)}{j\omega + \Omega \bar{k}_3 b_3} \tag{19.24}$$

which is the result that we desire.

19.2.2 Calculation of $\frac{\Delta C_i^0}{\Delta U}$

The flux density of interstitials is

$$J_i = -D_i \frac{\partial C_i}{\partial x} - \chi D_i K C_i \tag{19.25}$$

Here, D_i is the diffusion coefficient of the cation interstitials, $K = F\varepsilon/RT$, where ε is the electric field strength inside the barrier layer, and T is the temperature. The continuity equation is

$$\frac{\partial C_i}{\partial t} = D_i \frac{\partial^2 C_i}{\partial x^2} + \chi D_i K \frac{\partial C_i}{\partial x} \tag{19.26}$$

with the boundary conditions

$$-k_5 C_i = -D_i \frac{\partial C_i}{\partial x} - \chi D_i K C_i \quad \text{at } x = 0 \tag{19.27}$$

Substitution $C_i = \overline{C}_i + \Delta C_i e^{j\omega t}$ into Equations (19.25–19.27) and linearization of the boundary conditions relative to ΔU and ΔL yields:

$$j\omega \Delta C_i = \frac{\partial^2 \Delta C_i}{\partial x^2} + \chi D_i K \frac{\partial \Delta C_i}{\partial x} \tag{19.28}$$

or

$$-\overline{k}_5(\overline{C}_i^0 a_5 \Delta U + \Delta \overline{C}_i^0) = -D_i \left(\frac{\partial \Delta C_i}{\partial x} \right)_{x=0} - \chi D_i K \Delta C_i^0 \quad \text{at } x = 0 \tag{19.29}$$

$$-\overline{k}_2(a_2 \Delta U - b_2 \Delta L) = -D_i \left(\frac{\partial \Delta C_i}{\partial x} \right)_{x=L} - \chi D_i K \Delta C_i^L \quad \text{at } x = L \tag{19.30}$$

An analytical solution of the linear boundary problem [Equations (19.28–19.30)] can be easily obtained and the sought value $\Delta C_i^0 / \Delta U$ can be presented in the following form:

$$\frac{\Delta C_i^0}{\Delta U} = \frac{A + B}{\Delta U} = \Delta C_{iU}^0 + \Delta C_{iL}^0 \frac{\Delta L}{\Delta U} \tag{19.31}$$

where

$$\Delta C_{iU}^0 = \frac{b_{1U}(a_{22} - a_{21}) + b_{2U}(a_{11} - a_{12})}{a_{11}a_{22} - a_{12}a_{21}} \tag{19.32}$$

$$\Delta C_{iL}^0 = \frac{b_{2L}(a_{11} - a_{12})}{a_{11}a_{22} - a_{12}a_{21}} \tag{19.33}$$

$$r_{1,2} = \frac{-\chi K \pm \sqrt{\chi^2 K^2 + 4j\omega/D_i}}{2} \tag{19.34}$$

$$a_{11} = (r_1 + \chi K)D_i - \overline{k}_5, \quad a_{12} = (r_2 + \chi K)D_i - \overline{k}_5,$$
$$a_{21} = (r_1 + \chi K)D_i e^{r_1 L}, \quad a_{22} = (r_2 + \chi K)D_i e^{r_2 L} \tag{19.35}$$

$$b_{1U} = \overline{k}_5 a_5 \overline{C}_i^0, \quad b_{2U} = \overline{k}_2 a_2, \quad b_{2L} = -\overline{k}_2 b_2 \tag{19.36}$$

The reader should note that the expressions given above for cation interstitials are exactly the same for oxygen vacancies, with the oxidation number, χ, being replaced by 2, Subscript 2 being replaced by Subscript 3, and Subscript 5 being replaced by Subscript 6, so as to identify the correct reactions in Figure 19.1.

19.2.3   Calculation of $\frac{\Delta C_v^L}{\Delta U}$

By analogy it can be shown that:

$$\frac{\Delta C_v^L}{\Delta U} = \frac{A e^{r_1 L} + B e^{r_2 L}}{\Delta U} = \Delta C_{vU}^L + \Delta C_{vL}^L \frac{\Delta L}{\Delta U} \Delta L \tag{19.37}$$

where

$$\Delta C_{vU}^L = \frac{(b_{1U}a_{22} - b_{2U}a_{12})e^{r_1 L} + (b_{2U}a_{11} - b_{1U}a_{21})e^{r_2 L}}{a_{11}a_{22} - a_{12}a_{21}} \qquad (19.38)$$

$$\Delta C_{vL}^L = \frac{b_{2L}a_{11}e^{r_2 L} - b_{2L}a_{12}e^{r_1 L}}{a_{11}a_{22} - a_{12}a_{21}} \qquad (19.39)$$

$$r_{1,2} = \frac{\chi K \pm \sqrt{\chi^2 K^2 + 4j\omega/D_v}}{2} \qquad (19.40)$$

where

$$a_{11} = (r_1 - \chi K)D_v, \quad a_{12} = (r_2 - \chi K)D_v, $$
$$a_{21} = [(r_1 - \chi K)D_v + \bar{k}_1]e^{r_1 L}, \quad a_{22} = [(r_2 - \chi K)D_v + \bar{k}_1]e^{r_2 L} \qquad (19.41)$$

$$b_{1U} = -\bar{k}_4 a_4, \quad b_{2U} = -\bar{k}_1 a_1 \bar{C}_v^L, \quad b_{2L} = \bar{k}_1 \bar{C}_v^L b_1 \qquad (19.42)$$

By substituting Equations (19.24), (19.31) and (19.37) into Equation (19.18), we have the final result

$$Y_F^0 = I_U + I_L L_U + I_v^L (\Delta C_{vU}^L + \Delta C_{vL}^L L_U) + I_i^0 (\Delta C_{iU}^0 + \Delta C_{iL}^0 L_U) \qquad (19.43)$$

19.3   The Passivation of Copper in Sulfide-Containing Brine

The Scandinavian plan for the disposal of high-level nuclear waste (HLNW) calls for the encapsulation of spent nuclear fuel in a crystalline bedrock repository at a depth of about 500 m. The current plan calls for the emplacement of spent fuel in an inner cast-iron canister shielded with a 50 mm-thick outer layer made of copper [22–24]. After emplacement of the canisters in the boreholes in the floors of the drifts (tunnels), the remaining space will be backfilled with a compacted bentonite clay buffer [22]. The role of the inner cast-iron layer is to provide mechanical strength as well as radiation shielding, while the copper outer layer provides corrosion protection. In the anoxic groundwater environment, corrosion mechanisms involving sulfides have been identified to be important in controlling the canisters' lifetime [24]. Sulfide species, such as bisulfide ion (HS^-), are present in groundwater in the near-field environment, and hence in the vicinity of the copper canisters, thereby causing corrosion and reducing their lifetimes.

Figure 19.2 displays the potentiodynamic polarization of copper in a deaerated 0.1 M $NaCl + 2 \times 10^{-4}$ M $Na_2S \cdot 9H_2O$ solution at 25 °C, as measured in the present studies. A broad passive range of potential is seen, starting from −0.6 (the potential related to the formation of copper sulfide) and extending up to +0.15 V versus SHE. Four potentials within the passive region were selected for the impedance analysis, namely −0.495 V, −0.395 V, −0.205 V and −0.195 V versus SHE, with the electrode being controlled potentiostatically at these potentials for the entire time of the experiments.

A modified point defect model for the formation of the bi-layer $Cu/Cu_2S/CuS$ passive films on copper in the sulfide-containing solutions is proposed. The physico-chemical basis of the modified PDM is shown in Figure 19.3. In all previous systems this model was applied to the formation and dissolution of the passive oxide films on different metals, but

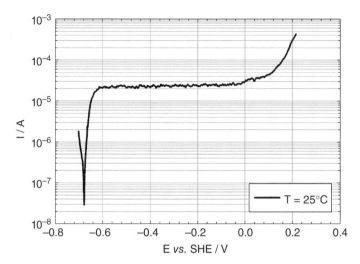

Figure 19.2 *Potentiodynamic polarization curves of Cu in a deaerated 0.1 M NaCl + 2 ×
10^{-4} M Na$_2$S·9H$_2$O solution at 25 °C. Scan rate = 1 mV s^{-1}.*

Metal / Barrier Oxide Layer	Outer Layer/Solution
(1) $Cu + V_{Cu}^{\chi'} \xrightarrow{k_1} Cu_{Cu} + v_{Cu} + \chi e'$	(4) $Cu_{Cu} \xrightarrow{k_4} Cu^{\delta} + V_{Cu}^{\chi'} + (\delta - \chi)e'$
$\|$	$\|$
(2) $Cu \xrightarrow{k_2} Cu_i^{\chi+} + v_{Cu} + \chi e'$	(5) $Cu_i^{\chi+} \xrightarrow{k_5} Cu^{\delta+} + (\delta - \chi)e'$
$\|$	$\|$
(3) $Cu \xrightarrow{k_3} Cu_{Cu} + \frac{x}{2}V_{S^-} + \chi e'$	(6) $V_{S^-} + HS^- \xrightarrow{k_6} S_S + H^+$
$\|$	$\|$
	(7) $CuS_{x/2} + \frac{x}{2}H^+ \xrightarrow{k_7} Cu^{\delta+} + \frac{x}{2}HS^- + (\delta - \chi)e'$
$\|$	$\|$
(x = L)	(x = 0)

Figure 19.3 *Interfacial defect generation/annihilation reactions that are postulated to occur
in the growth of anodic barrier sulfide films according to the Point Defect Model. $V_{Cu}^{\chi'} \equiv$
cation vacancy on the metal sublattice of the barrier layer; $Cu_i^{\chi+} \equiv$ cuprous cation interstitial;
$Cu_{Cu} \equiv$ cuprous cation in cation site on the metal sublattice of the Cu_2S barrier layer; $V_S^{\cdot\cdot} \equiv$
sulfur vacancy on the anion sublattice of the barrier layer; $S_S \equiv$ sulfur anion on the anion
sublattice of the barrier layer; $Cu^{\delta+} \equiv$ cuprous cation in solution.*

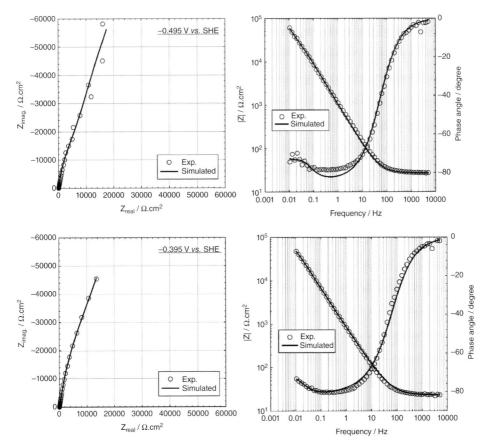

Figure 19.4 *Experimental and simulated Nyquist and Bode plots for copper in a deaerated 0.1 M NaCl + 2 × 10⁻⁴ M Na₂S·9H₂O, T = 25 °C as a function of applied potential. The solid lines show the best fit calculation according to the PDM.*

in the present study a model for the growth and dissolution of passive sulfide films was proposed for the first time.

Figures 19.4 and 19.5 show typical experimental electrochemical impedance spectra for the passive sulfide film on copper in a deaerated 0.1 M NaCl + 2 × 10⁻⁴ M Na₂S·9H₂O solution at 25 °C. The best fit results, calculated from the parameters obtained from optimization of the proposed mechanism based on the modified PDM (Figure 19.3), as listed in Tables 19.2 and 19.3, are also included in these figures as solid lines. It can be seen that the correlation between the experiment and the model is fairly good, indicating that the proposed model can provide a reasonable account of the observed experimental data. It should be noted that the obtained parameters should not only reproduce the experimental impedance spectra but also deliver values that are physically reasonable. The obtained kinetic parameters, such as the standard rate constants, transfer coefficients and defect diffusivities listed in Tables 19.2 and 19.3, show no systematic dependency on the applied

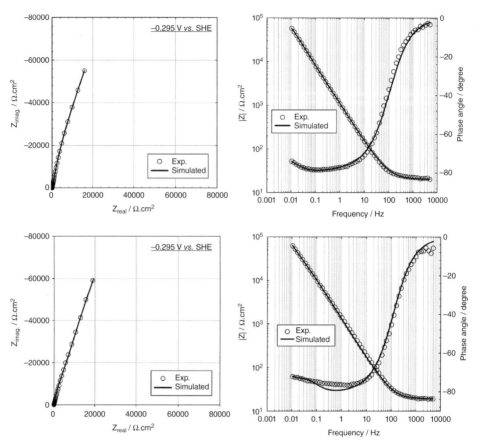

Figure 19.5 *Experimental and simulated Nyquist and Bode plots for copper in a deaerated 0.1 M NaCl + 2 × 10^{-4} M Na$_2$S·9H$_2$O, T = 25 °C as a function of applied potential. The solid lines show the best fit calculation according to the PDM.*

Table 19.2 *Averaged kinetic parameters obtained from PDM optimization on impedance data for copper in a deaerated 0.1 M NaCl + 2 × 10^{-4} M Na$_2$S·9H$_2$O, at 25 °C.*

Parameter	Value	Dimensions
α	0.26 ± 0.02	—
α_1	0.44 ± 0.02	—
α_3	0.21 ± 0.02	—
α_4	0.07 ± 0.03	—
k_1	$3.5 \pm 0.1 \times 10^{-07}$	(s^{-1})
k_2	$1.8 \pm 0.1 \times 10^{-16}$	(mol cm^2 s^{-1})
k_3	$2.4 \pm 0.2 \times 10^{-11}$	(mol cm^2 s^{-1})
k_4	$7.8 \pm 0.6 \times 10^{-13}$	(mol cm^2 s^{-1})
k_7	$7.3 \pm 0.2 \times 10^{-13}$	(mol cm^2 s^{-1})

Table 19.3 Other parameters obtained from the optimization of the PDM on impedance data for copper in a deaerated $0.1\ M\ NaCl + 2 \times 10^{-4}\ M\ Na_2S \cdot 9H_2O$, at 25 °C.

$E_{app.}$ (V vs. SHE)	−0.495	−0.395	−0.295	−0.195	Origin
δ	1	1	1	1	Assumed
χ	1	1	1	1	Assumed
n	−0.15	−0.15	−0.15	−0.15	1st stage optim
CPE-g (S s$^\varphi$ cm^{-2})	5.31×10^{-04}	4.90×10^{-04}	2.93×10^{-04}	2.71×10^{-04}	2nd stage optim
CPE-φ	0.82	0.78	0.81	0.83	2nd stage optim
pH	10.18	10.18	10.18	10.18	Measured
ε (V cm^{-1})	$1.94 \times 10^{+05}$	$2.10 \times 10^{+05}$	$2.10 \times 10^{+05}$	$2.20 \times 10^{+05}$	1st stage optim
C_{dl} (F cm^{-2})	1.60×10^{-04}	2.22×10^{-04}	1.43×10^{-04}	1.38×10^{-04}	2nd stage optim
R_{ct} (Ω cm^2)	$2.80 \times 10^{+05}$	$9.99 \times 10^{+10}$	$9.98 \times 10^{+10}$	$9.99 \times 10^{+10}$	2nd stage optim
R_s (Ω cm^2)	27	23	19	18	Estimated
D_v (cm^2 s^{-1})	7.44×10^{-14}	1.00×10^{-13}	1.00×10^{-13}	7.45×10^{-13}	2nd stage optim
R_{ol} (Ω cm^2)	$4.27 \times 10^{+04}$	$1.03 \times 10^{+05}$	$1.48 \times 10^{+05}$	$2.78 \times 10^{+04}$	2nd stage optim
C_{ol} (F cm^{-2})	2.35×10^{-04}	4.01×10^{-04}	4.66×10^{-04}	3.04×10^{-04}	2nd stage optim
σ (Ω cm^2 s$^{-0.5}$)	10	10	11	12	2nd stage optim
$R_{e,h}$ (Ω cm^2)	$9.11 \times 10^{+10}$	$9.94 \times 10^{+10}$	$1.00 \times 10^{+07}$	$1.00 \times 10^{+07}$	2nd stage optim
L_{SS} (cm)	3.86×10^{-07}	7.65×10^{-07}	1.14×10^{-06}	1.42×10^{-06}	Calculated
I_{SS} (A cm^{-2})	2.43×10^{-06}	2.44×10^{-06}	2.45×10^{-06}	2.46×10^{-06}	Calculated

potential as required by the fundamental electrochemical kinetic theory. This is a good test of viability of the proposed model and the obtained parameters.

As mentioned in Section 19.2, the rate of change of the barrier layer thickness that forms on a metal surface can be expressed as:

$$\frac{dL}{dt} = \Omega k_3^0 e^{a_3 V} e^{b_3 L} e^{C_3 pH} - \Omega k_7^0 \left(\frac{C_{H^+}}{C_{H^+}^0}\right)^n e^{a_7 V} e^{C_7 pH} \qquad (19.44)$$

where Ω is the molar volume of the barrier layer per cation, C_{H^+} is the concentration of hydrogen ions, $C_{H^+}^0$ is the standard state concentration, and n is the kinetic order of the barrier layer dissolution reaction with respect to H^+. Definitions of the other parameters are listed in Table 19.2. It should be mentioned that, since the pH of the solution was higher than the pH of zero charge for Cu_2S and CuS dissolution (PZC < 3.5 [25]), n should be a negative value as it is obtained in the optimization. Another point to be noted is that the rate of the dissolution reaction is potential-dependent if the oxidation state of copper in the barrier layer is different from its oxidation state in the solution. However, under an anoxic condition, the oxidation state of copper in both phases is +1, and therefore the rate of film dissolution is considered to be potential-independent.

Under steady-state conditions, $\frac{dL}{dt} = 0$ and the steady-state thickness of the barrier layer can be derived as

$$L_{ss} = \left[\frac{1-\alpha}{\varepsilon}\right] V + \left[\frac{2.303n}{\alpha_3 \varepsilon \chi \gamma} - \frac{\beta}{\varepsilon}\right] pH + \frac{1}{\alpha_3 \varepsilon \chi \gamma} \ln\left(\frac{k_3^0}{k_7^0}\right) \qquad (19.45)$$

Figure 19.6a shows a comparison of the calculated steady-state thickness of the barrier layer with the experimental results as a function of applied potential, while the steady-state current for passive copper in sulfide-containing sodium chloride solution calculated from

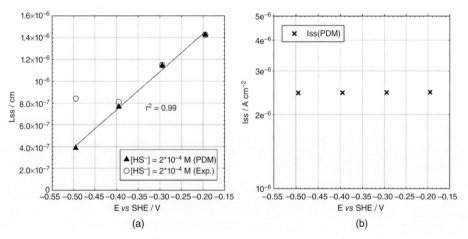

(a)   (b)

Figure 19.6 *Plots of the calculated and experimental steady-state barrier layer thickness as a function of potentials for copper in a deaerated 0.1 M NaCl + 2 × 10^{-4} M $Na_2S\cdot9H_2O$ at 25 °C.*

the parameters obtained from the PDM optimization is shown in Figure 19.6b. As seen from this figure, there is a linear dependence of $log(I_{ss})$ on the applied potential, which is consistent with the PDM diagnostic criteria for p-type passive films. The parameters obtained from the PDM optimization listed in Tables 19.2 and 19.3 were used to calculate, theoretically, the steady-state properties (thickness and passive current density) of the barrier layer using Equations (19.16) and (19.45). In order to calculate the experimental steady-state thickness, the well-known parallel plate capacitance formula was used [Equation (19.46)], assuming a value of the capacitance from the high-frequency (1 kHz) imaginary part of the experimental impedance data.

$$C = \frac{\tilde{\varepsilon}\varepsilon^0}{d} \tag{19.46}$$

where $\tilde{\varepsilon}$ is the dielectric constant (calibrated based on the obtained thickness, $\tilde{\varepsilon} = 724$), $\varepsilon^0 = 8.85 \times 10^{-14}$ (in F cm^{-1}) is the vacuum permittivity, d is the thickness of the film (in cm), and C is the capacitance (in F cm^{-2}). As can be seen, the thickness of the barrier layer increases with applied potential, as predicted by the PDM [26]. A good agreement was obtained between the calculated and experimental thickness, except at the lowest potential, which is closest to the active-to-passive transition (Figure 19.2); it is also possible that the barrier layer had not fully developed at that potential. This postulate is somewhat supported by the fact that instabilities were observed in the impedance measurements at that potential.

19.4   Summary and Conclusions

In this chapter, the application of the point defect model for analyzing impedance data for passive metal systems has been detailed. The feasibility of deriving an impedance version of the PDM, and of optimizing the model upon experimental impedance data to extract values for important model parameters, has also been demonstrated. These, in turn, were used to calculate the steady-state barrier layer thickness and passive current density as a function of voltage. These studies have provided a scientific basis for estimating the lifetimes of copper canisters in crystalline rock repositories in Sweden for the disposal of HLNW.

Acknowledgments

In these studies the authors gratefully acknowledge the support of Strålsäkerhetsmyndigheten (SSM) of Sweden.

References

(1) Abrantes, L., Fleischmann, M. and Peter, L. (1988) On the diffusional impedance of microdisc electrodes. *Journal of Electroanalytical Chemistry and Interfacial Electrochemistry*, **256**, 229–233.

(2) Fleischmann, M., Pons, S. and Daschbach, J. (1991) The ac impedance of spherical, cylindrical, disk, and ring microelectrodes. *Journal of Electroanalytical Chemistry*, **317**, 1–26.

(3) Chao, C.Y., Lin, L.F. and Macdonald, D.D. (1981) A point defect model for anodic passive films I. Film growth kinetics. *Journal of The Electrochemical Society*, **128**, 1187–1194.

(4) Lin, L.F., Chao, C.Y. and Macdonald, D.D. (1981) A point defect model for anodic passive films II. Chemical breakdown and pit initiation. *Journal of The Electrochemical Society*, **128**, 1194–1198.

(5) Chao, C.Y., Lin, L.F. and Macdonald, D.D. (1982) A point defect model for anodic passive films III. Impedance response. *Journal of The Electrochemical Society*, **129**, 1874–1879.

(6) Macdonald, D.D. (1999) Passivity – the key to our metals-based civilization. *Pure and Applied Chemistry*, **71**, 951–978.

(7) Macdonald, D.D. (2006) On the existence of our metals-based civilization. *Journal of The Electrochemical Society*, **153**, B213.

(8) Song, H. and Macdonald, D.D. (1991) Photoelectrochemical impedance spectroscopy I. Validation of the transfer function by Kramers–Kronig transformation. *Journal of The Electrochemical Society*, **138**, 1408–1410.

(9) Macdonald, D.D., Sikora, E., Balmas, M.W. and Alkire, R.C. (1996) The photo-inhibition of localized corrosion on stainless steel in neutral chloride solution. *Corrosion Science*, **38**, 97–103.

(10) Urquidi, M. and Macdonald, D.D. (1985) Solute–vacancy interaction model and the effect of minor alloying elements on the initiation of pitting corrosion. *Journal of The Electrochemical Society*, **132**, 555–558.

(11) Zhang, L. and Macdonald, D.D. (1998) Segregation of alloying elements in passive systems – I. XPS studies on the Ni–W system. *Electrochimica Acta*, **43**, 2661–2671.

(12) Macdonald, D.D. and Smedley, S.I. (1990) An electrochemical impedance analysis of passive films on nickel(111) in phosphate buffer solutions. *Electrochimica Acta*, **35**, 1949–1956.

(13) Macdonald, D.D. and Urquidi-Macdonald, M. (1985) Application of Kramers–Kronig transforms in the analysis of electrochemical systems I. Polarization resistance. *Journal of The Electrochemical Society*, **132**, 2316–2319.

(14) Urquidi-Macdonald, M., Real, S. and Macdonald, D.D. (1986) Application of Kramers–Kronig transforms in the analysis of electrochemical impedance data II. Transformations in the complex plane. *Journal of The Electrochemical Society*, **133**, 2018–2024.

(15) Ai, J., Chen, Y., Urquidi-Macdonald, M. and Macdonald, D.D. (2007) Electrochemical impedance spectroscopic study of passive zirconium. *Journal of The Electrochemical Society*, **154**, C52.

(16) Macdonald, D.D. and Sun, A. (2006) An electrochemical impedance spectroscopic study of the passive state on Alloy-22. *Electrochimica Acta*, **51**, 1767–1779.

(17) Geringer, J., Taylor, M.L. and Macdonald, D.D. (2012) Predicting the steady state thickness of passive films with the point defect model in fretting corrosion experiments. Proceedings, PRIME 2012; 222nd Electrochemical Society Meeting. Pacific Rim Meeting on Electrochemical and Solid State Science, 7–12 October, Honolulu, Hawaii.

(18) DataFit, Oakdale Engineering. www.oakdaleengr.com. Accessed 11 February 2014.

(19) Levenberg, K. (1944) A method for the solution of certain problems in least squares. *Quarterly of Applied Mathematics*, **2**, 164.

(20) Ellis 2: Complex curve fitting for one independent variable | IgorExchange (2012). Available at: http://www.igorexchange.com/project/gencurvefit.

(21) Bäck, T. and Schwefel, H.-P. (1993) An overview of evolutionary algorithms for parameter optimization. *Evolutionary Computation*, **1**, 1–23.

(22) King, F., Ahonen, L., Taxén, C. *et al.* (2002) Copper corrosion under expected conditions in a deep geologic repository. Swedish Nuclear Fuel Waste Management Company Report, SKB TR 01-23, 2001 Posiva Oy Rep. POSIVA 2002-01.

(23) King, F. (2013) Container materials for the storage and disposal of nuclear waste. *Corrosion*, **69**, 986–1011.

(24) King, F., Lilja, C. and Vähänen, M. (2013) Progress in the understanding of the long-term corrosion behaviour of copper canisters. *Journal of Nuclear Materials*, **438**, 228–237.

(25) Kosmulski, M. (2010) *Surface Charging and Points of Zero Charge*, CRC Press-Taylor & Francis Group, Boca Raton, FL.

(26) Macdonald, D.D., Biaggio, S.R. and Song, H. (1992) Steady-state passive films interfacial kinetic effects and diagnostic criteria. *Journal of The Electrochemical Society*, **139**, 170–177.

Index

References to tables are given in bold type. References to figures are given in italic type.

Developments in Electrochemistry: Science Inspired by Martin Fleischmann, First Edition.
Edited by Derek Pletcher, Zhong-Qun Tian and David E. Williams.
© 2014 John Wiley & Sons, Ltd. Published 2014 by John Wiley & Sons, Ltd.